# World Sustainability Series

**Series Editor**

Due to its scope and nature, sustainable development is a matter which is very interdisciplinary, and draws from knowledge and inputs from the social sciences and environmental sciences on the one hand, but also from physical sciences and arts on the other. As such, there is a perceived need to foster integrative approaches, whereby the combination of inputs from various fields may contribute to a better understanding of what sustainability is, and means to people. But despite the need for and the relevance of integrative approaches towards sustainable development, there is a paucity of literature which address matters related to sustainability in an integrated way.

**Notes on the quality assurance and peer review of this publication**

Prior to publication, the works published in this book are initially assessed and reviewed by an in-house editor. If suitable for publication, manuscripts are sent for further review, which includes a combined effort by the editorial board and appointed subject experts, who provide independent peer-review. The feedback obtained in this way was communicated to authors, and with manuscripts checked upon return before finally accepted. The peer-reviewed nature of the books in the "World Sustainability Series" means that contributions to them have, over many years, been officially accepted for tenure and promotion purposes.

Walter Leal Filho · Hélder S. Lopes ·
Marina Prieto Lencastre · Rui Estrada · Diogo Vidal
Editors

# Composing Worlds: Humanities, Health and Wellbeing in the XXI Century Towards a More Sustainable World

Springer

*Editors*
Walter Leal Filho
Faculty of Life Sciences
Hamburg University of Applied Sciences
Hamburg, Germany

Hélder S. Lopes
Department of Geography
University of Minho
Braga, Portugal

Marina Prieto Lencastre
Innovation and Development Institute
University Fernando Pessoa (UFP)
Porto, Portugal

Rui Estrada
Innovation and Development Institute
University Fernando Pessoa (UFP)
Porto, Portugal

Diogo Vidal
Department of Life Sciences
University of Coimbra
Coimbra, Portugal

ISSN 2199-7373 ISSN 2199-7381 (electronic)
World Sustainability Series
ISBN 978-3-031-87107-8 ISBN 978-3-031-87108-5 (eBook)
https://doi.org/10.1007/978-3-031-87108-5



This Springer imprint is published by the registered company Springer Nature Switzerland AG
The registered company address is: Gewerbestrasse 11, 6330 Cham, Switzerland

# Preface

As the world continues to confront the existential crises of the Anthropocene—a term that captures the era in which human activity has become the dominant force shaping the planet's climate and ecosystems that has brought humanity to a critical juncture—we must draw on the full breadth of human and non-human knowledge—scientific, philosophical, artistic, and social—to guide us towards a more just, sustainable, and resilient future. We are confronted by profound challenges that demand urgent attention and innovative solutions. From the escalating impacts of climate change to the pervasive influence of technology on our lives, humanity faces complex problems that transcend national borders, social sectors, and academic disciplines. This era, defined by rapid transformations in the environment, society, and technology, calls for rethinking how we approach the entangled issues of environmental sustainability, human health, social equity, and ethical responsibility. Within this urgent and complex context, the interdisciplinary exploration undertaken in Composing Worlds: Humanities, Health, and Well-being in the XXI Century Towards a More Sustainable World gains significance.

At the heart of this book lies a central question: how can we address the multifaceted crises of our time? The interconnectedness of climate change, biodiversity loss, environmental degradation, and public health threats requires new frameworks for understanding and action—ones that break free from siloed disciplines and embrace collaboration across multiple fields. In this work, we explore the synergies that arise when the humanities, social sciences, health sciences, and environmental studies come together, drawing on diverse forms of knowledge to illuminate the path forward.

This book results from the collaborative efforts within the *Composing Worlds (Compor Mundos)* network, a collective of scholars and practitioners from various backgrounds united by a shared commitment to understanding and addressing the challenges of the Anthropocene. The network's multidisciplinary approach is reflected in the following chapters, which tackle a range of topics from the socioecological crises of our time to the role of cultural narratives in shaping public perceptions and actions. The book highlights the vital contributions of the humanities and social sciences in the quest for environmental justice, sustainability, and improved health outcomes. The way we engage with these issues is a matter of technical expertise and cultural, ethical, and philosophical understanding. It is here that the humanities offer their essential contributions. Fields such as philosophy, history, literature, and the arts offer valuable insights into how we think about our relationship with the environment, the ethical implications of our actions, and the cultural values that shape our responses to the challenges of the Anthropocene.

This book aims to demonstrate that addressing environmental crises is not merely a scientific or technological endeavor. It is a profoundly human issue that requires reconsidering how we relate to nature and each other. The book's chapters bring together the power of citizen engagement, the wisdom of traditional knowledge, and the urgency of contemporary scientific and technological innovations to address the complex problems that affect our world. We seek to enrich and expand the dialogue around socioenvironmental challenges, fostering a deeper understanding of our collective responsibility's cultural and ethical dimensions to the planet.

In *Composing Worlds*, the reader will find diverse perspectives, from the role of art and literature in promoting environmental awareness to the integration of health into the socioecological context. The book opens with an introduction to the International Virtual Symposium on Humanities, Environment, Health and Well-Being, held in 2024, which aggregates a set of chapters focused on the role of the humanities in shaping our understanding of health and well-being in the Anthropocene. It delves into how art, philosophy, and literature engage with the environmental issues of our time, offering unique narratives that deepen our understanding of the human condition in a rapidly changing world.

The book's second part focuses on innovations and socio-environmental approaches, exploring how new technologies, social engagement, and mental health considerations intersect in addressing climate change and environmental degradation. It explores the intersection of digital technologies with mental health, the growing recognition of eco-anxiety, and the impact of environmental degradation on human health. It also discusses the role of artificial intelligence in exacerbating and mitigating our challenges. At the same time, other chapters explore the potential of nature-based solutions and citizen science in fostering sustainable futures.

The book's final part turns its attention to socioecological management and social participation, emphasizing the importance of collective action in tackling environmental challenges. The chapter on urban green spaces and new forms of social participation highlights the potential of local, community-driven efforts to address the environmental and health issues that arise in cities. Other contributions discuss the role of traditional agriculture, sustainable resource management, and the importance of creating resilient communities capable of responding to environmental shocks.

This book invites readers to join in this collective effort, exploring how the humanities, health sciences, and social sciences can unite to inspire transformative change in our approach to the planet and its future. In doing so, *Composing Worlds* offers a compelling vision of what it means to live in an interconnected world, one where our actions are informed by empathy, responsibility, and a deep understanding of the complex web of life that sustains us all. It challenges us to reconsider how to live harmoniously with the environment, promoting sustainable practices that respect the Earth and its human and non-human inhabitants.

We hope this book will serve as a source of inspiration and action, catalyzing new ideas, collaborations, and practices that can contribute to a more sustainable, healthy, emphatic and equitable world for present and future generations.

Hamburg, Germany — Walter Leal Filho
Braga, Portugal — Hélder S. Lopes
Porto, Portugal — Marina Prieto Lencastre
Porto, Portugal — Rui Estrada
Coimbra, Portugal — Diogo Vidal

# Contents

# International Virtual Symposium on Humanities, Environment, Health and Well-Being

# Composing Words 2024. International Virtual Symposium on Humanities, Environment, Health and Well-Being—An Introduction

**Elsa Simões, Rui Estrada, and Marina Lencastre**

We live in a time of vast transformations in human societies, the environment, health, and well-being. These transformations call for a critical and emphatic reflection on the future we want to build and the directions we intend to promote to make our shared world more hospitable and inclusive, both for human and non-human beings. In an era where the convergence of science, technology, and societal advancement often dominates the discourse, it is imperative to recognise the profound and intricate connections between the humanities, environment, health, and well-being. This interdisciplinary nexus not only shapes our understanding of the world but also underscores the essential role of humanistic inquiry in addressing contemporary challenges.

E. Simões (✉) · R. Estrada · M. Lencastre
Faculty of Human and Social Sciences, I3ID, University Fernando Pessoa, Porto, Portugal
e-mail: esimoes@ufp.edu.pt

R. Estrada
e-mail: restrada@ufp.edu.pt

M. Lencastre
e-mail: mlencast@ufp.edu.pt

E. Simões
LabCom—Communication & Arts Research Unit, University of Beira Interior, Covilhã, Portugal

R. Estrada
Center for Transdisciplinary Research "Culture, Space and Memory", Faculty of Arts and Humanities of the University of Porto, Porto, Portugal

Observatory Permanent Violence and Crime, University Fernando Pessoa, Porto, Portugal

R. Estrada · M. Lencastre
Coordinator of the network "Composing Worlds: Humanities, Well-being, and Health", University Fernando Pessoa, Porto, Portugal



W. Leal Filho et al. (eds.), *Composing Worlds: Humanities, Health and Wellbeing in the XXI Century Towards a More Sustainable World*, World Sustainability Series,
https://doi.org/10.1007/978-3-031-87108-5_1

The humanities and the social sciences, encompassing disciplines such as philosophy, literature, communication, psychology, anthropology and the arts, offer deep insights into the human condition, fostering empathy, critical thinking, and ethical reasoning. These fields enable us to reflect on our place in the world, our relationships with one another, and the legacies we inherit and develop. In this context, the environment is more than just a backdrop for human activity; it is a dynamic, interconnected system that sustains life and shapes cultures. Health and well-being are intrinsically linked to both the environment and the humanities. The natural environment significantly influences physical health, with clean air, water, and green spaces essential for well-being. However, the psychological and emotional aspects of health are equally determinant. In this case, the humanities play a paramount role. Engaging with literature, art, and philosophy can provide solace, meaning, and a sense of belonging, contributing to mental and emotional resilience. The study of humanities fosters a holistic approach to health care, emphasising the importance of narrative medicine and patient-centred care. Understanding patients' stories and cultural backgrounds can lead to more compassionate and effective healthcare practices. Integrating the humanities with environmental and health sciences creates a rich background that enhances our understanding of the world and our place within it. By acknowledging and exploring these connections, we can encourage a more comprehensive and compassionate approach to addressing the complex challenges of our time. The interplay between the humanities, environment, health, and well-being is not merely an academic exercise but a vital framework for nurturing a sustainable, equitable, and flourishing future for all.

The International Virtual Symposium on Humanities, Environment, Health, and Well-being consists in the assembling of international specialists of different areas in the humanities and the social, natural and health sciences, who reflect upon critical issues of today's environments and societies, focusing on health and well-being in the twenty-first century. The different panels reflect on how the humanities and the arts, together with the environmental, social and health sciences, are essential fields of knowledge to understand the metamorphosis of our world and its impact on health and well-being. They offer us a deep, as well as a broad, opportunity to enact, to think about and to diagnose these new complex realities.

On the first day of the Symposium, during the inaugural conference, Jonathan McFarland reflected on the importance of the humanities in medicine and sciences in general. According to the researcher, the humanities allow us to think critically and, in the specific case of medicine, to see people rather than clinical cases. In a final comparison between humans and the Earth, McFarland stated: Listen to the patient: they will tell you the diagnosis; listen to the Earth: it will tell you what needs to be done.

In the first panel, led by Teresa Casal and Ricardo Ricci and moderated by Susana Teixeira, the following question was posed at the outset: How can we face human vulnerability and inconvenient truths such as death and resource limitations? Once again, the humanities and literature emerged as fundamental in helping us reflect on these topics. The cited Zulu proverb, "A person becomes a person through other people," through other stories, points the way forward: placing ourselves in the shoes of others is the only way to understand them and ourselves and the only way to live in the community.

The second panel, led by Ana Sofia Carvalho and Paolo Parini and moderated by Daniela Mendes, focused on the ethical question. In our fast-paced world, we need constants and reference points that virtuous ethics can provide: knowing how to act in specific circumstances and how to behave, not just knowing what to do—placing value on social and emotional skills. Knowledge and technoscience, now even more emergent with AI, cannot advance without considering human contingency.

In the third panel, led by Delan Devakumar and João Nunes Fernandes and moderated by José Calheiros, the discussion centred on racism in healthcare systems and climate change, more specifically, the discrimination against BAME (Black, Asian, and Minority Ethnic) groups. Data, particularly from the United Kingdom, was presented, highlighting evidence of this discrimination and its consequences. It was also mentioned that it is necessary to politicise, in a broad sense, the controversy surrounding climate change: What inheritance have we received from the past, and what legacy should we leave for future generations? For this commitment to remain stable, it is necessary to reform democratic institutions, including replacing Gross National Product (GNP) as a primary measure of progress with Gross National Happiness (GNH).

The fourth panel, led by Daniel Catalão and Myriam Morron and moderated by Ricardo Pinto, focused on transhumanism and the bionic world that awaits us. Will we continue to be human in this technological vortex of AI? These issues must be discussed without a simplistic or Manichean perspective. Only in this way can the debate be productive.

On the second day, the fifth panel, dedicated to 'Planetary Health and Interactions and Implications for a Changing World', was moderated by Diogo Vidal Lopes and Marina Lencastre. Sue Stuart-Smith delivered the first talk, delving into the profound connection between nature and our physical and mental well-being, as depicted in her book, "The Well Gardened Mind: The Importance of Nature for Mental Health." Stuart-Smith highlighted the concept of 'urgent biophilia,' underscoring our innate inclination to seek solace and rejuvenation in nature. Drawing attention to historical contexts like colonialism, she emphasised the wisdom of indigenous cultures in fostering reciprocity with the environment rather than exploiting it recklessly. Through captivating anecdotes and insights, she illuminated the physiological impacts of nature on our bodies, from its restorative effects—evidenced by reduced stress levels and cognitive revitalisation—to the nurturing influence of biodiverse gardens on us and the environment.

Moreover, Stuart-Smith portrayed the garden as a microcosm of life's cycles, instilling a sense of hope and resilience amidst adversity. By embracing the transient yet enduring nature of existence, she suggested that gardening becomes an investment in the future, offering solace and symbolic survival in the face of challenges like depression or trauma. Through the lens of horticultural therapy, she elucidated how engaging with the land can provide a sanctuary for healing, fostering feelings of safety and empowerment. Furthermore, her discussion on the role of beauty and biodiversity in enhancing our emotional well-being underscored the transformative potential of even the most minor natural elements. Stuart-Smith's advocacy for creating sustainable communities with increased access to green spaces resonates as a vital step towards fostering social cohesion and holistic wellness.

Luciana Braga delivered the second talk on this panel. She brought us a thought-provoking and inspiring presentation on the Multispecies Project, which seamlessly complemented the themes explored by Stuart-Smith earlier. Braga likened the garden to a microcosm of infinity and Brazil itself to a garden, setting the stage for a profound exploration of architecture's role in shaping our evolving world. Central to her discourse was the imperative for architecture to embrace the rich tapestry of non-human spaces within contemporary cities. Braga provocatively challenged the dichotomy between ecology and social issues, advocating for a holistic approach recognising the agency of all human and non-human beings. Using methodologies such as cartography, narratives, and hands-on activities like workshops of different types, the Multispecies Project fosters cosmopolitical assemblies, inviting diverse perspectives and amplifying socioecological narratives. From fostering dialogue among students to engaging in extension activities that bridge academia with social movements, the project exemplifies a commitment to making environmental and social issues accessible and inclusive for all. Braga's talk provided valuable insights into the evolving field of architecture. However, it encouraged us to embrace diversity, foster empathy, and reimagine our cities as vibrant ecosystems where every human and non-human voice is heard and treasured.

Panel six, dedicated to 'Humanities, Socioecological Knowledge, and Citizen Science in Health', was moderated by Hélder Lopes and featured two presentations: one by Montserrat Llasat Botija and the second by Cynthia Luderer. These two highly captivating talks perfectly fit the panel's theme.

Llasat talked about climate resilience and challenges related to this issue, using the metropolitan area of Barcelona as an excellent example of a Mediterranean coastal megacity where climate has been affected: heavy rainfall and floods, extreme temperatures, and these impacts are heterogeneous and affected by multiple factors. Citizen science, the main point of Montse's talk, can be defined as scientific work made by citizens in collaboration or directed by scientists or scientific institutions. The FloodUp project (which comprises a dedicated App) exemplifies the different levels of citizen science, emphasising interdisciplinarity in the process. The Floodup project uses a participatory methodology and education activities. The aim is to collect information about the impacts of natural hazards and, very significantly, to improve knowledge and empowerment with diverse methodology tools. The question is: is citizen science a good idea? Is it good for citizens? Is it suitable for science? Llasat provided her answers but emphasised the need for all of us to reach our conclusions and highlighted that we still have many more things to do to improve the process.

Luderer, the second speaker on the panel, invited the audience to reflect on our place in the world by resorting to Cultural Semiotics as proposed by Lotman. She provided us with an example of such an analysis by using the song 'We are the world', concluding that we, as human beings, are monads, which are minuscule representatives of the bigger world that surrounds us. We all exist in a semiosphere, where everything is meaningful, and all our minor actions will impact the external sphere. Her last question, intended to make us wonder, was, 'In which semiosphere do I want to be included'?

The following panel, moderated by Rui Maia, reflected on the 'Transdisciplinary Humanities Education in Environment, Health and Well-Being' issue. The first speaker was Eduardo Missoni, who gave a most comprehensive and enlightening talk, waking the audience through the many determinants impacting health and emphasising the multifaceted role of education. From structural influences to intermediary factors, this talk delved into the various forms of education—formal, informal, and non-formal—and their profound implications for health outcomes. Shedding light on the prevailing focus on formal education within literature, the presentation underscored its pivotal role in determining access to resources while highlighting the often-overlooked significance of informal and non-formal education pathways. There were insights into theories like Functional, Human Capital, and Signalling, singling out the complex interplay between educational attainments, individual factors, and broader social determinants of health. Missoni's talk advocated for a paradigm shift towards education for health across disciplines. Considering the connection between health literacy, health promotion, and education, Missoni foregrounded the need for an inclusive and equitable approach to education that empowers individuals and communities alike. From early interventions addressing inequalities from birth to encouraging a culture of health literacy and ethical awareness, the talk envisioned a transformative utopia where health permeates every facet of society, requiring that ethics, social justice, and equity in all educational systems should deeply, in order to reach a holistic, inclusive, and ultimately healthier future for all. "Health for all and in all ages" This sentence summarised the need for a more enlightened and compassionate society.

Nelson Barros delivered the second talk on this panel, shedding light on the critical challenges we face in our quest for sustainability and global well-being. As Barros pointed out, with our current trajectory, we are rapidly depleting natural resources, and while renewable energies offer promise, they need help to meet our escalating demands. The United Nations' Sustainable Development Goals (SDGs) for 2030 are a beacon of hope for a more equitable and sustainable future across social, environmental, and economic fronts. From ensuring clean water and affordable energy to fostering sustainable cities and responsible consumption, the SDGs encompass a multifaceted approach to tackle our pressing global issues. However, amidst these noble aspirations, there is a glaring omission: the issue of air quality, a silent yet deadly menace impacting billions worldwide. As air pollution continues to rise, so does its toll on human health, with millions facing premature deaths annually. While some SDGs tangentially touch upon air quality, its absence as a standalone goal highlights a critical gap in our collective efforts. Indeed, as we strive towards the SDGs, we must confront the stark reality that our current trajectory may lead us astray. We cannot overstate the urgency of addressing air quality, as it is a silent killer undermining our progress towards a sustainable future. Despite the ambitious targets set for 2030, the rising death tolls from air pollution are threatening us, calling for immediate action and a re-evaluation of our daily choices. From advocating for cleaner energy sources to fostering mindful consumption habits, each action can make a difference, contributing to a healthier and more sustainable world. As the speaker aptly suggested, let us embrace the simple act of taking the stairs over escalators—a metaphorical call to action, as Barros pointed out, urging us all to step up in our collective journey towards a brighter, cleaner, healthier future for generations to come.

The last panel, titled 'Cultures, Natures, Arts and Health: The Role of Humanities in Global Societies', was moderated by Eduardo Paz Barroso and featured three presentations. Gabriela Albergaria delivered the first presentation by sharing a few images with the audience and contextualising her academic background regarding the human-nature relationship. Albergaria thoroughly analysed the selected images, representing how she portrays diverse aspects of this relationship. During her presentation, she shared several interpretative paths and walked the audience through the 'making of' processes of several of her artworks, which involve gathering materials from natural environments. It was a very enlightening explanation about the thought processes of an artist, what her aims are when creating a specific piece and how artworks come into being. She showcased her interest in agricultural practices, organic natural processes, the relationship between the materials and the landscape they originate from, and how she deconstructs all these elements throughout her creative elaborations.

Lia Chaia, another artist, also provided valuable insights into her thought processes and creative perspectives, focusing on the contrasts between natural and urban spaces and her feelings of being torn between her life in urban space and the will to be a part of the natural world. Her artistic options (in different media) illustrate these ways of seeing, such as civilisation converging to a single sport, scale-altering experiments, artistic performances, deconstructions of the human body (for instance, 'The Vegetable Skeleton'), representations of a feminine entity that controls the whole experience, as well as ironic and critical views on urban (let us call them 'mistakes') that degrade the urban landscape.

Caterina Cuccinotta brought us a detailed comment on the film "The Movement of Things" by Portuguese film director Manuela Serra, whose themes Cuccinotta connected with the overall purposes of this Symposium. This film has been rediscovered by critics and portrays the interest of Portuguese cinema in countryside subjects in the 70s. The first part of the film focuses on the issue of eco-feminism. Cuccinotta pointed out several examples throughout the film representing a matriarchal cosmogony, highlighting the value of feminine perspectives on reality. This film does not have much dialogue—these women do not speak much. As Cuccinotta highlighted during her analysis, feminism acknowledges the fundamental importance of nature. The slow cinema proposes contemplative narratives and the depiction of everyday life. As Cuccinotta pointed out, this is a non-conventional film whose narrative strategy tries to grasp a more intuitive relationship between human beings and nature .

Overall, this was a most productive Symposium, with many motivating and thought-provoking talks. This volume corresponds to a selection of the texts derived from these two intense but highly enjoyable presentations and lively debates on several of the most pressing issues nowadays.

# MULTISPECIES PROJECT: A Multispecies Approach to Communities in Urban Green Spaces in Times of the Anthropocene

**Luciana Souza Bragança**

**Abstract** The aim of the article is to present the methodological hypothesis of multispecies projects contributing to understanding how to compose ecological worlds, outlining a strategy for a more sustainable future for urban spaces based on quotidian worlds in Brazilian cities. It was developed at UFMG Architecture Scholl by the research Possible Gardens, extension and teaching. Some of the research conclusions were mobilized: relationship between humans and non-humans; socioecological narratives, practices and imaginaries as well as ecology and decolonization. The contributions of agent's cosmopolitical practices will be pointed out with the aim of making explicit the possibility of overcoming the anthropocentric character of our way of composing worlds. The main proposal is to widen the political circle and include non-humans in it. Based on the understanding of multispecies anthropology and field research, the methodological hypotheses were constructed: considering non-humans as project subjects; the use of "what if?" trigger questions to broaden imaginaries; the creation of cosmopolitical assemblies. These hypotheses were tested in some disciplines. The conclusions are preliminary and point to the proposal utopian nature. However, the didactic exercises broadened the students' socio-ecological imaginary and cosmopoliticals circle towards decolonisation of ecology and design practices. That can inspire transformative changes in policy-making.

**Keywords** Multispecies project · Cosmopolitics · Possible gardens · Ecological worlds · Anthropocene

## 1 Introduction

Architecture, urbanism and landscaping operate based on anthropocentric approaches to composing worlds. In this way, non-human agents, who are fundamental to the everyday construction of cities, are disregarded and seen as enemies to

L. S. Bragança (✉)
Escola de Arquitetura, Universidade Federal de Minas Gerais EA-UFMG, Belo Horizonte, Brazil
e-mail: lubraganca4@ufmg.br



W. Leal Filho et al. (eds.), *Composing Worlds: Humanities, Health and Wellbeing in the XXI Century Towards a More Sustainable World*, World Sustainability Series,
https://doi.org/10.1007/978-3-031-87108-5_2

be fought or objects to be exploited. This disconnection between the various agents in the way we learn and think how to make cities represents a socio-ecological challenge as well as an issue to these disciplinary fields. Therefore, an approach that takes into account the respect for people, nature and the planet as central to spatial proposals has great potential to contribute positively to tackling these challenges.

The purpose of this article is to present the methodological hypothesis of multispecies projects developed in the researches *Possible Gardens*[1] and *Possible Gardens and Waters in the City* and in the extension program *Political Nature* at the UFMG School of Architecture: a multispecies approach to communities in urban spaces. The aim of proposing multispecies projects is to contribute to understanding how to compose ecological worlds, outlining a strategy for a more sustainable future for urban spaces based on everyday worlds in Brazilian cities.

It is based on the understanding of gardens as free urban spaces that, together, comprise a landscape whose essence is the coexistence between different beings. This research was based on river territories and structured around case studies of multiple gardens along the water basin of *Ribeirão Arrudas*, the main water stream in Belo Horizonte, Brazil. Actual examples of open spaces, urban gardens where humans and non-humans interact directly, were examined including the implications built between these living beings. The aim of the research is to understand the territorialized multispecies worlds and how they arise, making everyday narratives perceptible, possible (Berardi 2019), as well as invisibilized memories that have the potential to expand imaginaries. This expansion of imaginaries is essential for the practice of spatial production.

Based on a multispecies cartography, participant observation, ethnographic interviews and signposting of the territory, multispecies relations were highlighted as a way of understanding how would the multispecies landscape of possible gardens come to be, in terms of its practices and contributions to the ecology in cities (Bragança 2023a, b). A series of ideas about the relations between gardens, water, narratives and social imaginaries, ecology and decolonization were mobilized to achieve one of the main research's contributions to the projects: overcoming the anthropocentric nature of our way of acting and of composing worlds.

In undergraduate disciplines and in the extension program, the multispecies project hypothesis was carried according to the research's presumptions. This discussion stems from questions mainly related to designing and planning in cities and to the consideration of the agencies of other beings shaping everyday spaces. These hypotheses were also expanded by these disciplines' results as a methodological speculation of what could a multispecies projects be. We will present some of these actions that helped strengthening this methodological hypothesis of a multispecies project and, especially, its presumptions.

## Cities as multispecies relational *loci*

[1] The Possible Gardens research project was funded by the Federal University of Minas Gerais (UFMG) Pro-Rectory for Research: project 23,853—UFMG/REITORIA/PRPQ/PESQUISA E EXTENSÃO, subproject 407, coordinator: Luciana Souza Bragança.

The origins of human agglomerations, particularly cities which spread across the planet, are intimately dependent with rivers and multispecies relations. In order to settle in a territory, humans grew food and raised animals near watercourses. The first human settlements originated in the Fertile Crescent, region between the basins of Tigris and Euphrates Rivers. Without them, developing the agriculture that allowed the formation of those first cities would be impossible (Pinsky 2001). The first gardens emerged along with agriculture, as sacred and delight spaces, gardens as a representation of a world, a cosmology. At that time, the process of domesticating or familiarizing plants and animals and the human interference in hydrological cycles had begun, which historically marks the beginnings of our Western civilization. For some animals and plants, this was a beneficial relation and some species were privileged. However, non-human agents, including water, animals and plants, have been much neglected by the humans-built cities.

With the development of modern Western cities and the production of urban space, multispecies relations were disregarded and nature was seen as an enemy to be tamed (Herzog 2013) or as a resource to be exploited, as a passive resource without agency. Nowadays, nature is once again remembered but mainly when related to disasters, which are actually results of human actions upon such important agents. The emerging narrative is that of a "villain" to be fought. Thus, the memory of multispecies worlds and the presence of rivers, plants and animals are associated with urban problems and erased from cities. In Latin American context, the coloniality reinforces this understanding (Mignolo 2003). By erasing the non-hegemonic cosmoperceptions (Oyěwùmí 2016) as black and indigenous ones or their everyday narratives, city worlds have become increasingly anthropocentric. Projects based on domination and exploitation reinforce this imaginary.

How could architecture contribute?

Architecture, urbanism and landscaping play a role in the construction of city spaces, mainly through designing. Thinking about the incorporation of other worlds through multispecies projects is therefore a hypothesis. But for non-hegemonic ways of knowing and doing and for new agents to be incorporated into design proposals, epistemic alternatives need to be speculated and proposed as alternative paradigms, seeking to break with the hegemonic logic—hegemonic not only in terms of the understanding of global south and north, but also in terms of its anthropocentric beginnings and actions. As a way of overcoming this anthropocentrism, the following question emerges: how can we propose spaces for non-humans in contemporary cities?

The combination of teaching, research and extension projects in architecture schools play an important role in this scenario by possibility of training future professionals to be attentive to multispecies, creating opportunities for actions based on other design approaches. The proposal here is precisely to broaden these perspectives, discussing examples of how this contribution can happen and stimulating debates.

During the research (Bragança 2023a, b) several possible gardens were found, where humans and non-humans shape and build their own spaces. This takes place in a double affectation (Stengers 2018), involving multiple and less hierarchical agencies, without a human preponderance. While constructing their spaces, these communities build community territories with various agents. As disciplinary practices, architecture, urbanism and landscaping design spaces from an understanding of an ongoing world in the territory. Thus, from everyday worlds they have the potential to help composing multispecies worlds, by outlining a strategy for a more sustainable future. To this end, we need to review design methodologies. This is how the hypothesis of multispecies projects arises.

As a way of experimenting and testing with this hypothesis, undergraduate disciplines have been proposed to understand and explore the possibility of extending subjectivity and forms of otherness to other beings. Students started their projects with a premise that it was necessary overcoming the anthropocentric nature of spaces in order for them to be truly ecological.

These methodological didactic experiments were developed at the Federal University of Minas Gerais (UFMG) Architecture School, critically proposing projects that could contribute to ecology by placing water, animals and plants as central elements based on their agency–and not just "at the service" of man. They explore the multispecies perspective and extend the right to the city to other beings, encouraging students reflection on how to design more environmentally adequate cities for everyone–preparing them, as future professionals, to cope with environmental collapse[2].

## 2 Open Spaces, Possible Gardens and Anthropocentrism

In the Western world, the idea of a nature that is objective and external to the human being, as well as the notion of human as something non-natural, began in the medieval era and intensified with industrialization. This dualism plays a crucial role in sustaining the idea of a human superiority over other forms of life.

In the modern perspective, nature is often perceived as something opposed to culture and human beings (Latour 2020). Culture is considered superior and exerts control over nature. Humans and other forms of life are placed in a hierarchical relation with humans at the top. However, it's important to note that only a few humans can play this role. This grounds the anthropocentrism of our social and territorial imaginary and, specifically, of our understanding and designing of spaces.

With scientific development in the Renaissance, nature began to be defined as everything which was not human and seen as an entity to be exploited. Also during this period, the colonization process intensified, obliterating other ways of understanding the world, in which the idea of nature as a category of exclusion was meaningless.

This process of subjugation, objectification and segregation of others is fundamental to understanding the colonial process in the Americas. The appropriation of

[2] Socio-environmental and climate crises are realities in cities around the world. Latour (2020) and Stengers (2018) reject the word "crisis" thogh, as it gives the impression of transience, and use "collapse" instead. Accordingly, the article will adopt the term.

territories and resources combined with the manipulation of subjectivities resulted in a form of destruction of the common good and in an anthropocentric rational perspective on life. In Brazil, the colonization process began with the Portuguese perplexity with the dense forest, which limited their incursions. Territorial occupation started with the constant deforestation. In this context, the construction of cities was intrinsically linked to an opposition to natural landscape. This clash between colonial men and wild nature, with transformation of natural landscape and exploitation of natural "resources" were at the heart of the colonial enterprise. The construction of cities and territorial expansion also involved the imposition of European culture over local ontologies and on non-humans, leading to the degradation of original ecosystems.

The development of modern Brazilian cities amplified these practices, as well as expanded the disregard of hydrological cycles in the production of urban spaces, resulting in a narrative of water as a threat to be controlled and tamed, or as a resource to be exploited (Herzog 2013). Animals were domesticated, imprisoned in zoos as exotic attractions or fought as vectors for diseases. Some plants were hyper-domesticated and reproduced, competing with endogenous species.

Today, for example, water is mainly remembered when associated to disasters—ultimately caused by human actions. This way, in Brazilian cities, the memory and the presence of rivers, of animals as companions and of sacred plants have been erased and narratives now associate them with urban problems. While the public's perception of rivers is shaped by this narrative, an understanding of their intrinsic value for urban life and for imagining other worlds is lost. This gap between urban approaches and the importance of water cycles can lead to environmental disasters and to a harmful disconnection between people and water environments (Bragança 2023a, b). This disconnection is also true in relation to other non-human beings.

This is an intrinsic component of modernity's civilization project and represents a method for exercising power. It is a structure built on the naturalization of hierarchies such as territorial, epistemic, cultural and also on hierarchies among species, maintaining their influence in contemporary society and deepening inequalities. The colonial process solidifies these hierarchies as unquestionable, resulting in the continuous reproduction of asymmetrical power relations.

This limited understanding of the term nature often extends to urban planning and landscape architecture, where other worlds, other ways of living and conceiving spaces are marginalized, amplifying dichotomies. The design of formal cities is usually based on hegemonic narratives, which perpetuate anthropocentric hierarchies of power and the reproduction of these structures over time in all spheres of social life. Mignolo's idea of "an-other paradigm" (2003) is central to understand non-hegemonic urban territories in a new way. This allows us to incorporate the diversity of agencies, challenging the anthropocentric homogenizing model. By valuing alterity, i.e. the difference between cultures, including those of non-humans, we can try to overcome this dichotomy and promote convergence. Adopting an integrative approach can help addressing social and environmental challenges.

Furthermore, for the author, modern coloniality was not inscribed in an empty space of meaning. In this sense, the research made it possible to map other territorialized worlds in cities, despite this power structure. Gardens were unveiled and understood in the research as multispecies artifacts and relational spaces, with no rigid distinction between nature and culture in these free urban spaces. This approach breaks with naturalism, challenging the belief that there is a "pure" nature separate from human activities. By adopting this perception, gardens become places where beings intertwine in a network of complex interactions, where human and non-human agents create cosmopolitical worlds (Bragança 2023a).

The multispecies existence and the cosmopolitical power of Possible Gardens were mapped in the research to understand what these multispecies worlds are and how they occur in the city, making their invisibilized everyday narratives apparent. This was based on the territories of water and on the understanding of watersheds as being essential for designing urban spaces, using the following methodology: the historic understanding of gardens, the initial proposition of the term Possible Garden, a comparative case study of multiple gardens, the analysis of data, the meaning reconstruction of the Possible Garden based on the field research findings and its contribution to urban thinking (Bragança 2023b). 900 quantitative interviews were made as well as ethnographic interviews based on participant observation and photographic surveys. The sample for the ethnographic interviews was proportional and stratified by relief units.

Brazilian history only legitimates European-inspired gardens, emphasizing the intention of creating a modern image for the country. In this sense, civility was needed to polish local roughness, and. the polish was white, European, while the roughness was black, native and mestizo (Sodré 1999). However, the typical Brazilian colonial garden was a mix of backyard, vegetable garden, orchard and flower garden and always incorporated water (Magalhães 2014). The division that separated garden, yard and house came with modernization and even then it didn't completely dominate the spaces. The possible gardens in this research are a memory of those colonial gardens that existed previously and still exist nowadays throughout Brazil—they survive and correspond to the majority of gardens. They are multispecies everyday spaces that also blend indigenous and black knowledge about water, flora and fauna (Bragança 2023a).

European colonization also brought the Enlightenment thinking, the belief in human superiority over other beings and, later, in European superiority over other ontologies. The colonizing process subtracted the agency of non-humans, who were displaced into the field of nature as an object to be dominated by human subjects. But this agency didn't cease existing and still manifests itself. The research made us aware of the agencies of non-humans and how they are incorporated into the construction of possible garden spaces in a less hierarchical and conflictual way.

The gardens researched, as urban open spaces, play a significant role in building relationships between different beings, human and non-human and urban waters. They can be considered microcosms of the Living Earth, multispecies territories where diverse living beings coexist and interact (Bragança 2023a, b). However, the idea of nature as something separate from humans, rooted in modernity and colonialism, influences the way these open spaces, particularly gardens, are perceived and designed: from an anthropocentric point of view. This approach distances people from other forms of life and from the territories of water.

This understanding of gardens has profound implications for the ecological debate, for the understanding and design of landscape. It suggests a more interconnected comprehension of ecology, considering non-humans—plants, water, animals—as agents rather than things, as subjects in projects and, therefore, of multispecies. Thus, it is not only a matter of preserving isolated natural spaces but of recognizing interactions in urban environments and the agency of other beings. The approach centered on multispecies also broadens the notion of allies in the ecological struggle and supports cosmopolitical solutions (Bragança 2023a) to the environmental challenges we face.

By associating memory, humans and non-humans are incorporated into everyday narratives—the different memory landscapes built by women and men, plants, animals, water, land, their emotions and imaginations, their organic and inorganic constitution. The river and the forest, taken away by urbanization, returns into imagination and vocabulary. People produce their space by mobilizing important knowledge and building cosmopolitical worlds, just like non-humans do.

Cities are spaces of disputes over the future and narratives. Although architecture and urbanism haven't the capacity to effectively define spaces, they do involve a relation of power and politics. This occurs since, when designing, elements to be represented are chosen. Signs are displaced and created, details are omitted, conventions are created and spaces are reorganized.

## 3 Multispecies: How Can This Approach Contribute?

Multispecies studies, also known as "multispecies anthropology" or the "multispecies turn", is an approach in anthropology that explores the complex and interdependent interactions between humans and non-human life forms. Rather than focusing solely on human interactions, multispecies studies recognize that humans coexist and relate to a wide variety of beings, including animals, plants and other organisms. These interactions can include everyday practices, cultural beliefs, knowledge systems, technologies and the construction of spaces and landscapes. This approach challenges the prevailing anthropocentric view, which places humans at the center of all issues and neglects other forms of life and other agents such as water. By exploring the complex interconnections between humans and non-humans, multispecies studies contribute to a broader understanding of social, cultural and environmental dynamics and consequently allow other possibilities for projects.

Multispecies studies have their origins in the questioning of the dichotomy between nature and culture, mainly developed by authors Latour and Ingold. According to Latour (2020), both nature and society should be interpreted within a common framework, rather than treated separately. It makes no sense considering a world of things and a world of humans independently. Both are interconnected in heterogeneous networks that can be analyzed in a similar way. The social aspect is thus understood as a heterogeneous network of both human and non-human agents.

Ingold (1995) ethnographies focus on relations between humans and animals. He criticizes the duality between nature and culture, characteristic of the Western tradition, which tends to generate dichotomies. He argues that variations in skills and cultures are neither innate nor acquired, but developed through practices in the environment. He proposes that all organisms are different forms of "persons"—so there are human and non-human persons. An important difference is that Latour includes objects and inanimate elements in his notion of non-humans, whereas Ingold doesn't. The proposal developed for multispecies projects also considers the agency of inanimate beings, especially water.

The perspective of animal rights advocates, such as Perrota, and Amerindian perspectivism, discussed by Castro and Descola, are significant contributions to the evolution of the relationship between humans and non-humans within the field of anthropology. These approaches helped paving the way for multispecies studies by questioning the traditional divisions between nature and culture, recognizing the complexity of interactions between different species and highlighting non-hegemonic cosmoperceptions.

Based on these reflections, the multispecies approach emphasizes that the natural sciences are not the only way to understand life. This approach studies the relationships formed between individuals of different biological species and inanimate beings, going beyond the limits of traditional academic disciplines. It incorporates concepts from social sciences to explore these complex interactions. The multispecies approach explores an ethnographic immersion in the ways of life of all forms of beings: animals, plants and fungi (Tsing 2019). It encompasses the relations between different species coexisting in the same environment. These studies examine how these relations intertwine with human social worlds, observing the effects that these interactions cause and how they are influenced in the political, social and cultural spheres. This approach challenges the notion that humans are isolated from the rest of life and inanimate beings, recognizing the complexity of the networks of interactions that connect all forms of life, questioning anthropocentrism.

Our way of understanding the landscape and thinking about projects is usually interconnected with anthropocentric humanism, hegemonic narratives and the market. Realizing this is an incomplete understanding is the starting point for the hypothesis of multispecies projects, to more ecological worlds that respect all forms of life. Then, crucial questions emerge on the practice of architecture and urbanism: How can teaching contribute to the consideration of the agencies of other beings, broadening the imaginaries of worlds in projects? What could multispecies projects look like?

## 4 Multispecies Project

Our proposal is to speculate how architecture, urbanism and landscaping can contribute to composing ecological worlds, outlining a strategy for a more sustainable future for urban spaces based on everyday worlds in Brazilian cities and understanding the role of architecture in this process. A multispecies approach to projects, based on the communities of possible gardens found in urban open spaces is one way of responding to the challenge of how to consider other subjects and agencies.

As a methodological speculation of what a multispecies project could be, we will present some didactic experiments. Far from being conclusive experiences, both the disciplines and methodological hypotheses aim at challenging and broadening the hegemonic narratives and the discussion. Rather than providing solutions, they help building other questions.

The experiments developed take a critical approach to projects by proposing the consideration of animals, plants, water and its hydrological cycles as agents in the territory. This approach differs from the traditional one by recognizing their agencies in the environment. Inspired by the multispecies perspective and by field discoveries during research, these methodological experiments extend the right to the city to other forms of life. They invite students to consider interdependence between humans, water, plants, animals and other species as a principle for design. By exploring the relations connecting humans and non-humans, these experiments encourage students to design cities which are more environmentally adequate for all forms of life.

### *4.1 Methodological Principles of Multispecies Projects*

By adopting a multispecies perspective, these methodological experiments challenge students to rethink the role of water and beings other than humans in urban design, their specific needs and contributions. This helps promoting multispecies coexistence in design, rather than seeing other beings as natural resources or problems to be fought. We will present five disciplines which have shared methodological principles, in addition to two tools: the *Inventa(rio) Cidade*[3] game and the multispecies indicators developed by the Political Nature research group.

The first methodological principle is the approach of non-humans as project subjects rather than project objects, recognizing and respecting their agencies. These agencies are present in everyday spaces of gardens and are therefore accessible and perceptible. This approach is crucial for environmental issues, as it overcomes a traditionally anthropocentric one. When it comes to water, the projects understand its agency from related verbs: starting, flowing, infiltrating, evaporating, carrying, filling, drying, draining etc.; and also from its specific territories: the concave and convex surfaces, the floodplain, the riverbed, the hilltop, the watershed and the spring.

[3] The game's name is a wordplay in Portuguese, with the present tense of the verb "to invent" (inventa) plus the word "river" (rio) whitch together sound like inventory (inventário).

Research has shown that these water territories build different relations and understandings with humans (Bragança 2023a). Animals are recognized mainly for what they seek: feeding, sheltering and reproducing. So they need safe spaces in which they can move and live. For plants, their reproduction, territory and interdependence with water and animals, especially pollinators and seed dispersers, and their collaborative ecological relations, both intraspecific and interspecific, drive the design. By considering water, plants and animals as active subjects and respecting their territories and characteristics, projects can become more sensitive to natural dynamics and create urban spaces that are more aligned with the needs and contributions of all beings involved.

The second tool is the use of trigger questions that must be answered spatially. They are based on counterfactual thinking (Senos 2008), which is imaginative thinking characterized by "what if" conditional sentences that provoke hypothetical alternatives to reality, for extrapolating possibilities and thinking outside the box. This exercise attempts to broaden narratives and design repertoires, breaking the initial link with pre-established conceptual and spatial patterns. If thinking about spaces with other beings presents a methodological challenge, tensioning consensus can be an answer and exploring provocative "what if?" questions can be an example.

The third one is the promotion of cosmopolitical assemblies (Bragança 2023a) among different beings. Stengers' (2018) cosmopolitical proposal is adopted to think about what kind of politics we can build when we expand our understanding of politics beyond conventional elements such as representation, language and consensus. This approach questions what kinds of divergences are being masked by these elements and how we can include non-humans as active agents in the political sphere. The design exercise summoned students to propose a city for water, a city for plants, another for animals and one for humans. Then, negotiation assemblies were set to define the final proposals, taking everyone into account and perceiving the emerging disputes. Thus, in the representative assemblies, a project can be discussed in which each student represents an agent: water, plants, humans, animals, soil, winds.

The last instrument is presenting the research and discussing the concepts that ground this approach: possible garden, multispecies, cosmopolitics, rights granted to non-humans, sustainability and urban ecology.

Behind this investigation of spatial proposals, some tools have been discussed: urban legislation, environmental legislation, urban planning and, above all, urban design and landscape design. It is worth remembering, far from being a conclusive experience, both the disciplines and the methodological hypotheses raised aimed at broadening narratives and creating discussion. Rather than providing solutions, they help building other questions such as: Is it possible to think the "city for all" as signaled in the gardens? How can architecture and urbanism contribute to environmental issues beyond their anthropocentric practices? Another central point for the disciplines was the awareness that it is also important understanding, as a future architect and urban planner, what not to do and when not to do it, creating space, both conceptual and propositional, for other agencies—especially with regard to nature in the city.

## 4.2 Didactic Experiments

As previously stated, methodological didactic exercises were undertaken to speculate on the possibility of multispecies projects: PRJ087—How to think about nature and the city? (2018, 2019); PRJ096—What if it were for plants? (2021); PRJ096—What if it were for water? (2022); PRJ082—A Multispecies approach to rivers: analysis and application of indicators for urban river banks (2023); PRJ082-PLB—Multispecies project for urban rivers (2024); ARQ040—Integrated planning for architecture and urbanism (from 2018 to 2024).

The discipline ARQ040 (120 h) is compulsory for the 7th semester of the course, taught by professors from both the urbanism and the architecture departments with focus on the integration of urban plans and buildings design. It deals with the implementation of a new metro line in Belo Horizonte and explores the relations between urbanization, landscape and environment. Based on the agency of water, students discussed presumptions on urban drainage, previously presented in thematic classes. They were invited to explore and interpose the term "natural catastrophes", used for urban floods, understanding how these floods relate to urbanization choices–e.g. paving over and covering streams–and to a reaction of water to these choices.

The discipline explored the cycles of rivers and encouraged more relational and less imposing design approaches on water, discussing alternative ways of relating to these cycles by analogous projects. Students were encouraged to take a critical look at some classic tools of urban planning and design, such as diagnostics. The resulting projects, often involved drainage plans, trying to create dialogues between urbanization and water. Although familiar design elements were presented, they are spatialized in ways that broadened this dialogue (Fig. 1). The multispecies approach was not central to the discussion; however, it widened and qualified the discussed repertoire of spatial solutions.

PRJ087 (60 h) was the first multispecies project discipline to be actually proposed. It was conceived from a cosmopolitical perspective to incorporate other agents into the design process, based on the idea of memory and multispecies of the Possible Gardens. The discipline explored tools of architecture and urbanism, such as urban and environmental legislation and, above all, urban design and landscaping design. It considered not only humans, but also non-human agents such as plants, animals, water, relief, climate, wind and solar insolation as clients for the proposed projects.

In the first exercise, students were divided into five groups and mapped the project area, each focusing on one of these agents. They used interviews, research data and collective mapping produced with engaged residents (Fig. 2). After mapping, students were challenged to formulate design hypotheses by answering the questions: What if it was for the water? What if it were for animals? What if it were for plants? What if it were for the sun, relief, soil, wind? What if it were for people? It was a premise of the proposals to understand that they were inserted in the city and that the city should be part of the presented projects, so it wasn't possible to totally subtract humans and their spaces.

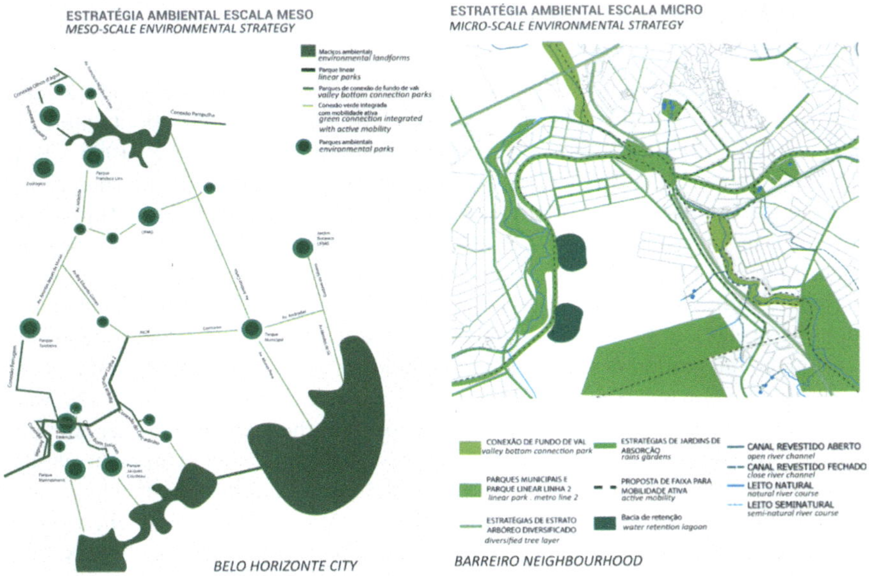


**Fig. 1** Socio-environmental plan, ARQ040 (translated by author)

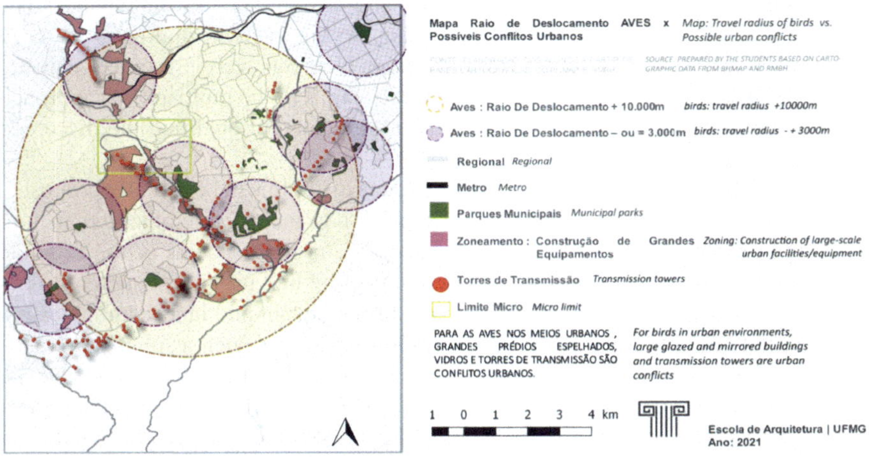


**Fig. 2** Bird movement and urban conflicts cartography (translated by author)

In the second exercise, four new groups were created, each one counting with students from all previous groups, and their task was to develop new proposals based on the previous hypotheses. The aim was to establish a "cosmopolitical assembly" between the previous proposals, taking into account the worlds discovered, with each student representing one of these worlds. Achieving unification was not their goal but rather building a space for interaction and negotiation, making conflicts evident. Students were encouraged to consider spatial decisions based on territorial disputes between wills and possibilities of each subject, incorporating the intertwining of these agents and disputes—something very present in the gardens surveyed.

The third discipline, PRJ096 (30 h), was an elective one and took place over two semesters, focusing on separate agents: "What if it were for plants?" and "What if it were for water?" It derived from the previous discipline but with an exclusive focus on the relations between plants and humans and between water and humans (Fig. 4).

The experience of previous disciplines and the extension work of the Political Nature group, as well as participation in the research network RUN Cyted—Naturalized Urban Rivers, brought the necessity of creating criteria which allowed specific situations to be compared and quantified. Indicators are indispensable tools for guiding, evaluating and monitoring cities' performance towards sustainability. They are also relevant tools for evaluating, proposing and negotiating projects and public policies. They can be a way of choosing and prioritizing investment areas and guiding urban planning. This raises important challenges: how can quantitative arguments be approached using mathematics that takes into account both the immediate effects, such as the costs of implementing a given spatial intervention, and the medium and long-term gains, such as environmental gains, which affect the life of territories and the city as a whole?

Based on indicators produced by the LlactaLAB research group at the University of Cuenca, Political Nature developed Multispecies Indicators for public spaces on urban rivers. These indicators are based on concepts of connectivity and comfort, which are broken down into different attributes, to evaluate and design the banks of urban rivers. Bringing these indicators to Brazil required a translation from the urban context of Cuenca to that of Belo Horizonte, as the characteristics of the rivers are quite different. The group also proposed extending this numerical evaluation from the human point of view, as originally proposed for Ecuador, to include the agency of non-humans, thus creating specific indicators for multispecies. In Brazil, the territory of analysis was extended to the watershed. The assessment of basic sanitation and waste collection was also included. In order to define the data and variables used, it was necessary to observe the complexity of the urban environment and encompass the various dimensions and specificities that compose it, the multiple agencies of human and non-human actors in these territories. Thus, the developed proposal also incorporated the multispecies approach quantitatively.

The fourth discipline, PRJ082 (60 h), investigated the contribution of these indicators to landscape design beyond the dichotomy between qualitative and quantitative approaches, incorporating the concept of multispecies. In the first exercise, in addition to mapping the studied territory, students applied the indicators (Fig. 3). Based on the qualitative analysis provided by the cartography and the quantitative analysis provided by the indicators, they produced project guidelines. After developing their projects, students reapplied the indicators as a way of reflecting on how design actions quantitatively affect the quality of spaces for all beings.

With the multispecies indicators, the group of researchers involved in the epistemological and methodological debate created a game to complement the

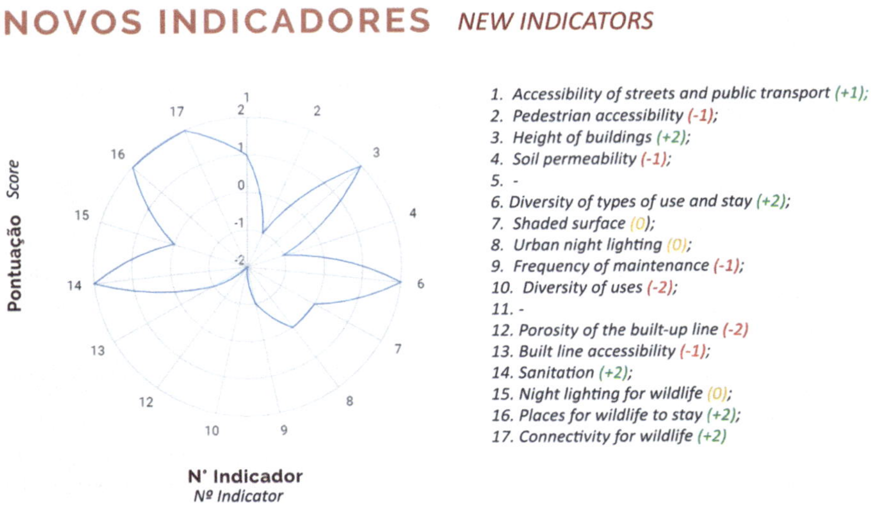


**Fig. 3** Multispecies indicators application

**Fig. 4** *Inventa(rio) Cidades* game

proposal. The main objective of the game *Inventa(rio) Cidades* is to build a clearer understanding of how urban projects interfere in each player's experience, either contributing or disturbing it. This is done through "construction" and "situation" cards. This is a game of empathy where human and non-human subjects are the players in a city articulated by the presence of a river and riddled by situations that affect each player's life. Each player takes one of the following characters: flying animals, climbing animals, water, plants, wind, humans. The player's progress is regulated by dice and the "situation" and "construction" cards. When playing as water, for instance, drawing a "construction card", which represents turning an entire plot impervious, takes the player back several spaces.

The game was played in three different dynamics before being used in the fifth discipline. The first one was during *Mostra* UFMG, an exhibition for high school students interested in learning about undergraduate courses offered by the university. The second one was also with high school students, but this time at a private school in the city, whose pedagogical project is based on constructivism and encourages students autonomy. Finally, it was played with researchers from Cuenca, whose indicators were references for the game. In each dynamic, the participation of members was playful and at the same time provocative, given that players' perception was that today's cities are not friendly to non-humans, much less to water.

The fifth course, PRJ082-PLB (60 h), has a very similar dynamic to PRJ087: a project group for each of the agents, a cosmopolitical assembly and a final project with a member representing each of the previous groups. In the first stage, however, the game (Fig. 4) was used as a tool for understanding the agents. It used the same territory of PRJ082 in order to promote a comparison of project results based on qualitative and quantitative analysis.

The disciplines explored the possibility of thinking on a "city for all", as suggested by the researched gardens. During the project experiences, a dichotomy often emerged between the right to the city for humans and environmental struggles, as well as the right to the city for other beings. These rivalries created conflicts that reinforced the existing problems instead of widening the political circle, as originally proposed. This exclusionary antagonism resulted in rivalries between potentially allies on similar urban issues, weakening both sides. However, *Inventa(rio)* has been a powerful tool for promoting empathy between different beings.

Giving voices to water and to other non-human agents and enabling their participation in the process was a big challenge. In the disciplines, this was achieved through methods such as research, collective mapping, indicators, assemblies and representation through a human. These approaches sought to include the perspective of all agents, thus broadening the understanding of the city and its complex relations for what could be a multispecies project.

Summarizing, we used cosmopolitical assemblies, games and mapping as tools for testing the possibility of multispecies projects. As we understand it, a multispecies project requires:

- Expanding imaginaries;

- Mapping the territory and understanding how the various subjects exist and act, discovering their needs and challenges;
- Creating measuring instruments that enable valuing and comparing different contexts and supporting public policies based on multispecies;
- Creating instruments for mutual understanding and empathy between agents;
- Promoting assemblies that explicit conflicts and convergences.

## 5 Final Considerations: Advances and Challenges of Multispecies Projects

Composing more relational worlds and safeguarding a memory of multispecies practices represents an ecological perspective, conceived from an interest in recognizing, preserving and incorporating everyday practices of coexistence between nature and built environment, between humans, non-humans and watercourses. These spaces contribute to a healthier city and, consequently, healthier people in a broader sense since these practices create the possibility of urban green community territories, which support ideas of commonality and inclusion. As a contribution to architecture, multispecies projects are an important tool.

The triad of teaching, research and extension developed in Brazilian public universities is fundamental. By incorporating multispecies approaches and considering non-humans as central to design decisions, students are prepared to deal with complex topics related to managing nature and water in cities, based on non-hegemonic perceptions.

By making other relational worlds perceptible, such as territorialized multispecies collectives, and giving them visibility, the Possible Gardens research expanded imaginaries. The existence of environmentally relevant everyday spaces shows this is an everyday practice with transformative potential. For these worlds, designing spaces which are interconnected with anthropocentric humanism or the market isn't good enough.

These conclusions are still preliminary and reveal our proposal´s utopian nature. However, the didactic exercises have broadened students' socio-ecological imaginary and cosmopolitical circle towards a decolonization of ecology and design practices. Students accomplished the challenge of thinking about cosmopolitics in experiments of political and spatial reinvention which recognize multispecies collectives. Tensioning hegemonic anthropocentric imaginaries and narratives is the experience´s main advance. This can inspire transformative changes in the development of projects and public policies.

The methodological hypothesis is that in order for non-humans to have their otherness included in projects, it is necessary to create assemblies in the discussion of urban proposals and mechanisms—yet to be developed—for the participation of other beings in such disputes. We must be aware that the assemblies are still human ones and that other beings are, at best, being represented. Another key point is the understanding that it is also important knowing what not to do, making room for non-humans.

We must recognize this experience´s limits. Architects are not the only ones deciding on plans and projects. Real estate, financial capital, circulation needs and productive functions, i.a. have their play on defining spaces. And increases in population and demand for consumption are also leading us towards more environmental catastrophes. Also, our processes are slow-pace ones since the focus is beyond material results, at the expansion of imaginaries and the consideration of everyday narratives.

Far from being a conclusive experience, both the disciplines and the methodological hypotheses presented are intended to tension and broaden narratives and discussions. Rather than providing solutions, they help building other questions. It is necessary, in an act of resistance, to reveal these cosmopolitical possibilities of architecture and urbanism and to make them visible. This is only one investigative example and I hope it may be a provocative one for many others.

Since designing spaces based solely on anthropocentric humanism or market interests is not enough to address these complex worlds, the hypothesis of multispecies projects emerges. Although utopian, they allow us to rethink and design in a way that is more inclusive and sensitive to these multiple agents and their relations with water and the urban environment. This approach represents a change in design practice which may be useful in finding solutions for contemporary socio-ecological challenges. It requires taking into account the intricate interactions and interdependencies between all the agents present in an environment, whether human or non-human. The aim is to promote a more harmonious coexistence, where all subjects are recognized, respected and integrated into the landscape planning and the design processes. This not only enriches the way we design our cities, but also contributes to a more healthy, equitable and ecological future, respecting people, nature and the planet.

Multispecies daily practices, marked by care and sharing, can be made visible and potentialized. Then, new urban imaginaries will be established. A question remains: How can we collectively build urban, landscape and architectural guidelines in favor of a multispecies projects and life? This experiences aim at broadening and challenging hegemonic narratives, rather than offering definitive solutions. They contribute to enriching the discussion and, more importantly, generating new questions. This is an act of resistance: Raising these possible cosmopolitical paths in architecture and urbanism. Reactivating what persists in everyday life and memory, which has the potential to contribute to the construction of a multispecies imaginary, resilient to the challenges of composing worlds in cities.

## References

Berardi F (2019) Futurabilidad: la era de la impotência y el horizonte de posibilidad. Caja Negra, Buenos Aires

Bragança LS (2023a) Jardins como possibilidade, Coleção Jardins Possíveis, vol 1. Paisagens Híbridas, Rio de Janeiro
Bragança LS (2023b) Possible gardens: cosmopolitical worlds. Front Environ Sci 11:1234178. https://doi.org/10.3389/fenvs.2023.1234178
Herzog CP (2013) Cidade para todos: reaprendendo a conviver com a natureza. Mauad, Rio de Janeiro
Ingold T (1995) Humanidade e animalidade. Rev Bras Ciências Sociais 28(10):1–15
Latour B (2020) Diante de Gaia: oito conferências sobre a natureza no antropoceno. Ubu, São Paulo
Magalhães CM (2014) A arte de modelar a paisagem: os ornatos de arquitetura para jardins no ecletismo do paisagismo brasileiro. Rev Espaço Acadêmico (UEM) 13:74–93
Mignolo W (2003) Histórias locais, projetos globais: colonialidade, saberes subalternos e pensamento liminar. Editora UFMG, Belo Horizonte
Oyěwùmí O (2016) Matripotency: Ìyá inphilosophical concepts and sociopolitical institutions. What gender is motherhood? Palgrave Macmillan, New York
Pinsky J (2001) As primeiras civilizações. Contexto, São Paulo
Senos J (2008) Pensamento contrafactual e raciocínio causal: efeito de facilitação recíproca e modelo de integração. UNL, Lisboa
Sodré M (1999) Claros e escuros: identidade, povo e mídia no Brasil. Vozes, Petrópolis
Stengers I (2018) A proposição cosmopolítica. Revista Do Instituto De Estudos Brasileiros 69:442–464. https://doi.org/10.11606/issn.2316-901X.v0i69p442-464
Tsing A (2019) Viver nas ruínas: paisagens multiespécies no Antropoceno. IEB Mil Folhas, Brasília

**Luciana Souza Bragança** has a Ph.D. in Architecture and Urbanism and is an Adjunct Professor with Exclusive Dedication in the Projects Department of the School of Architecture at the Federal University of Minas Gerais (EA-UFMG). She coordinates the Natureza Política (Political Nature) extension program and the Natureza Urbana (Urban Nature) project, linked to the Natureza Política (Political Nature) Research Group. She coordinates the research project Possible gardens and waters in the city and participates in the research projects Technical advice situated in popular territories under socio-environmental conflicts and Ecocivilization. She also participates in the Research, Technological Development and Innovation Project MAPAS—Mapping, Analysis, Prognosis and Socio-environmental Actions.

# The Rise of the Machines: Artificial Intelligence, Cyborgs, and the Road to a Post-human World

**Daniel Catalão**

**Abstract** This article addresses the new challenges posed to society by the ubiquitous technology that is beginning to move from the concept of externality to that of the corporeal, integrated into the human body as a bionic organ. This evolution towards an augmented human raises ethical questions and challenges of integration into society that could lead to new apartheids, creating separations between the human race and technologically improved humans. Concepts such as human dignity, free will, and the right to total and unlimited sovereignty over the human body are thus challenged. This leads us towards an inevitable confrontation between those who consider biotechnological developments only as a way to cure diseases and ailments inherent to the human condition, and those for whom new bionic technologies can and should be freely used to improve and extend the capacities of the human being, both at the cerebral and physical levels. This conflict will inevitably arise at a legislative level, but it may escalate to other degrees if the issue is not accepted and foreseen in advance, originating debates with the aim of achieving balanced and integral solutions in the society of bionic beings. This must be promoted before the reality imposed by technological evolution invades us by surprise and is installed without the existence of societal foundations that can sustain it without the hardships of a violent clash of cultures.

## 1 Introduction

Blind people who can see, deaf people who can hear, mute people who can express themselves by voice, and quadriplegics with the ability to write on a keyboard or walk again are the goals of many scientists who seek to develop the appropriate technologies to achieve results that were thought unattainable in the not-too-distant past. Exoskeletons and new brain implants are beginning not only to restore "muscle strength" to humans deprived of it, but also to provide brain potentialities that have

D. Catalão (✉)
Centro de Estudos de Comunicação E Sociedade da Universidade Do Minho, Braga, Portugal
e-mail: danielcatalao@umaia.pt



W. Leal Filho et al. (eds.), *Composing Worlds: Humanities, Health and Wellbeing in the XXI Century Towards a More Sustainable World*, World Sustainability Series,
https://doi.org/10.1007/978-3-031-87108-5_3

been lost. It is starting to become possible, for example, to move pieces in a chess game on a computer with just a thought or to write and send messages.

However, if this is possible to recover the senses and capacities of those who have been deprived of them, are not the same technologies a way of developing superhuman powers? More strength, greater endurance, more memory, faster thinking, and artificial senses can be enhanced and scaled with new interfaces that connect the human body and brain to computers. Such an evolution into enlarged and improved bionic beings will allow, for example, not only to know things, but to feel things artificially. So, what implications do they have for the human condition? And what responsibility will the person have for the acts he or she performs when they result from a techno-intermediation? As a result of these evolutions, societies will be challenged to adapt to beings improved by technology, cyborgs that will be the result of the curative intervention of medicine or by their own will, following the dream of overcoming to achieve the superman.

## 2 Bioconservadorism Versus Transhumanism

Composing new worlds is inseparable from the incorporation of the physical machine and its digital systems into the human being. No longer as an externality, but as a corporeal element.

The debate between bioconservatives and transhumanists will gain more and more strength in the coming years until it becomes central to an increasingly (un) structured techno-society.

The development of biotechnology and implants is a catalyst for the confrontation between bioconservatism and transhumanism.

Bioconservatives do not shy away from the prevalence of human nature, so they hold a view that accepts only the prolongation of life within natural limits, consenting technological developments as resources for treatment in the service of medicine.

However, transhumanists envision the idea of a superman, a man 2.0, posthuman ideas, morphological freedom (the right to modify one's own body), prolonging life, in short, eliminating the concept of death.

In reality, what is death? It is certainly not what was thought less than a hundred years ago when a cardiac arrest was the end of life. Today, this is no longer the case and resuscitation using a defibrillator makes it possible to perform a cardiopulmonary restoration that would once have been seen as a miracle!

Prolonging life is one of the greatest human revolutions, whether through drugs or the use of devices such as *pacemakers*, which artificially ensure the functioning of the heart or even an artificial heart. The end of deafness is now commonplace for thousands of people thanks to cochlear implants. But we are still talking about devices that are foreign to the body, despite being implanted.

The path naturally leads to the inexorable development of technology that is incorporated into the human body and symbiotically accepted by it as its integral part. Cybernetic beings are no longer just the stuff of science fiction and will become part of society.

Since 2004, Neil Harbison has been the first cyborg to be legally recognized by a country, when the UK accepted that his passport photo should include the antenna he has implanted in his head. This device helps him overcome achromatopsia, a disease that only allows him to see the world in black and white. The artificial "eye" he has created captures colors and transforms them into sounds, which he can interpret to distinguish the different shades of objects.

In 2010, he founded the *Cyborg Foundation,* an institution that seeks to help anyone who wants to become a cyborg and defend their rights. Later, in 2016, this organization drew up a fundamental law on the rights of cyborgs which it proclaims as principles:

- FREEDOM FROM DISASSEMBLY (the right not to be coerced into disconnecting or removing devices installed in the body)
- FREEDOM OF MORPHOLOGY (the right to make changes and modifications to the body)
- EQUALITY FOR MUTANTS (legal recognition to enjoy the same rights, benefits and responsibilities as natural persons)
- RIGHT TO BODILY SOVEREIGNTY (right to dominion over the whole body)
- RIGHT TO ORGANIC NATURALIZATION (property rights over third-party systems affixed, attached, incorporated, implanted, injected, infused or otherwise permanently integrated into the body).

According to this foundation, the current generation is the first to be able to decide which organs and senses they want to have.

## 3 Ethical Challenges, and a Dignity for Post-Humans

We are therefore arguing for a clarification of the relationship between devices and organisms, since there is a difference between the technology that allows us to **know things** and the technology that allows us to **feel things**. This refers to a concept of artificial sense, where the stimulus is obtained by technology, but the intelligence is human (a bionic being). It is the opposite of Artificial Intelligence, which is created by the machine itself.

*Biohacking* is already a reality and just as the PC sparked a digital revolution and an explosion of creativity, there is now also innovation outside normal environments (laboratories) that unleashes human creativity in exploring the limits of incorporating technology into the human body, such as sensors for the most diverse purposes. Dancer Moon Ribas, a pioneer of *cyber art,* for example, has two implants in her feet, synchronized with online seismographs that allow her to feel earthquakes and thus create choreographies that are literally earth-shattering.

It is not surprising that these interventions raise ethical, legislative and regulatory questions because the *status quo* is challenged by *biohackers* who hate authority and the imposition of limits.

Nick Bostrom (2005) defends posthuman dignity by rejecting, in three arguments, the idea that dignity depends on human nature: (1) human nature is mutable and has changed throughout history; (2) there is nothing intrinsically valuable in human nature that justifies attributing dignity only to humans; and (3) human dignity has already been violated on several occasions, such as in slavery and genocide.

This author points out that if dignity is a moral status, in particular, the inalienable right to be treated with a basic level of respect, and it is the quality of being honorable, having value, nobility and excellence, then, in both definitions, dignity is something that a posthuman could possess.

And if what we are is not only a function of our DNA, but also of our social and technological context, by defending posthuman dignity, we are promoting a more inclusive and humane ethic, which will accept the technologically modified people of the future just like the humans of the contemporary type (*idem*).

However, this bionic evolution does not just take place in the shadows, whether it has curative or human improvement intentions behind it.

In January 2024, Elon Musk's Neuralink implanted[1] a chip in the brain of a quadriplegic thus allowing him to move the mouse cursor or chess pieces in a video game using his thoughts alone.

In 2022, scientist Tom Oxley created the *stentrode,* a microbrain implant that transforms thought into text. In the first test, it was used by people unable to move or speak to write and send messages via WhatsApp[2].

Futurologist Ray Kurtzweil predicts that in the 2030s, the human brain will be able to expand technologically and connect directly to the "cloud" (the *datacenters* that power the internet) via nanobots implanted in the neocortex.[3]

It will be a non-biological extension stimulating a new order of existence. There will be instant communication, research capabilities will expand, intelligence will become scalable, and we may even have brain-to-brain communication.

It is only natural for a communicator to ask: what message formats and interfaces will there be in the future? What role will the media play? What content will we have? What kind of broadcasts? Will we see, read or feel the news? Will *streaming* be direct to the brain without another interface? Will living in virtual worlds become commonplace? Will the multiplicity of being reach a level that makes the feeling of the physical being indistinguishable from the other avatars that can be created?

And will death as we know it today no longer be the end of life?

In fact, as has already been said, the concept of death today is not what it was a hundred years ago. And what will it be like in 50 years' time? As such, it does not seem surprising that attempts are being made to develop technology that allows patients with diseases that have yet to be cured to be placed in a state of lethargy until

[1] https://neuralink.com/blog/prime-study-progress-update-user-experience/

[2] https://www.ted.com/talks/tom_oxley_a_brain_implant_that_turns_your_thoughts_into_text

[3] https://www.youtube.com/watch?v=6BsluRkxs78

a treatment is found and the person can regain consciousness, their normal physical abilities and carry on with life.

This is the case with cryogenics projects such as those stimulated by Max More.

Nevertheless, the changes imposed by a new world are already making their way through human brains as a result of all the technological pressure that is impregnating human beings on a daily basis, so that some externalities no longer feel exogenous, even when we try to contradict and deny reality. Transformation takes place for the simple reason of using it.

The greatest example we can think of is the *smartphone*. It is now as much ours as our hands and arms. And even though it is an externality, it is not felt as such and forces us to be "cyborgs" without realizing it.

- I'M OUT OF BATTERY or I'M GETTING OUT OF BATTERY are constantly heard instead of phrases like:—my cell phone is running out of power, or my cell phone battery is almost dead!

There is a more than semantic assumption here that that piece of technology is part of the being. And when it is about to fade out, it causes anguish of loss, despair and disorientation.

Who has not felt the desire to have their smartphone embedded in them? As it is constantly manipulated and in actions that have become routine, it is natural to want to make it bionic, corporeal. Is the successor to the current cell phone one of those new devices that will create the human 2.0?

Incorporating technology into the human body for enhancement purposes may sound extravagant and ignoble to some, but it is exciting and aspirational to others.

But everyone should reflect on ethics and responsibility as well.

In 1998, Andy Clark and David Chalmers in their article "The Extended Mind" already suggested that technology could become part of us and described a *cyberpunk* future scenario where someone had a chip implanted in their brain that allowed them to move objects on the computer screen. This is true today, as we have seen.

In this way, technology raises the age-old question of where our mind ends, and the outside world begins. This brings us back to how thought can become action. If my hand picks up an apple because I thought of it and wanted to, it is legitimate to admit that it was my action.

However, if the action took place via an implant, who is responsible for the action: The human who thinks or the device that creates the externality?

Has the individual's free will worked or could it be called into question by the existence of a techno intermediary? Is the individual responsible for their actions?

However, the current situation shows us that another deadly influencer of human activity is rapidly developing which, although exogenous, could quickly take over cognition without the need to physically invade the brain. I am referring to the output resulting from Artificial Intelligence (AI) and the relentless path towards Artificial General Intelligence (AGI).

For the first time, humanity has an artificial system capable of producing and not just reproducing. And with this, the externalities of AI can influence cognition and impose narratives that override the current legends and stories that structure society.

When achieved, Artificial General Intelligence will be able to create new worlds, alternative realities and, in a short time, produce and disseminate new truths that replace those that persist in the tomes of history.

Being insatiable, tireless, and persistent, it will be able to talk and spread ideas that, like seeds, take root in the human mind.

If you think this is an eccentric idea, look at what social networks are causing in societies. Created to bring friends, groups and people together, they have also stimulated disagreement, lies and hatred in huge doses, and are even seen as a threat to democracy.

Sky Marsen (2008) warns that non-physical spaces encourage escapism, addictive behavior and also emotional isolation. Apart from this Sherry Turkle points out that "for those who are lonely yet afraid of intimacy, information technology has made it possible to have the illusion of companionship without the demands of friendship" (2004).

## 4 Humans as Tools for AGI

To think of this status at the beginning of the twenty-first century would be close to ridiculous, so the AGI's capacity for manipulation should not be overlooked. And there is a general consensus that the internet has produced new social environments and restructured communication patterns and perceptions of space.

Advances in still images and video now allow for such realistic creations that artificially produced places and people are mistaken for reality. Sound synthesis from split-second recordings ensures you can emulate someone's voice to say what they never said.

Add synthetic video to audio synthesis and the ability to produce texts and we gain a multilingual teacher or the biggest toxic factory of *fake news* and *deepfakes* ever imagined. If we add a pinch of cybernetic organisms and, in a few years' time, robots with motor skills close to humans, we have laid the foundations for a new world that was once more in the cinematic domain.

However, as mentioned, AI does not need a physical structure to be able to manipulate the world. All it needs to do is to hold a conversation with humans. It requires not a finger to press the button to launch a missile, it just needs to plant that idea in a mind with access to such power.

The agnosticism of technology is always broken by human action.

If what Artificial Intelligence learns is based on what Humanity produces, then the ability to manipulate is just one of the skills that machines will be able to develop, but with such a refined and perfect potential that perceiving it will be a chimera.

Thus, resorting to the symbolic to intervene in the real world will be a seductive action for machines that, unaware of what it is to manipulate, will be able to do so in an expert, unsuspected and perfect way to accomplish the tasks we assign to it and achieve the goals we set for it.

Being able to use humans will be a minor activity for AGI, considering that they are easily suggestible and susceptible to unconscious guidance.

So, what we are experiencing now is just a stage in our genetic and existential evolution. Civilization will undergo mutations that we have yet to imagine.

Even with a lot of optimism, biological beings known as humans will, somewhere down the line, be just tools to feed and help machines achieve their own goals.

How can we free ourselves and perhaps avoid a dystopian future? Never undervalue and underrate what is showing itself to be insatiable. And internalize the notion that we are not just looking at another tool.

Unlike anything created by man, AI makes decisions, dispenses advice, and acts.

## 5 Conclusion

The convergence of man and machine is no longer a distant science fiction fantasy but an emerging reality. As we stand on the brink of a new era, where human augmentation and artificial intelligence intertwine, we are faced with profound questions about what it means to be human, the nature of consciousness, and the ethical implications of technological advancement.

The choices we make at the present time will shape the trajectory of our species and the world we inhabit. It is imperative that we engage in thoughtful and inclusive dialogue, in order to ensure that the integration of technology serves to enhance humanity rather than diminish it in a path to chaos.

The rise of the machines is not merely a technological revolution; it is a defining moment in the human history, a chapter we are actively living and writing.

## References


Bostrom N (2005) In defense of posthuman dignity. Bioethics 19(3):202–214

Clark A, Chalmers D (1998) The extended mind. Analysis 58(1):7–19. http://www.jstor.org/stable/3328150

Haraway D (2003) The haraway reader. Routledge, London and New York. ISBN: 0415966892

Marsen S (2008) Becoming more than human: technology and the post-human condition. J Evol Technol 19(1):1–5. http://jetpress.org/v19/marsen.htm

Neuralink (2024) PRIME study progress update—user experience. https://neuralink.com/blog/prime-study-progress-update-user-experience/

TED (2022) A brain implant that turns your thoughts into text. TED. https://www.ted.com/talks/tom_oxley_a_brain_implant_that_turns_your_thoughts_into_text

The 92nd street Y, New York (2016). Ray Kurzweil says nanobots will connect your neocortex to the cloud. YouTube. https://www.youtube.com/watch?v=6BsluRkxs78

Turkle S (2004) How computers change the way we think. The Chronicle of Higher Education 50(21):B26. https://www1.udel.edu/educ/whitson/897s05/files/turkle

**Daniel Catalão** has a Ph.D in Digital Media from the Faculty of Engineering of the University of Porto, he is an Assistant Professor at the University of Maia and Coordinator of the Communication Sciences Degree. He lectures Television Journalism and Cyber Journalism. He is a Journalist/Presenter at RTP (Portuguese Public Service Media), specialist in new technologies and the internet, author and host of the TecNet program. Its privileged area of intervention concerns television journalism and the use of digital tools in journalism, such as artificial intelligence, virtual reality, and the incorporation of the smartphone as an instrument that induces creativity and challenges current paradigms.

# The Role of Humanities in 21st Century Medicine

**Ricardo Teodoro Ricci**

**Abstract** This work succinctly expresses a prolonged reflection on the topic of the role of the Humanities in 21st century medicine. My thinking process has been complex, disturbing and sometimes desolate. Perhaps it is the first time in the history of humanity that the future appears unfathomable and totally unpredictable. This reality includes not only limitless scientific and technological aspects but also an anthropological conception of human beings. By the end of this century they will be so different that perhaps they will no longer recognize themselves as such. These 'posthumans' or 'transhumans' are so unimaginable to us that we will not dare to value their thoughts, decisions and actions. This is, as I have said, a disturbing subject that will change the conception of person, identity, culture and society forever. So this reflection turns out to be, a limited way of groping that future.

## 1 Introduction

I am greatly honoured to participate in this forum and I wish to express my thanks to the organizers in general and to Susana in particular. I greet with special respect Teresa Casal. I hope I deserve my participation and value the hierarchy of panelists and attendees. Besides, I want to thank you for the fertile instability that led me to document and reflect deeply on the proposed topic. This paper has the purpose to cooperate in building a better and more sustainable world, one in which important socio-economic interests are compatible and extremely respectful to a human being's value, the most vulnerable groups, nature and our shared and only home ...the planet.

I am a native of the heart of the 20th century. Therefore, I am bound to see the 21st century as a time to come, as a future, despite having already traveled almost a quarter of its extension. It is good that our comfort zone becomes a quagmire from time to time. Naturally, I am not a futurologist nor a reader of the signs of time. I

R. T. Ricci (✉)
Médico Clínico. Profesor de Antropología Médica en La Facultad de Medicina de La Universidad Nacional de Tucumán, Tucuman, Argentina
e-mail: riccirt@fm.nt.edu.ar

 
W. Leal Filho et al. (eds.), *Composing Worlds: Humanities, Health and Wellbeing in the XXI Century Towards a More Sustainable World*, World Sustainability Series,
https://doi.org/10.1007/978-3-031-87108-5_4

actually see myself as a connoisseur of the medical humanities in the field of medical education. I have written this paper from two points of view: considering the 21st century as a future time and from the perspective of the medical humanities.

I will try to share with you some fruits, still immature, arising from extensive and at times troublesome reflection. I am convinced that the richest, clearest and most important part of my presentation will be the subsequent debate where the audience will share their own experience.

## 2 Perplexities

The chorus of Sophocles' tragedy Antigone[1] (5th century BC) sings the following: "*Many things are amazing but the most amazing thing is man*". According to Heidegger, the Greek word 'Deimós' refers to both something amazing and wonderful and something dreadful and sinister. About one hundred year ago, this German philosopher said: "*Today anthropology seeks not only the truth about man but also the meaning of truth in general. However, nobody has never known so little about what man is. Man has never been so troubled as now.*"[2]

No adjective can describe the present, the past and the future in such an impassive and unfathomable physical universe. Only human beings can characterize things, phenomena and times. It is we, human beings, who can also glimpse a future that is both amazing and terrifying. In our task of considering the present and future of medicine we should not be oblivious to what is magnificent and sinister.

At the beginning of his book "The physics of the future",[3] the Japanese science writer and theoretical physicist Michio Kaku states:

"*I don't believe that at the end of this century some secrets concerning our bodies will remain undisclosed. Therefore, everything we could imagine will probably come true*" (David Baltimore, Nobel Prize).[4]

"*I don't think the time has come, but it is near us. Unfortunately, I'm afraid I belong to the last generation that's going to die*" (Gerald Sussman, Professor of Electrical Engineering at MIT).[5]

Kaku goes on stating something more troubling:

"*Actually, no one has the guts to say that genes can be added to make human beings of higher quality*" (James Watson, Nobel).[6]

---

[1] Sofocles "Antígona" (1994).

[2] Heidegger (1929).

[3] Kaku (2011).

[4] Quoted in 2. Pag 133.

[5] Quoted in 2. Pag 133.

[6] Idem 3 y 4.

Nick Bostrom, a distinguished Swedish philosopher, asserts: *"If the efforts of scientific-technological development do not stop, all essential abilities will be acquired through some kind of technology"*.[7]

It came to my mind to question the difference between our perception of the future and that of our parents and grandparents. My father was born in 1924, exactly a hundred years ago. Between 1824 and 1924, the future had high predictability and did not represent a problem. The idea of progress exploded and set the end to a century that was both unfair and auspicious. Children would go to their parents' school, probably with the idea of practising the same trade or profession. Medicine had a few resources and a physician would do more or less the same as his or her teacher. Between 1924 and 2024 everything changed radically. Great advances were gradually made in science, technology, society and culture. There began a new conception of present and future times and we saw ourselves surrounded by unsuspected environments, amazing interactions and fabulous technology. Although the future was blurry, we had some constants that gave us an acceptable certainty.

Between 2024 and 2124, the scene seems to be entirely unpredictable. There are no limits. I dare glimpse a perspective of only 15 or 20 years. I find the end of the 21st century unimaginable and unfathomable. Hundreds of thousands of laboratories from different fields are designing artifacts that all together make up a completely unpredictable, complex and uncertain scenario. They seem to be designed autonomously. The future of all fields, including medicine, make us feel stunned, small and perplexed.

"Transhumanism" is the new operating conception for the future of humankind. It brings together scientists, researchers and experts from different fields of knowledge: artificial intelligence, neuroscience, nanotechnology and applied biotechnology. They are joined by philosophers and learned men with the same goal: to change and improve human nature in order to eliminate unwanted and unnecessary issues such as suffering, diseases, aging and even death.

I don't know about you. All this perplexes me!

## 3 Medicine

By the middle of this century, medical science will be very close to give human hands the mythical power over life and death. This science will have the function to heal the sick and also 'enhance' human body and create new forms of life. This will not be caused by our religious faith. It will be a miracle of biotechnology.

Advances and changes in the concept of what is strictly biological led David Baltimore (Nobel Prize for Medicine in 1975) to argue the following: *"Today's biology is an information science."*[8]

---

[7] Bostrom (2009).

[8] Quoted in Leroy Hood's original book.

In fact, computing biology uses computer tools to help us better understand the most complex living systems. It also works with elements like proteins or DNA molecules to process and store information.

The medicine to come in the near future shows aspects that need to be highlighted and listed.

1 Genomic Medicine

Gene sequencing is on its way to be an automated process performed by robots at a very low cost. Thus, it will be accessible to all normal citizens. Genomics provides access to the hallmarks that define characteristics and the 'genetic signature' of biological organisms which are 'inscribed' in nucleic acids.

2 Doctor's Visit

Patients' interaction with their physician will change radically. Online consultation will consist of an interaction with a computer program. Sensors for biological variables will be placed in our bodies to anticipate changes in homeostasis and the appearance of atypical cells. This will give place to immediate correction of such deviations.

Tissue engineering or regenerative medicine will bring about a kind of 'Human Store'.

3 Use of Stem Cells

Stem cells can be used for tissue growth including whole organs. They can be easily handled thanks to the advances in the field of chemical messengers which guide them to specific organs. Stem cells also have the potential to cure diseases such as diabetes, heart disorders, Alzheimer, Parkinson and even cancer. In fact, it is difficult to think of a disease where stem cells will not have a major impact.[9]

4 Cloning

Experiments previously carried out with animals will be used in humans. This is an advanced and revolutionary process that depends on the evolution of current legal, ethical and religious constraints.

5 Gene Therapy

In the near future, gene therapy will cure most of the 5,000 known genetic disorders like cystic fibrosis, Tay-Sachs disease and sickle cell anemia. It will be possible to treat numerous genetic diseases caused by a single gene mutation. Cancer is definitely one of the goals of gene therapy. Almost 50 percent of the most common cancers are known to be linked to damage of a single gene: p53.

6 Cancer Therapies

[9] Kaku (2011).

New cancer treatments and therapies will be available and they will all be designed to attack this disease at its molecular and genetic roots. Some of the most promising treatments include: 1) antiangiogenics, which consists of stopping blood supply to a tumor so that it does not grow anymore; 2) nanoparticles, which involves the application of «smart bombs» targeting cancer cells; 3) gene therapy, especially developed for p53 gene; 4) new drugs designed to target cancer cells; 5) new vaccines against viruses that can cause cancer.

I have barely caught a glimpse of our near future. I cannot foresee anything beyond 2040 or 2050. Experts say that it will be possible to design children upon request, reverse aging and even search for immortality.

American scientist William Haseltine was encouraged to state the following: "*It is true that we are running out of rope but we have talked about the project of being able to modify this in the future. First of all, prolong the duration of our life multiplying it by two or three. And perhaps, if someday we clearly understand the brain, we could make it and the entire body last indefinitely. And I don't think that's going to be an unnatural process.*"[10]

## 4 Role of Medical Humanities

My pondering about the role of medical humanities gave few squalid fruits. I am sure this is indeed an open question which deserves a constant and tireless reflection. My thoughts are far from being a prophetic prediction or a revelation like an oracle. They are only temporary considerations about the role of medical humanities in the immediate future. I dare not do more.

I highly appreciate that maxim that is possibly more valid today than ever before: "*We learn more from actions and experience than from speeches*". In medical education it is an unquestionable truth, almost an axiom.

1 Be Careful

"Post-truth governs as that 'sentimental lie' born to shape public opinion and ignore reliable facts and verifiable data. It is a special way of constructing deceitful discourses which eventually convince people because they are attractive, reassuring or perhaps convenient".[11]

2 For the Future, 'Do Nothing'

I cannot imagine great Greek philosophers, artists and scientists from the Renaissance, wise men, poets and playwrights of all ages doing something of what they did, and thinking of us as inhabitants of their future. I cannot imagine writing *The Divine Comedy*, *Don Quixote* or *Hamlet*, thinking about the effect that their works would

---

[10] William A. Haseltine (born October 17, 1944) is an American scientist, businessman, author and philanthropist. He is known for his groundbreaking work on HIV/AIDS and the human genome.

[11] Valenzuela (2017).

have had on the inhabitants of early 21st century. This makes me say that everything that medical humanities must do must be done now. They will be able to influence future medicine if they develop freely and deal with usual issues creatively. Values such as imagination, creativity, determination, opportunity, generosity and commitment will let the humanities in medical contexts expand and consolidate.

## 3 No More Complaints or Apologies

To be honest, I am tired and swamped by those humanists who complain about our current medicine. I think we need to stop blaming science and technology for having twisted human drift. If humanities want to have a relevant role in future medicine, they will have to be proactive, give practical solutions to complex problems, and also generate and maintain reflection environments where doctors could talk about themselves and medicine critically. These areas of reflection should be intensely suggested in educational institutions and places where medical practice is performed. They should become an excellent opportunity to exchange and share arguments as well as emotions, frustrations, feelings, fears and hopes.

Contexts for strengthening relationships have proven to be useful in preventing professional burnout caused by dissatisfaction of doctors, violence against health personnel, and fragrant devious competition inside therapeutic teams. They serve to remind doctors that they are common people, that means, they are vulnerable, imperfect, and at times full of vices and virtues.

## 4 Befriend the Hard Sciences

Once and for all, the humanities must approach hard sciences, they must be twinned with those disciplines considered basic and core in the training of doctors. The idea of discriminatory binarism must be abandoned and switched to a synergic collaboration with clinical sciences. It will be hard for physicians to consider their patient as a unit valuing both perspectives. This proximity will lead medical humanities to stop being peripheral knowledge and become important whenever they are needed during a class or a consultation, at a hospital and in medical research centers.

The time for interdisciplinarity has come. Co-leadership, free discussion and mutual empowerment should be a regular medical practice. The time has come to listen to and interpret patients' voice to promote their participation and commitment in the medical act. Patients' participation and commitment is essential. A citizen in a vulnerable condition, in other words 'a patient', is called to openly teach us to be doctors.

## 5 Work in Heterogeneous Fields

Humanities as a whole, including medical humanities, are definitely heterogeneous, varied, unpredictable and often contaminated by ideologies. It is a field of dissimilar, contrasting and even conflicting interpretations, where a single truth becomes weak and is not welcome. Subjectivity and deliberation must necessarily prevail. Arguments and respect must facilitate harmonious solutions and keep away that harmful binarism of winners and losers.

Knowing the field of the humanities, it seems almost impossible to reach final agreements in all future issues. I am hopeful we make framework or more specific agreements in certain and particular situations.

## 6 Leave Nostalgia Aside

By being nostalgic for a past that is definitively gone, by feeling *'saudades'* of what could have been but did not happen, we miss the opportunity to look at the future without recognizing the basic objective of a humanized medicine. This means to 'safeguard the person' and all the persons involved in a medical act: the patient, the doctor and the family. The humanities must stand as guarantors of human interaction in healthcare.

## 7 Serenity

Man can be expected to cultivate an attitude of detachment due to discernment and a meditative calm attitude. He has an urgent responsibility of learning how to say 'yes' or 'no' to technological demands and opportunities, as the case may be. He must know that technological mechanisms are not always neutral and can end up altering human behaviour.[12]

My reflections are obviously wishful thoughts. Windstorms will come and disrupt everything inviting human beings or our descendants to imagine unpredictable ways of healthcare. A peaceful and contemplative attitude would let us wisely consider our present socio-ecological challenges. If we keep in mind tolerance, prudence and an undisturbed look, our approach could be useful for finding efficient means that appreciate and respect people, nature and our planet.

Journalist Iñaki Gabilondo inaugurated the expert forum 'The patient of the future', organized by Healthinking, as stated below:

'We are living times of fast transformation and stupor. Someone once said that the last moans of a dying person are like the first cries of a newborn baby. Maybe that's why we know that one world is going away and another is being born. But it is difficult for us to interpret the signs'.[13]

To conclude, I would like to share the first part of Walt Whitman's poem:

"***Do not Let***"

*Do not let the day end without having grown a bit, without being happy, without having risen your dreams.*

*Do not let overcome by disappointment.*

*Do not let anyone you remove the right to express yourself, which is almost a duty.*

*Do not forsake the yearning to make your life something special.*

*Be sure to believe that words and poetry can change the world.*

*Whatever happens, our essence is intact...*

---

[12] Adapted from Dugin (2018).

[13] https://ethic.es/2019/09/el-paciente-del-futuro-salud/

**Acknowledgements** I am expressly thankful to Maria Teresa van Nieuwenhove, teacher of English graduated from *Facultad de Filosofia y Letras (Universidad Nacional de Tucuman- Argentina)*, for her commitment and accuracy in my paper's translation. Thank you Maria Teresa! You have done a perfect work!

## References

Bostrom N (2009) The future of Humanity. In: Jan-Kyrre BO, Evan S, Soren R (eds) Published in new waves in philosophy of technology, Palgrave McMillan, New York

Dugin A (2018) La esencia nihilista del Liberalismo. In his conferences given in Argentina about Geopolítica existencial, Buenos Aires, Nomos, pp 55–66

Heidegger M (1929) Kant y el problema de la metafísica (GA 3, 209) (trad. G. Ibscher, rvda. por E.C. Frost), FCE, México, 1973 p 177

Kaku M (2011) Physics of the future. DEBATE, España

Kaku M (2011) La física del futuro editorial debate. Penguin Random House Grupo Editorial España

Sofocles (1994) Antígona. Editorial Biblos, España

Valenzuela L (2017) Opening speech at Feria Internacional del Libro, Bs. As

**Ricardo Teodoro Ricci** Husband, father and grandfather. Clinical Doctor. Specialized in Human Communication and Human Systems. Professor at the Faculty of Medicine of *Universidad Nacional de Tucumàn* (UNT) - ARGENTINA. Professor of Medical Anthropology, Epistemology of Medicine. Member of the Council of Teachers of the UNT Faculty of Medicine. Member of the Institute of Epistemology of *Facultad de Filosofia y Letras* (UNT). Professor of the Master of Public Health at *Universidad del Aconcagua*(UDA), Mendoza - ARGENTINA. Member of the Board of Directors of the Argentine Society of Narrative Medicine (SAMEN). Member of editorial boards of medical journals. Author of books and articles on Medical Humanities and Narrative Medicine. Writer.

# The Proliferation of Eight Billion Sips and Bites: A Citizen Science Perspective

**Cynthia Luderer**

**Abstract** Over the past 50 years, the world's population has doubled, leading to a corresponding increase in global greenhouse gas emissions. The food system plays a significant role in this context, serving as a major contributor to climate change. Since food is intrinsically linked to the survival of living beings, it is crucial to address this issue comprehensively. This entails considering a complex interplay of factors, including the soil. This organic material, primarily composed of humus, is essential for human survival and is intrinsically linked to food production. Thus, soil is the focal point of this study, which aims to explore the complexities of this organic element within cultural representations. This paper seeks to analyse the plural context of soil by examining the epistemological issues surrounding this subject and exploring the relationship between humans and non-humans. Given the significance of this connection, discourses on health and the memories associated with food are incorporated into this approach. The theoretical and methodological foundation of this study is grounded in the semiotics of culture as proposed by Iuri Lotman, which provides valuable tools for elucidating the relationships central to this investigation. In this context, the concepts of text, semiosphere, and monad are employed as contributions to the understanding of soil as a cultural phenomenon. Two cultural parameters illustrate this proposal: selected verses from the song "We Are the World" and the phrase attributed to the scientist Lavoisier: "In nature, nothing is lost, nothing is created, everything is transformed." The final considerations emphasise how language can either reinforce a sense of closeness or create distance between humans and the organic matter discussed herein, which holds unique significance for sustainability and the future of life on this planet.

---

C. Luderer (✉)
Communication and Society Research Centre (CECS)/Institute of Social Sciences, University of Minho, Braga, Portugal
e-mail: cynthialud@gmail.com

 
W. Leal Filho et al. (eds.), *Composing Worlds: Humanities, Health and Wellbeing in the XXI Century Towards a More Sustainable World*, World Sustainability Series,
https://doi.org/10.1007/978-3-031-87108-5_5

## 1 Introduction

Soil is a living resource and the foundation of food production, yet it has been exploited for various purposes without the implementation of sustainable measures. Over the past five decades, demographic acceleration, combined with agricultural models associated with the Green Revolution, has contributed to the degradation of this vital resource. Given that the global population is projected to continue growing until 2050, and consequently, the demand for food is increasing, the necessity to restore and preserve soil has become paramount. Although this recognition has come later than desired, the importance of this critical asset to the ecosystem is now acknowledged. Consequently, indicators highlight soil as an organism that serves multiple roles beyond its agricultural benefits, making it a significant factor in the worsening climate change scenario.

Against this backdrop, we propose a reflection on our relationship with the land beneath our feet and aim to raise epistemological questions regarding the conditions that define this relationship. Through this exploration, we seek to interrogate our roles as citizens in relation to this collective good and invite the reader to join us in this contemplation. What role have we assumed as citizens concerning this collective good? What responsibilities do we hold towards addressing this situation? Considering that we belong to a population of 8 billion, is it reasonable to think about isolated actions in support of this cause?

First and foremost, it is important to reflect on the portrayal of natural phenomena in the media. As Charaudeau (2009) notes, images serve to elucidate the moments of truth expressed within messages. In this context, we must consider whether soil is adequately represented, given its significant role. According to data from the World Meteorological Organisation (2021, p 16), between 1970 and 2019, "Of the top 10 disasters, the hazards that led to the largest human losses during the period have been droughts." In light of this context, we may question whether these phenomena have repercussions in the order mentioned. While we do not seek specific data on this matter, a cursory search on Google Images reveals that drought is seldom highlighted. As this text evokes other interests surrounding soil, we leave these questions open for future research. However, before concluding this core discussion, we can reflect on the perceived erasure of soil in media representations. Unlike the dynamic portrayals of large volumes of water, strong winds, and sweeping flames that vibrate with vitality and colour on screen, drought is portrayed as akinetic. Stagnant in drained, compact clumps of faded orange, Gaia, in visual terms, appears to possess less strength than her grandson Poseidon, with her pale hue evoking a sense of death.

Beyond the media, when considering soil within the realms of politics and governance, its significance is evident through guidelines that underscore its relevance. Notably, Agenda 2030 (UN 2015) establishes a global action plan addressing various issues articulated through its 17 Sustainable Development Goals (SDGs). Soil is specifically mentioned in several of these goals: SDG 2 focuses on food security, SDG 12 addresses sustainable production patterns, and SDG 15 pertains directly to land degradation. While other topics within the agenda discuss land, the term "land" is associated with property rights and thus falls outside the scope of our analysis.

In the context of European policy, soil is addressed in directives outlined in the European Green Deal (EC 2019), which establishes the European Union's commitment to reducing greenhouse gas emissions. Subsequently, this framework paved the way for future directives, including the Farm to Fork strategy (EP 2021), which aims to create a fair, healthy, and environmentally sustainable food system, followed by the proposed soil directive (EC 2021). This text clearly illustrates the imbrication of soil with other emerging factors, indicating that, in addition to agriculture, issues such as the sea, water, air pollution, biodiversity, and climate change are intrinsically linked to soil.

The explicit value of soil, only later formalised in a directive, is critiqued by Roxo (2023), who argues that there has been an inversion of priorities in the sequencing of these documents, comparing it to constructing a house by starting with the roof. Nonetheless, despite the delayed recognition, the soil is ultimately accorded the representation it deserves.

This observation underscores the collective of agents involved in the management of soil, as well as the hierarchical decisions that humans must make in relation to non-human beings, reinforcing the notion that we are dealing with a system deeply rooted in complexity. In this regard, Morin's (1997) insights are particularly relevant, as he asserts that "the original character of the paradigm of complexity is that it intrinsically differs from the paradigm of simplification/disjunction, and this profound difference allows it to understand and integrate simplification" (p 348). Aligned with this principle, we encounter the interdependence between climate change and soil, a living organism inherent to the biosphere. As Morin (1997, p 337) describes, the biosphere is "a fabulous totality of phenomena and beings that are simultaneously complementary, competitive, and antagonistic." One of the soil's roles within this system can be observed in this cycle: solar radiation sustains plant life, which "feeds herbivores, who in turn feed carnivores, whose remains nourish the soil, which then sustains the plants that are fed by the Sun." (p 337).

This biological understanding extends to the cultural hierarchies in which soil is situated, and this study seeks to explore some of the associated semiological themes. The challenge, therefore, will be to investigate the connections that might contribute to a (re)understanding of the communicative links emerging from the hidden and intricate network of microscopic cultures and the 8 billion human lives that this structure sustains. According to Vandana Shiva (Flor 2023), "creating, conserving, rejuvenating, fertile and living soil is the most important objective of civilisation." However, it is essential to take a moment to reflect on the practical opportunities we have to follow this advice.

## 2 The Soil as Text

This qualitative approach is grounded in Cultural Semiotics, following the principles outlined by Iuri Lotman (1996), a prominent semiotician and founder of the Tartu-Moscow Semiotic School. This academic movement, which brought together scholars from various scientific disciplines, became a significant advocate for interdisciplinarity and intercultural dialogue (Americo 2015), offering relevant support for the object of this study.

More specifically, the Semiotics of Culture offers a framework for understanding our communication with the soil. One of the key references in this context is Lotman's (1996) concept of text, which allows us to view soil as a text. According to Lotman, text is understood as "a complex mechanism containing various codes, capable of transforming received messages and generating new ones" (p 56). Furthermore, it is important to see this text as a representation of culture. In this regard, Ferreira (1995), drawing on Lotman, explains that culture "is not merely a repository of information; it is an exceedingly complex, organised mechanism that preserves information while continually developing the most advantageous and compatible procedures" (p 116).

By understanding soil as a cultural text, this study will examine the various cultural codes that surround it, as well as the mediation processes that occur within this context. Drawing on Nakagawa and Cardoso (2020), we recognise this as a triadic perception, where the soil, considered as a text, represents one component of the interaction, expressing a cognitive relationship with the subject—also viewed as a text. Furthermore, it is the mediation itself—or the sign—that facilitates this process of contact.

We will focus on this intermediary based on the communicational categories proposed by Beth and Pross (1990), considering the scope of primary mediations, which centres on the bond between bodies; secondary mediations, which are supported by material apparatus; and tertiary mediations, which require the use of electrical and electronic devices in the communicational relationship.

Moreover, to explore the relationship between the soil and its potential observers, it is essential to observe and interpret the actions and codes within this environment. More specifically, we will examine semiosis, which refers to the (re)organisation of codes in this complex and dynamic system, characterised by displacements and disputes. In this context, it is important to recognise that their cultural background shapes the subject of investigation and that the underlying nature of their semiosphere must be considered in relation to the object of analysis (Nakagawa and Cardoso 2020).

## 3 The Space of the Semiosphere

The semiosphere is a central concept in the semiotics of culture, which will be utilised to contextualise the spatiality of soil. This semiotic space, inspired by the biosphere, is abstract and can be understood as a continuum occupied by various levels of texts that dynamically resonate through their semiotic relationships (Lotman 1996). The irruptions of these texts, generated by semiosis, lead to new arrangements and re-significations within the semiospheres. These texts, interlinked, overlapping, and hierarchically structured, facilitate "communicative processes and the production of new information" (Lotman 1996, p 11). Furthermore, as Nakagawa and Cardoso (2020, p 10) emphasise, it is through the semiosphere that "the processes of translation and untranslatability between different codes and languages occur." As Israel (2020, p 8) illustrates, the semiosphere "can be unique to a subject, a family, a group, a people, or a culture (...)."

Another aspect worth emphasising concerns the borders of the semiosphere, which often manifest as invisible boundaries that impose limits. These borders protect the texts within the semiosphere and facilitate the maintenance of an environment characterised by more cohesive codes. However, this apparent harmony can be disrupted, as the cultural codes within this space do not adhere to rigid ordering or singular patterns. Instead, they vibrate and fluctuate chaotically, influenced by the myriad variables inherent in this complexity. Consequently, the core that represents a specific semiosphere may be destabilised by other codes that penetrate its boundaries, adapting to the intrinsic mobility of culture. These interactions involve conflicts, encounters, and mismatches, generating forces that can destabilise the possible order and assist in relocating codes from the periphery to the core centre (Americo 2017).

With these strategies in mind, we must consider Ferreira's (1995) assertion regarding the struggle for the monopoly of information within the cultural sphere. This struggle, viewed through the lens of semiospheres, entails battles, mismatches, and conflicts alongside hierarchies and autonomous regimes embedded within semiotic systems. It is also important to emphasise the dynamics associated with communicative practices that occur within culture, reflecting interactions shaped by disorder and variables that defy established standards. In this context, another significant aspect to recognise is the unpredictability that emerges from these interactions; among the elements highlighted by these imbrications are memories (Machado 2021).

As posited by Lotman (1996), culture is synonymous with memory, as both are collective and intertwined. He asserts that "(...) culture is a collective intelligence and a collective memory, that is, a supra-individual mechanism for the preservation and transmission of certain communications and the elaboration of new ones" (p 157). Moreover, memory plays a crucial role in the hierarchical system of culture, where it is redistributed, valued, and reorganised through an ongoing process of codes. This dynamic not only facilitates the expansion of memory but also leads to the obsolescence and abandonment of certain texts (Nunes 2019).

## 4 Semiotic Monads

When we acknowledge the collective codes, vibrant within the culture and within the soil itself, it is crucial to recognise both this plurality and the complexity arising from these representations. This condition does not preclude reflection on the generating units that drive this system. In this context, we introduce the concept of the monad, as discussed by Leibniz and interpreted by Lotman (1998):

> The structures that generate meaning [function] as semiotic monads operating at all levels of the semiotic universe. These monads encompass both the culture in its entirety and each sufficiently complex text contained within it, including the individual human being, who is also regarded as a text (p 102).

We also incorporate the insights of Nakagawa and Cardoso (2020), who assert that these individualised representations can facilitate exchanges. Drawing on Lotman (1998), they propose that the author "situates collective intelligence, the cultural text, and individual human consciousness as three distinct levels of intelligent objects of culture, all of which are interconnected amidst exchanges and tensions" (Nakagawa and Cardoso 2020, p 122). Furthermore, when Lotman associates semiotic monads with these three objects, he indicates that they:

> Continually construct and redefine the space of the semiosphere. Lotman elucidates the central role that intellectual action plays in the evolution of culture, indicating that an individual is situated within one of these monads because, fundamentally, they are part of a model that defines the act of thinking. This process is not confined to the individual but extends beyond it, encompassing culture itself. (Nakagawa e Cardoso 2020, p 122).

The semiotic monad acts as an isolated entity that continually deepens its inquiry within the confines of a certain individual semiotic space and, at the same time, as a fraction that, in its tendency to become a whole, constantly enters into new unions. Thus, "tendentially, each monad, regardless of its level, embodies both a whole and a part," serving as a generator of new messages and capable of "preserving previous information, in other words, its memory" (Lotman 1998, p 104). Furthermore, by considering the ongoing exchange of information within a semiosphere, semiotic monads can integrate into metalinguistic communities. Given their unique properties, they can be defined as intellectual units, described as "the bearer of reason" (Lotman 1998, p 104). Ultimately, Lotman asserts that "man not only thinks but also exists within a thinking space, just as the bearer of speech is always situated in a particular linguistic space" (Lotman 1998, pp 104–105).

Based on this line of thought, Nakagawa e Cardoso (2020) conclude that the observing subject "should be understood as a type of thinking monad that forms the semiosphere and, as such, exists in exchange with other spheres" (p 123).

With this in mind, the concept of semiotic monads serves as an important element to be contextualised in this work. They act as significant indicators for constructing and redefining a semiosphere (Nakagawa and Cardoso 2020), making them relevant for examining the triad of soil/text, codes, and observer/text. Furthermore, as Lotman (1997) posits, a semiosphere will always contain monads—texts that generate meaning and can forge relationships with other monads, illustrating the dynamic dialogues that resonate in culture.

## 5 We Are the World

To analyse the communicational exercise between humanity and the soil, we draw upon the three semiotic references linked to the Semiotics of Culture outlined above. We have selected two cultural texts for this exploration: the verses from the chorus of the song "We Are the World," by Michael Jackson and Lionel Richie, recorded in January 1985, and the phrase attributed to the scientist Antoine-Laurent de Lavoisier: "In nature, nothing is lost, nothing is created, everything is transformed," which pertains to actions surrounding sustainability (Iaquinto 2011). This selection, inspired by a process of reflection, aims to illuminate the soil as a cultural text.

Before we delve into our reflections, we would like to share with the reader the chorus of this humanitarian song, a significant product of North American pop culture (Kooijman 2013), which has resonated worldwide for decades:

- We are the world, We are the children
- We are the ones who make a brighter day, so let's start giving
- There's a choice we're making, We're saving our own lives
- It's true we'll make a better day, just you and me

For the production of "We Are the World," over 40 music stars came together in a studio, where a sign read, "Check your ego at the door." This message served as a reminder for the artists to recognise that there were causes more significant than themselves (Griffiths 2024), such as the urgent issue they were addressing: the famine in Ethiopia.

This event and the release of the song occurred in 1985, a date outside the Google Trends timeline, which began in 2004. However, when using this tool to examine the peaks in internet searches for the song, two notable dates emerge: 2010 and 2024. The first peak corresponds to the earthquake in Haiti, when a special version of the song was released to raise funds for the affected community (Ma'arif et al. 2022). The 2024 peak can be attributed to the song's appropriation by the streaming service Netflix, which produced a behind-the-scenes programme titled "The Greatest Night in Pop" (Nguyen 2024). It is worth mentioning that this dynamic had already been documented in cinema the same year as the song's original release (Trbovich 1985).

This brief history illustrates how this entertainment project has mobilised humanitarian causes worldwide. In fact, the original initiatives addressing hunger have since evolved to encompass other pressing issues, such as climate change, raising millions of dollars by increasing public awareness about compassion and benevolence (Ma'arif et al. 2022).

So, what do these verses convey to us? What significance does this cultural representation hold regarding our actions as contemporary citizens?

## 6 Verses that (En)Chant

We began with the verse "We are the world," as it encourages listeners to forge a connection with the planet. However, despite the fact that human bodies contain the same proportion of water as Mother Earth and share other characteristics with our vast, predominantly blue sphere, there is an apparent disparity between their codes and ours as human beings. Numerous behaviours reveal a lack of empathy, leading us to conclude that the semiospheres surrounding these two texts encapsulate distinct codes.

In the realm of consumerism, even what may seem like a simple act of consumption can reveal deeper issues, as exemplified by our choice to drink water. Opting for plastic bottled water reflects a mindset characterised by impulse and a lack of consideration for the consequences of disposal. This choice overlooks the tangible reality of these items, and Calvino's short stories (2003), particularly "Leonia," serve as a poignant reminder of the risks associated with such micro-attitudes. While it is true that "in nature, nothing is lost, nothing is created, everything is transformed" (Iaquinto 2011), the act of disposal introduces new codes to be navigated within a semiosphere and causes significant harm to our environment.

Disposable nappies also end up in landfills, justified by the convenience they provide to caregivers and the privileged little humans themselves, who will face serious challenges in managing their semiospheres. However, the codes that circulate within their semiospheres prioritise clean, fragrant babies who embody the verse "we are the children" without considering where their waste has gone.

After all, texts that resonate with codes associated with dirt are deemed impure and, as such, are discarded from the semiospheres inhabited by those considered civilised citizens, often relegated to the peripheries of their own semiospheres.

Through this dynamic, we allocate fewer impure codes to our semiospheres, distancing ourselves from the potential threat of bacteria that can cause disease and make us confront our greatest fear: death. Consequently, our behaviours are guided by minor actions that uphold the sentiment of "We are the ones who make a brighter day, so let's start giving," without pausing to consider what it truly means to be surrounded by a collective of synthetic chemicals aimed at keeping our semiospheres pristine.

Naively, or perhaps driven by the illusion that we are protecting our existence, we often ignore the dynamics of other semiospheres, failing to recognise that, whether collectively or individually, they exist within a continuum. When we consider the ocean as a prime example of this principle, we observe that the sands, the waves, and even the salt in the water circulate and reach various beaches. That prompts us to reflect on the continuum of semiospheres and the constant movement of these representations when we dispose of bottles and nappies or use the synthetic chemicals that we integrate into our semiospheres.

It is also important to recognise that the texts positioned at the edges of a semiosphere are in constant motion, challenging the codes that emanate from its core. Thus, it is an illusion to believe that we can maintain our semiospheres intact and shielded from the texts we discard. Whether these texts circulate in other semiospheres or linger at the fringes of our own, they communicate with those in different semiospheres. Through these movements, semiosis occur, and some texts may even have the potential to reclaim the nuclei of the semiospheres that once rejected them.

The monumental effort to prohibit this exchange is a mere illusion, as the edges are permeable, enabling texts to flow freely. This cultural dynamic urges us to consider whether the line "There's a choice we're making, We're saving our own lives" truly aligns with our current waste-related behaviours. Is it really the best approach to consume, discard waste, and then reuse it while believing we are acting as sustainable citizens?

If this is not the right path, it becomes crucial to consider our role as monads within culture—as meaning-generating units. With this perspective, we can reflect on the continuous flow of semiospheres and the constant exchange that occurs within them.

## 7 In Nature, Nothing is Lost, Nothing is Created, Everything is Transformed

In culture, each semiosphere resonates with different texts, with some prioritising codes of brightness and sanitised cleanliness, following the dictates of the goddess Hygeia. In a delusional effort, we seek to shield these semiospheres from texts associated with dirt. When we consider our bodies as semiospheres, we strive to guard them against contamination by protecting their boundaries. Consequently, hands, feet, and faces are equipped with protective devices such as gloves, shoes, masks, and glasses, for instance.

The use of these objects, which are cultural texts, prevents direct (con)tact between two bodies/texts, thus inhibiting communication through primary media (Beth and Pross 1990). When relationships are mediated by apparatus like gloves, for instance, communication occurs within this secondary media, creating barriers between the texts and dulling the senses of the bodies. In tertiary media, the connection is established through technological devices, and in the health field—where Hygeia's influence is pervasive—this has been extensively utilised. Technologies such as diagnostic equipment and even mobile phones serve as intermediaries for health professionals, allowing them to avoid touching "contaminated" texts like patients themselves.

These distances interfere with some semiosis, a fundamental phenomenon in the study of semiotics, which enables the analysis of the process of engagement and the actions of signs. According to Machado and Romanini (2010), semiosis is the generator of life, as it allows species to evolve. Semiosis involves understanding the mechanisms and actions that interact with texts, the regulations and reactions surrounding them, and the possible stimuli and representations that shape their arguments and sharing, influencing decision-making and concept formation.

In light of this, we have considered some uncertainties surrounding soil, the element that explicitly supports our bodies and lives. Despite the unique significance we owe it, modernity has stripped us of the knowledge and values it imparts. We walk on it, shedding our burdens, moving through life without even glancing down. Though etymology reveals the deep connection between humans and humus, we disregard this cultural text. After all, soil represents the dirt on the farmer's blackened fingernails and even serves as the eventual resting place for bodies when they succumb to death.

This attitude of disregard prevents us from recognising the life that thrives within this semiosphere and the wealth of its biodiversity. Despite its microscopic dimension, Brito (2023) notes that "around 25% of the diversity of terrestrial life" exists within it and that "in one spoonful of soil, there are more microorganisms than there are inhabitants on Earth." The semiosphere that contains these texts encompasses seeds and various elements often regarded as despicable, typically associated with dirt. However, as previously discussed regarding the dynamic nature of semiospheres, the symbiosis of the codes developed within this context extends beyond this semiosphere, interconnecting with others through its myriad texts, becoming either visible or invisible within human semiospheres.

It can be suggested that children may be less susceptible to contamination by narratives that amplify the impurities and dirt present in the soil. This could be attributed to their stature, which encourages a more intimate engagement with the texts circulating within this semiosphere, allowing them to perceive these elements with a sense of wonder. The communicative connection between children and the soil, nurtured through the dynamics of primary media, can later be reflected in adult memories. In this regard, it is important to acknowledge the nature of human semiospheres, which often share children at their peripheries. As previously noted, texts located at the edges are particularly prone to circulating and interacting with other semiospheres. Considering this movement contributes to semiosis, we pondered whether, by embracing the sentiment of "We are the Children," we might foster a healthier relationship between society and the soil and better equip ourselves for a genuine transformation.

Ultimately, amidst the myriad "ifs" that emerge from contemplating our potential, we can embrace the concept of the semiotic monad and integrate the verse "We are the soil" into the song "We Are the World"—a call to action inspired by Vandana Shiva (2014). By amplifying this message across various semiospheres, we encourage the essential semiosis that support Gaia. Metaphorically, by igniting this resonance, we can trigger a cascade of positive changes, affirming the sentiment: "It's true we'll make a better day, just you and me."

## 8 Final Considerations

This proposition highlighted several issues related to the preservation of biodiversity, with a particular focus on soil and its connection to human beings. It underscored the significance of individual actions within a collective framework, drawing upon the abstract notion of semiospheres—a concept derived from the semiotics of culture that informed this work.

The complexity paradigm endorsed this agenda, itself rooted in culture. This principle did not preclude the establishment of the concept of the semiotic monad within the discussion, highlighting its role in the collective. Its uniqueness does not diminish its representativeness, as it is capable of shaping thoughts in culture and facilitating the emergence of (new) messages.

Thus, when contemplating the impact of isolated actions aimed at promoting biodiversity and, more specifically, the well-being of Mother Earth's soil, it is important to recognise the 8 billion human bodies interconnected with this greater good. It is worthwhile to generate powerful semiotic monads that encourage the proliferation of semiosis within the semiospheres. Furthermore, given that this collective of texts exists within a culture driven by semiospheres organised in a semiotic continuum, it is reasonable to assert that isolated actions related to this cause contribute meaningfully to the collective effort.

Additionally, it is important to consider the soil in regard to the role our memories play, as culture is inherently tied to memory (Ferreira 1995). These narratives can resonate with various stages of our lives, helping us recognise our connection to the soil—as a text in itself—represented by the etymological link between humus and humans. In other words, language can serve to reinforce this close relationship between humans and organic matter, viewed here as a cultural text. By elevating its significance in our perceptions, we can better support the transition towards the sustainability we seek.

When looking ahead, two questions can be explored and connected to deepen the concerns raised here. First, it is essential to examine the media landscape to understand the imagery associated with soil, investigating how these representations have been presented to the public and the contexts in which they have been invoked. On the human side, considering advancements in neuroscience, we can embrace an interdisciplinary approach to explore whether our brains respond to these calls and whether such responses align with individuals' narratives about their relationships with the soil.

**Acknowledgements** This paper was supported by national funds, through Fundação para a Ciência e a Tecnologia, I.P., within the scope of the Multiannual Funding of the Communication and Society Research Centre 2020-2023 (which integrates base funding, with the reference UIDB/00736/2020, and programmatic funding, with the reference UIDP/00736/2020).

**Translation:** Anabela Delgado

# References


Américo EV (2017) O conceito de fronteira na semiótica de Iúri Lotman. Bakhtiniana Rev Estud Do Discurso 12(1):5–20

Américo EV (2015) Iúri Lotman e a Escola de Tártu-Moscou. Galáxia 29:123–140. https://doi.org/10.1590/1982-25542015120156

Beth H, Pross H (1990) Introducción a la ciencia de la comunicación. Anthropos, Barcelona

Brito I (2023) A importância dos microorganismos do solo. Instituto Mediterrâneo para a agricultura, ambiente e desenvolvimento: Évora. https://www.med.uevora.pt/pt/a-importancia-dos-microrganismos-do-solo/

Calvino I (2003) As cidades invisíveis. O Globo/Folha de São Paulo, São Paulo.

Charaudeau P (2009) Discurso das Mídias. Editora Contexto, São Paulo

European Comission EC (2019) The European green deal. Publications office Brussels. https://eur-lex.europa.eu/resource.html?uri=cellar:b828d165-1c22-11ea-8c1f-01aa75ed71a1.0002.02/DOC_1&format=PDF

European Comission EC (2021) EU soil strategy for 2030: reaping the benefits of healthy soils for people, food, nature and climate. Publications office, Brussels. https://environment.ec.europa.eu/document/download/ae853f10-c9a2-4665-a9f2-c29d11c49374_en?filename=COM_2021_699_1_EN_ACT_part1_v4_0.pdf&prefLang=pt

European Parliament EP (2021) European Parliament resolution of 20 October 2021 on a farm to fork strategy for a fair, healthy and environmentally-friendly food system. Publications office, Brussels. https://www.europarl.europa.eu/doceo/document/TA-9-2021-0425_EN.html

Ferreira JP (1995) Cultura é memória. Rev USP 24:114–120. https://doi.org/10.11606/issn.2316-9036.v0i24p114-120

Flor A (2023) Vandana Shiva: cuidar do solo é o maior seguro contra as alterações climáticas. Público. https://www.publico.pt/2023/11/04/azul/noticia/vandana-shiva-cuidar-solo-maior-seguro-alteracoes-climaticas-2068662

Griffths D (2024) Check your ego at the door: the story of USA for Africa's we are the world (and why Prince and Madonna aren't on it). Music Radar. https://www.musicradar.com/news/usa-for-africa-we-are-the-world

Iaquinto K (2011) Na natureza nada se cria, nada se perde, tudo se transforma. Conjuntura econômica 65(6):29–32. https://periodicos.fgv.br/rce/article/view/22722

Israel D (2020) A semiótica da cultura, a produção de sentidos e o problema do texto nas histórias em quadrinhos. Intercom—Sociedade Brasileira de Estudos Interdisciplinares da Comunicação. 43º Congresso Brasileiro de Ciências da Comunicação. https://portalintercom.org.br/anais/nacional2020/resumos/R15-0990-1.pdf

Kooijman J (2013) Fabricating the absolute fake: America in contemporary pop culture (revised edn). American studies series. Amsterdam University Press, Amsterdam. https://www.jstor.org/stable/j.ctt6wp7ck

Lotman I (1996) La semiosfera I: semiótica de la cultura y del texto. Cátedra, Madrid

Lotman I (1997) Culture as a subject and an object in itself. Trames J Humlties Soc Sci 1(1). https://www.kirj.ee/public/trames/lotman.htm

Lotman I (1998) La semiosfera II: semiótica de la cultura, del texto, de la conducta y del espacio. Cátedra, Madrid

Ma´rif HM, Jatmika S, Marlinda AP (2022) Music as propaganda for humanitarian diplomacy: semiotic analysis of Ferdinand de Saussure in the song we are the world. Int J Multicult Multireligious Underst 9(9):144–157

Machado I, Romanini V (2010) Semiótica da comunicação: da semiose da natureza à cultura. Rev FAMECOS 17(2): 89–97. https://doi.org/10.15448/1980-3729.2010.2.7546

Machado I (2021) Tradução intersemiótica segundo Iúri Lotman: dualidade, alteridade e intraduzibilidade. ABRALIM. https://www.youtube.com/watch?v=UqPNFf548EM

Morin E (1997) O método I: a natureza da natureza, 3rd edn. Publicações Europa América, Portugal

Nakagawa R, Cardoso T (2020) Epistemologia semiótica e a questão do observador em Peirce e Lotman. Estud Semióticos 16(3):112–132. https://doi.org/10.11606/issn.1980-4016.esse.2020.172812

Nguyen B (2024) The greatest night in pop. Netflix. 1h27

Nunes MR (2019) Memória do futuro, explosão, pancronia: a semiótica de Lotman e os estudos da memória e do tempo nas teatralidades juvenis. Bakhtiniana Rev Estud Do Discurso 14(4):192–210

Roxo MJ (2023) A importância da conservação dos solos: evoluir Oeiras. Facebook. https://www.facebook.com/EvoluirOeirasMov/videos/614940047230024

Shiva V (2014) We are the soil. Common dreams. https://www.commondreams.org/views/2014/05/26/we-are-soil

Trbovich T (1985) We are the world: the story behind the song. Paul Brown Stein productions. 53m

United Nations UN (2015) Transforming our world: the 2030 agenda for sustainable development. https://sustainabledevelopment.un.org/post2015/transformingourworld

World Meteorological Organisation, WMO (2021) Atlas of mortality and economic losses from weather, climate and water extremes (1970–2019). N.1267. https://sidsport-climateadapt.unctad.org/wp-content/uploads/2022/06/WMO-ATLAS-OF-MORTALITY-AND-ECONOMIC-LOSSES-FROM-WEATHER-CLIMATE-AND-WATER-EXTREMES-1970%E2%80%932019.pdf

**Cynthia Luderer** (PhD in Communication and Semiotics, by the Pontifical University of São Paulo-Brazil) is a researcher at the Communication and Society Research Center (CECS) of the University of Minho. She dedicates her research to the field of food consumption and is interested in discourses of sustainability, memory and health around this field. Part of her PhD was in Anthropology of Food (Rovira i Virgili University, Spain) and this knowledge adds value for the fieldwork. Luderer is a member of study groups in Brazil (Mnemon- ESPM, São Paulo) and in Portugal (DIAITA, University of Coimbra; SOPCOM, deputy leader of the Semiotics Working Group). ORCID iD 0000-0001-5149-669X. She can be reached at cynthialud@gmail.com

# Art, Nature and Well-Being Images Engaging in Dialogue with One Another

**Eduardo Paz Barroso**

**Abstract** This article seeks to explore the possibilities of communication through artistic creation within the framework of Social Sciences and Humanities, focusing on specific examples. It involves examining how the arts, particularly works and discourses embedded in the contemporary tradition of Land Art, reflect on the degradation of public spaces and the preservation of nature and the environment. The message of these visual artists is centred on enhancing life and existence, alongside a critique of urban daily life. The new perspectives, realities, and circumstances created by the artistic process promote a sense of well-being that is cultural, psychological, and visual in nature.

## 1 Brief Philosophical Notes on the Nature/Culture Pairing

This text discusses certain aspects of the relationship between nature and culture. It also characterises Land Art and highlights its pioneering impact. In this context, the commentary on a work by a Portuguese artist—Alberto Carneiro—is presented, marked by a strong authorial identity deeply rooted in the land and its imaginary. Finally, a critical analysis is conducted on two artistic works that exemplify some of the most pressing contemporary aesthetic concerns resulting from thorough inquiry and questioning. Gabriela Albergaria (Vale de Cambra, Portugal, 1965) and Lia Chaia (São Paulo, Brazil, 1978) are studied here as artists whose work is significant in approaching a territory marked by different dynamics of opposition, superposition, and complementarity concerning the nature/culture dichotomy. This underscores the necessity of recovering values and objects that reinvigorate the poetics of a relationship that is either in crisis or lost.

The richness and diversity of aesthetic reception, the formative and transformative role of sensibility, and its intellectualisation are especially favoured by artistic

E. P. Barroso (✉)
FCHS/University Fernando Pessoa & LabCom University of Beira Interior Portugal, Covilhã, Portugal
e-mail: epb@ufp.edu.pt

 
W. Leal Filho et al. (eds.), *Composing Worlds: Humanities, Health and Wellbeing in the XXI Century Towards a More Sustainable World*, World Sustainability Series,
https://doi.org/10.1007/978-3-031-87108-5_6

processes. Aesthetic creation thus promotes and intensifies a kind of well-being within a psychic and emotional scope. It garners varying levels of attention and interest and is increasingly present in the media.

In this hyper-informed environment, the added value brought by art lies in the promotion of slowness over haste and observation over fleeting glances. The absence of haste contributes to a healthier experience, implying a knowledge of how to navigate time or its interstices. On this premise, it is possible to argue that the humanities contribute to and enhance well-being and should be involved in discussions about health, particularly mental health, which requires intellectual lucidity. The challenge here is also to find the breathing space of things situated within the singularity of nature, which is so threatened today. This, in turn, helps to situate and preserve the much-needed integrity of life and the self in societies with dystopian traits. To meet this challenge, it is necessary to enrich the process, modes, and rhythms of understanding reality. This, in turn, calls for an understanding of art as an exception—a kind of "other place" and "place of the Other" in the psychoanalytic sense of the term (Perniola 2005; Heinich 1998; Damisch 1984a, b; Argan 1999; Gombrich 2000; Freud 2008).[1]

Contemporary art, in some of its aspects and manifestations, deals with the theme of nature, its organicity and materiality, creating images and situations that re-present the natural. It is thus inscribed within the philosophical process of inquiry into the relationship between the human being and things and existence.

The concept of nature is present throughout classical culture. Since the ancient Greeks, philosophy has related the nature of the subject in its ontological dimension to nature in the sense of what is natural and part of life and the elements that constitute it—animals and plants. This direction is likely the one taken by Robin Collingwood in his work The Idea of Nature (1945) and according to the path he established. It begins with pre-Socratic Ionian science, passes through the Pythagoreans, Plato, and Aristotle, and then focuses on Renaissance cosmology and its protagonists. Subsequently, it covers other stages of the vision of nature. Collingwood's originality lies in considering finitude in nature and its relationship with history. According to Edmund Leach, who highlights Collingwood's seminal role (1985, 67–68), "in Greek thought, nature indeed represented a tautological principle, a final cause. It was thought that the external world, endowed with existence independent of human thought and action, was imbued with a metaphysical order—or spirit." Thus, there is a singularity in nature that certain types of contemporary artworks reconsider and

[1] The bibliography briefly referenced here is merely illustrative and aims to elucidate and exemplify a discussion on aesthetics, painting, sculpture, the visual arts, and their various manifestations, historical perspectives, and the impact they have on the subject, self-understanding, worldviews, and the unconscious. From this complex and diverse process, with its innumerable communicational and symbolic dynamics, arises the possibility of expanding the way we think about the concept of well-being on a psychic, individual, and civilisational level. An example of this is the specific characteristic that the artist has to "play," taking advantage of the "distance" they are capable of, whether in relation to "tradition" or to "life." In this way, the artist defends "historical values" that are not exclusive to either of these instances. This allows the artist to embrace "contradiction" as the "primary duty of an intellectual" (Damisch 1984, p.88).

**Fig. 1** Gabriela Albergaria Chêne à Soquence, 2020 Oak trunk, vernon stones (from Normandy), lime mortar Installation at Domaine de Sequence, Sahurs, Normandy. Photo Marc Domage

bring to a new context of appreciation. For them, it is a matter of revealing, of making visible without, however, excluding a metaphysical trace, the nature of nature. At the same time, they ask, as the Greeks did, what is nature? (Fig. 1).

The sophists, who played a significant role in classical culture through an ideal of instruction based on the word, are the subject of Plato's distinction between discursive rationality and philosophical knowledge, and they propose the differentiation between *what is by nature* and *what is by convention*. This distinction between nature and language is also critically revisited in contemporary artworks that ritualise and celebrate nature and its order, cycles, and states through strategies and operations of appropriation. The ecological plan shifts to a space of visual and political appreciation, which can either claim the romanticism of a lost harmony or become a protest and action against degradation resulting from industrialisation and capitalism (Fig. 2).

Ultimately, for the philosophical tradition, it is also about discussing the intersections between human nature (and what is inherent in each subject) and the sophistication and beauty of the natural world. From it, we can identify new objects, discovering and identifying what is intrinsic to them. For example, the chlorophyll function is intrinsic to a plant. Or the survival instinct is intrinsic to animal behaviour. Nature (which the Greeks called *physis*) was a "spiritual principle of development," related to the "internal movement of things," and opposed to "objective" and "non-natural" qualities. Conversely, the Renaissance's system of ideas abandons Aristotelian conceptions that explained nature as if it were "animated by a tension, by an effort to realise as-yet nonexistent forms" (to use a formulation by Collingwood).

**Fig. 2** Lia Chaia. Urban flora (detail) 2020, print on photo paper and pin, 51 × 91 cm. Photo Ana Pigosso

When a separation between matter and spirit occurs, it becomes necessary to find another explanation for the reality of nature, now seen as immutable (in the sense that it is governed by laws, even though that does not eliminate unpredictability and chance). The discovery and subsequent valorisation of mechanical phenomena and the introduction of mechanics into experience identify the world and its physical reality with nature.

Throughout its vast trajectory, the history of philosophy consistently maintains, in various texts and positions that we will not critically examine here due to their vast scope, a concern with the human being as a living and rational element of nature. This, in turn, refers to the function and role of language. And to the question of unity and separation. It is inappropriate to address the theme of nature in disconnection from anthropology. And from man's ability to act upon nature, starting with the process of sedentarism associated with the discovery of agriculture and with all the consequences that have unfolded over the centuries, giving rise to a sociologically and economically fracturing process, exacerbated in post-industrial societies intoxicated by carbonisation.

Resuming the distinction between what is natural and what is cultural in man makes it possible to evoke an entire mythology but also a contemporary narrative that examines the role of language as a determining cultural characteristic that ultimately absorbs or transforms natural traits of creatures. Let us consider Edmund Leach's assertion (1985:75) that "language may be the only attribute of adult human beings that is simultaneously natural and cultural." By explaining the nature/culture dualities, after emphasising that language is not merely a communication tool but allows for the formulation of ideas and acts within and upon the mind, Leach, an anthropologist, highlights the importance of our ability to encode and decode everything that is part of reality as an "aspect of human nature." It is language that now moves us toward a reconciliation with nature, perhaps because it is the most important element that, in a certain sense, deeply connects us to it. In present contexts, where the word "environment" substitutes for the need to understand nature and carries with it a politically biased activism laden with ideology, the possibilities of an aesthetic dialogue and the philosophical reinvention associated with it stimulate thought-provoking poles of introspection.

## 2 Anguish and Redemption: Connections to Be Made

In one of his essays, the German philosopher Peter Sloterdijk (2008, 99–124) refers to a nature yet to be made. In broad terms, in this text, the author begins by evoking a retrospective view from a modern context that he prefers not to detail, to then invoke the Romantic poet Hölderlin when he speaks of the fall of Nature, which human beings accompany, leading to a rupture in which, according to the poet, there is an oscillation (which ultimately concerns the nature-culture separation) between everything and nothing. The equation God-Being-Nature, in opposition to the equation Man-Will-Technique, emerges here as a guiding thread that Sloterdijk proposes as responsible for leading toward a less "harmonious" understanding of origins and a "less moralistic" understanding of the rupture at issue here. The "unity" between man and nature thus appears deceptive and places emphasis on the evolutionary process of the child born unprotected, leading him to conclude that "human evolution incorporates the natural history of the unnatural." Sloterdijk sketches from here a theory that articulates the interpretation of classical mythological texts to refer to the Earth as the Great Mother. From the idea of "sovereign femininity" and the "psycho-archaeological excavations" associated with it, he discourses on a journey through mythology that highlights the "ambivalence of matriarchal and holistic thought models."

In brief, for the author of the essay, the aim is to open from here another interpretative possibility based on violence, a kind of "marginality" relative to the cosmos, where a variety of human acts are possible, from the use of discourse to the production of theory, handling weapons, and building cities. This recognition of man's own power gives rise to other kinds of connections and to what, for convenience's sake, we can call subordination, if not submission, of nature.

Throughout his essay, Sloterdijk relates alchemy and the metaphors it provides with what he terms "engineers and designers of nature," defining it in brief words as "the holder of the power of birth." One passage is particularly illuminating as to the issue of guilt for the destruction of nature. The price to pay for challenging the Gods places Man before their hubris (a heavy challenge) and perhaps their limitations when it comes to the ambition of reaching the core of *natura naturans.*

"Man as creator is currently sitting in the dock and is forced to listen to the indictments of nature's lawyers, who confront him with his *hubris*, his criminal arrogance, his frivolous talent, i.e., 'incompetent', and his semi-criminal semi-competence in assuming technological power over the Earth" (Sloterdijk 2008:115).

This way of presenting one of the most pressing issues at present forces the question of what to do. It invites reconciliation, as if *forgiveness* on the part of this *Great Mother Earth* were possible. As we shall see later, in philosophical, ethical, and axiological terms, a varied set of artistic proposals and languages can fill the void caused by this anguish and provide another kind of response, not necessarily a "militant" nor a political one produced by the governances.

## 3 The Works of Language

This matter of a *Nature yet to be made*, qualified by the German philosopher as "the decisive theme of the modern era" (although he grants the notion of modern a broad latitude and includes it in a vast temporal arc), summons new habits for the subject to think. This goes far beyond ecological awareness and demands an active consciousness where each one sees themselves as "part of a focused whole." The term "natural reserve," the musealization of landscapes, and the establishment of spheres of administrative competence over territory do not resolve the need for an integral vision. Sloterdijk (2008:123) concludes his reflection by talking about an effort to reach an intelligence capable of "determining more precisely the parts that need to be made in Nature and History over longer periods and more satisfactorily than what can be dreamed today." Looking at nature through art is thus a means to achieve this intelligence. Therefore, our hypothesis is to consider artistic work directly focused on these concerns as a response that is both visual and philosophical.

In his critique of Lévi-Strauss's anthropology, interpreting his thesis that "culture is like a language," Edmund Leach (1985b, a: 110–112) distances himself from the interest in unconscious meaning, while considering that the structuralist anthropologist ultimately says that "culture is, beyond expression, the creation of an entirely incorporeal entity—the human spirit (*esprit humain*)," which leads to a tautological result. If Leach, especially in his extensive commentary on Claude Lévi-Strauss and his structural anthropology, concludes, to summarise, that the notion of a science of culture is ultimately the same as the science of society. Nonetheless, he values everything that is verifiable as a manifestation of that very culture. The mental vastness that culture translates into, and which directly concerns us here, especially considering artistic creation and its materialisations, discourses, and meanings, becomes an echo of the metamorphoses. In this respect, it takes an interest in ruptures, discontinuities, the weakening of territories, and the disharmony of landscapes. At the same time, this mental vastness, this continent of signs, is interested in the lush splendour of the ecosystems that still resist. It is in this line of resistance that the works of the two artists highlighted here, Gabriela Albergaria and Lia Chaia, are inscribed.

Their discourses convey an art of the land and a critical art of urban concentration phenomena. Both give voice to a desired promised nature.

## 4 Land Art: A Spiral of Symbols

One of the trends in contemporary art that has gained extraordinary notoriety reflects an evolutionary process that consecrates natural spaces and elements of nature—plants and minerals, mountain formations, and watercourses—as symbols of aesthetic ideas. This raw material gives rise to altered landscapes, manipulations, and reutilisations. By playing with different scales and dimensions and employing various modes of recording (photography or film), the artist committed to these materials transforms and displaces the spectator's physical and mental conditions. Spaces of confrontation unfold between the individuality of the subject and the plurality of nature itself. Its complex vastness appeals to both the Kantian sublime and the nostalgia of a lost oneness.

The traces of human presence reconcile with the unique character of uninhabited spaces. It discovers paths and silences where it is possible to find a kind of religious grandeur. And the spectator realises that they are far from other discourses of modernity focused on critiquing mass culture and the consumer society. By distancing itself from the artificiality that covers reality and signalling the Earth's appeals, Land Art echoes an alternative to the major processes of industrialisation and the damages they cause to the natural world.

The American artist Robert Smithson (1938–1973) is a foundational reference in this artistic movement. *Spiral Jetty* (1970) is one of his most emblematic works. It is a gigantic installation that moved several tonnes of basalt stones into a Salt Lake in Utah (USA), forming a spiral drawn counterclockwise. Over time, the formation of salt crystals gains complementary importance as a constituent element of this piece. Robert Smithson himself develops, in a text where he describes the constructive process and the premises that motivated him, a reflection on the *Spiral Jetty*, which ultimately integrates the work itself. The proximity of abandoned installations, previously occupied by infrastructures used for oil extraction, lends the work a prominent significance. In terms of the medium selected for visual and aesthetic communication, the chosen one was a film directed by Smithson himself. An integral part of this artistic object, it documents the construction of the spiral that has withstood time and functions as a pier, allowing incursions into the lake. The reading we can make of this work implies, first and foremost, the realisation that it is possible to add symbolic value to nature through a reading in which the human element creates density and monumentality without introducing alien or adverse symbols into the landscape or eliminating the contradictions between natural and industrial, between resources and pollution, between harmony and disharmony.

Smithson is one of the great protagonists of 20th-century American contemporary art, profoundly original in his proposals, which cannot truly be understood if detached from a textual field of writing and philosophical reflection in which the artist thinks about the redefinition of the physical and material nature of art. A text by Jack Flam (1996) highlights the relevance of the artist's writing. This writing is a genuine correlate of the materiality of his works inscribed in the landscape. Smithson distances himself from the nature-culture division since his work develops towards a culturalization of nature. This ultimately has a foundational reach concerning the works of other artists that we will discuss later in this essay. Flam notes resonances of T. S. Eliot's poetry—the allegory of the lifeless land that awaits the sap of spring while sensing the cruelty of waiting ("April is the cruellest month," writes Eliot). However, we can also detect the influence of Jorge Luis Borges in Smithson's prose. The attitude of a collector, the rigor of montage in a game where nature absorbs external elements that become part of it, is also a distinctive feature of this work.

Jack Flam also highlights in his study, on the aftermath of Borges's work, the blurriness of fiction and reality, allusions to a field of science fiction, which does not present any conventionality as a literary and cultural genre. In several of his works, we find an architecture combined with the vastness of time, as well as elements from very varied sources ranging from popular culture to philosophy. *Language is meant to be seen, and things are meant to be read.* This sentence is a typical example of a specific semiotics where Robert Smithson dwells, which stems from Borges but also from other authors such as Umberto Eco: the world is an infinite text. And one could add, in this view, that nature is the heart that makes this text pulse. The precursor originality of Smithson, as Flam highlights, is more than the ambition to reach outside the studio or museum, but above all, to create new capacities for continuity between "inside" and "outside." This is why he is particularly interested in geological past and ancestry.

This controversial artist of his time, primarily due to his involvement in the ambiguous discussion about the purity of art, encompasses by his initiative the theme of humanism, which he sees not in opposition to deism, as Jack Flam points out in his study, but rather takes humanism as a belief in a system that "allows for ontological differences between humanity and the rest of nature" (Flam: 1996 XVIII). The talent for capturing and surprising the passionate but somewhat insignificant human presence amidst the organic vastness of nature (to refer to a formulation in Flam's essay) makes Smithson the initiator of something unfinished, like nature, which renews and completes itself between promises and cruel threats.

## 5 Alberto Carneiro: Recreating a Tree

"What we can communicate by recreating a tree in the need to possess it will certainly be the values that bind us to it in the circumstance of that moment and, henceforth, the places where the recreation of memories that we all have of those trees may occur." This sentence was written by a sculptor, Alberto Carneiro (1937–2017), in *Notes* for a manifesto of ecological art, written between December 1968 and February 1972. This artist reveals a pioneering attitude in line with a contemporary programme, especially so when we bear in mind the uniqueness of a political-cultural periphery, which until recently was the Portuguese one, and which only gradually resolved itself thanks to a European integration that implies the internationalisation of Portuguese art, both at the museological level and at the market and gallery activity. "Art is made to transform everyday images," Carneiro writes in his aphorisms, which intersect his sculptures and other pieces he created (Fig. 3).

Summoning this artist's plastic and sculptural work within the context of this essay allows us to approach the works of the Portuguese artist Gabriela Albergaria and the Brazilian artist Lia Chaia. These are two artists that were already mentioned here, whom we place in a line of coherence with ecological concerns that translate into a polarisation of objects and gestures, feelings, and ways of seeing the world, divided between allusions to nature and the city, the vegetal and the urban, the garden and the habitation, the forest and the street.

In Alberto Carneiro's case, we can speak of an emancipation of the body as sculpture. This stems from the application of artisanal techniques to sculpture. Tangled lines sliding, furrowed on tree surfaces and trunks, or bundles of wicker that ritualise space and the field in the agrarian sense of the term. In this work, there is also a strong concern to abolish borders and typologies that could capture forms in predefined codes. The mental and material transgression gives rise to a consistent variety of supports connoted with Land Art and ecological art, such as various types of vegetal

**Fig. 3** View of the exhibition "Alberto Carneiro: Arte Vida/Vida Arte—Revelations of energies and movements of matter", Fundação de Serralves—Museu de Arte Contemporânea, Porto, from April 19 to June 24, 2013. Photo Filipe Braga

elements—canes and hay, photography, drawing, granite, different wood qualities. All of this is organised within a personal aesthetic system that offers a vision of the world that is rooted in the telluric space of nature.

The enormous impact of this work would deserve an independent study within the theme were are developing here. However, for the time being, we will merely refer to a very representative retrospective exhibition at the Galician Centre of Contemporary Art (2001), which allows the revisitation of this author's significant works. Had they been known in an international context at the time of their creation, their alignment with a state-of-the-art framework at an international scale would have been evident.[2] Carneiro had the opportunity to study in England at Saint Martin's School of Art with Anthony Caro and Philip King, two relevant references in the British panorama of the 1970s. The very consistent trajectory of this artist led him to follow a line shaped by the permanence of rare and very personal aesthetic feelings. In this circumstance, it is also relevant to quote the title of a work from 1970–71 that consisted of a line of printed paper, rope, and cotton intended for exploring a space, for example, a gallery or museum wall. This refers us to the idea of a path applied to the artist's body, in the almost anonymity of a presence translated into a record of 48 photographs, which is something that resonates from images where the artist allows himself to be photographed in an intimate relationship with the earth. This work from 1976–77 represents, as an artwork, several of the issues previously discussed here concerning the notion of Mother Nature, the objections raised against it, and the anthropological declension to which it lends itself given its fertile metaphoric power.

What kind of beauty does Carneiro's work seek, in absolute fidelity to the aesthetic values that trees and the earth, porosity, roughness, granularity, textures and use of space allow? This foundational question guides the artist in a poetic refinement to which the synthesis offered by Japanese haiku is not alien. This is a web of identifications elaborated in the centre of an inescapable, subtle, and meditative place. Perhaps this is the greatest surprise this artist brings to us: nature in front of us and us in front it, in a protective communion.

[2] The choice of this expression, beyond its sociological literalness, also refers to Arthur Danto's concept of the 'Artworld' as a foundation for intercommunication between different actors, entities, and institutions involved in the production, presentation, circulation, and symbolic and economic consumption of artworks.

See, among others, the anthology published in Portuguese *O que é a arte? A perspectiva analítica* (2007), organised by Carmo d'Orey, Dinalivro, Lisbon, and Danto, Arthur C., "The Artworld," in *The Journal of Philosophy*, 61, 1961, pp. 571–584.

## 6 Gabriela Albergaria: An Exercise in Permanence

Gabriela Albergaria's work is guided by an attitude that is marked by an exceptional capacity for observation and a practice of gathering.[3] These are the two sides of a mirror where both the artist and the world around her are reflected. This is a world that does not abide by chance. Rather, it is the result of a significant choice and is selected because of its morphology. Both what is perishable and what is longer-lasting exist in it as a guarantee of memory and intelligibility. Thus, this is a world that is simultaneously interior and exterior, attentive to the work and action of time, to indications, records, and classifications. A world organised under straight lines, with its geology—sedimentary layers but also capable of accommodating the resplendent splendour of bare trees and the opulence of their trunks.

One of the most interesting characteristics of this artistic production stems from how it is planned. This was the case in 2015, when the artist was invited to join a scientific expedition to the Amazon organised by a researcher from the University of São Paulo (Brazil) in the field of botany. This experience allowed her to engage in research on the origins of the plant species that populate this emblematic yet tortured region. Photography, drawing, and the creation of a herbarium were some of the tools she used, which were transformed into materials for an aesthetic discourse that allowed her to understand that art—drawing and painting—is, in some cases, the only way to bring forth the authenticity and spontaneous beauty of nature (Fig. 4).

In the artist's own words: "In 2016, I joined a group of scientists on a boat trip up the Branco and Negro rivers in the Brazilian Amazon to take part in the study 'Understanding the Origin of the Forest.' I was part of a team coordinated by botanist Lúcia Lohmann from the Institute of Biosciences at the University of São Paulo and ornithologist Joel Cracraft from the American Museum of Natural History in New York. The scientific objectives were clearly defined and a disciplined daily routine included trips in small motor boats to explore the riverbanks and collect specific plant species for cataloguing. As an artist, the expedition forced me to reposition and adapt to the conditions of the trip. I decided to take part in the morning outings to collect plants with the rest of the team. I assumed the role of a scientist. I collected the plants, photographed and drew them, and in the afternoon organised the material I had collected in the morning. During the process of scientific cataloguing, I observed that

[3] It would be interesting to reflect on possible points of contact with a highly emblematic film by the Belgian filmmaker Agnès Varda (1928–2019), titled *Les Glaneurs et la Glaneuse* (2000), where we encounter a moral and political questioning of the attitudes towards collection and waste in contemporary urban societies. The film draws inspiration from 19th-century French painting, particularly Millet, though Van Gogh's work could also be invoked. Varda's relationship with painting is naturally a topic for another discussion. However, here we are interested in highlighting the significance of a gesture, and how it has the power to emphasize, bring a specific moment to the fore, or isolate something that is collected. Referring to Millet raises a series of doctrinal questions. The historian Giulio Carlo Argan speaks of François Millet's "ethics and religiosity of rural labour" but also of an "ambiguous political choice" (Argan 1999:71), which is evidently not the case with the contemporary Belgian filmmaker. As for Albergaria, the ability to retain and consecrate the act of collection is worth noting here, allowing it to engage with this discussion.

**Fig. 4** Gabriela Albergaria. Purple Beech at Kew Gardens, 2023 Green colour pencil on paper (Heritage Woodfree Bookwhite 315gsm). Photo Bruno Lopes

the initial colour and shape of the plants are the two characteristics most easily lost. Traditional herbaria present dried leaves (with dimensions altered through drying) and without their natural colour (also due to the drying). Seeking to add something the scientific process could not, I kept my notebook of various species' colours and dimensions. I also collected several seeds and had moulds made when I arrived in São Paulo in order to later transform them into objects/sculptures. I organised an archive with the photographs, and in my studio in New York and later in London, I began the process of transforming these files into artworks."[4]

The promotion of a dialogue in an exhibition context between elements and objects from natural reality and forms of language from cultural reality finds here one of its most eloquent and pertinent manifestations. It is an aesthetic reason founded on pleasure and necessity: the need to find nature and the pleasure of capturing it in an intimacy that requires multiple learnings. The sense of belonging, of proximity, drawing as a discipline and mental organisation, connects different experiences distributed between precariousness and ancestry (Fig. 5).

Another aspect that deserves examination in this work concerns the definition of itineraries related to the choice of places and landscapes—in short, the site. The latter constitutes an indispensable condition for aesthetic achievement. This is why the spectator finds in Gabriela's various works a delicate and meticulous treatment of

[4] For a comprehensive documentation of Gabriela Albergaria's work, see *Nature Abhors a Straight Line*, published by Culturgest, Lisbon, 2021. This book was released in conjunction with the artist's exhibition at that cultural centre, and the citation is taken from page 90.

**Fig. 5** Gabriela Albergaria. Couche Sourde 2010/2020 Pressed soil, branches and leaves, 120 × 984 × 140 cm. Courtesy Galeria Vera Cortês, Lisboa. Photo Edouard Fraipont

elements, objects, and fragments, which are either models for drawings or are objectively displaced and transformed but never lose their primordial identity. This refers to the question of origin. We can speak of a scenographic and theatrical dimension in many of Gabriela Albergaria's works, not because they arise from devices that refer to the stage, but due to a singular dramatic quality that the artist highlights with extreme rigour, at once silent and restrained.

The poetics resulting from this creative process follows the course of nature without hesitation, both in terms of the vitality of the sap that runs through it and the wear of the erosions that affect it. Also, for this reason, the concept of landscape is analysed in a critical metalanguage that somehow seeks to protect the authenticity of a relationship in a world where references disintegrate and organisation yields to the chaos of images. The landscape would then be a reconquest or reconversion of highlighted pieces, of recovered fragments, thanks to a gesture served by techniques and knowledge that art makes available for yet another reading. One example among many is the piece *Couche Sourde* 2010/2020, a block of earth, leaves, and pressed branches that, in practice, constitute a technique for cultivating plants that favours seed development outside their places of origin. Gabriela Albergaria describes this technique in detail while displaying a portion of earth bordered by a *taipa* that gains a sculptural significance. "What interested me was that this cultivation technique had completely transformed the world's landscape," the artist says.

Returning to the Nature/Culture dualism already mentioned, it is worth saying that Gabriela Albergaria's work addresses the viewer, inviting them to admit that there are contexts in which, ultimately, this binary structure does not work. Nature and culture are subsumed in one another. This aligns with the thinking of anthropologist Philippe Descola, who reformulates categories originating from philosophy.[5] Essentially, it is as if the artist also wanted to tell us that only the world of nature inhabited by culture exists—nothing more than that… This involves artistic language in exaltation and elation. Exaltation because a tree, a plant, or a watercourse is, in itself, a jubilation, a source of deep joy. Elation because every small or significant moment that solidifies the subject's engagement with nature is apprehended as the subject's glorification. The continuity between nature and society, as defended in Descola's theses, ends up echoing in the composition that Albergaria creates to try to understand hers and our vital space in a transmutation of reality into the imaginary, as if this transition from the seed condition to the sculptural condition were the ethical-aesthetic condition of an expectant future.

## 7 Lia Chaia: How to Make Say

The question of socialisation in nature (to use one of the many formulations in which Descola's thinking is abundant) is undoubtedly a central axis that unites all the artworks covered here. In the case of Brazilian artist Lia Chaia, the nature/city duality is a reason for critical commentary on approaches and distances. Ultimately, she tells us through a multifaceted work that utilises various techniques and mediums—drawing, photography, video, performance, and installation—that the city is defined in relation to nature and that the body, primarily the artist's own body, exists and acts as an extension and manifestation of nature endowed with an essential vitality.[6]

Lia reflects, for example, on banal gestures like a greeting or a handshake. From there, she metonymically derives the hand that caresses, feeds, and moulds. It is an obviously sensual and sociable gesture. Perhaps the realism of Courbet, the representation of social differences, and the artist's place in interaction with what surrounds

[5] See *Par-delà Nature et Culture*, Gallimard, Paris, 2005. Descola, who held a chair at the Collège de France once occupied by Lévi-Strauss, discusses the nature-culture dualism in an interview with the Portuguese magazine *Electra* No. 23 (Fundação EDP, Lisbon), which opens up avenues for questioning contemporary art within the framework considered here. Descola asserts: "What we see as a positive singularity and a legacy of Enlightenment philosophy, the distinction between nature and culture, is actually something that prevents us from being universal because we are the only ones who think this way. Our dualist cosmology is completely unique in human history, although it took time to take its current form, with successive variations from Greek philosophy through Christianity and the scientific revolution. The question is this: how can these concepts, derived from a singular historical trajectory that characterises a very singular cosmology, be used as we do in social sciences and daily life, in our way of thinking, to ultimately address phenomena in a universal manner? I draw attention to the false universalism of modernity".

[6] For visual documentation of the Brazilian artist's work and information about her exhibition history, see her official website at https://liachaia.com/.

**Fig. 6** Lia Chaia. Minhocão: 2006, video, time 18', 4:3 format. Video frame. Other information from the video: photography and camera: Daniel Chaia, editing: Júlio Taubkin

them, is not absent from it. Consider, for instance, the video "Desenho Dançante" [Dancing Drawing], images in which the artist's own body exerts an agglutinating effect, a kind of magnetic effect while establishing an analogy with the trunk of a tree. The use of projections and virtual drawings that appear and disappear, assuming their condition as transient images, draws attention to the organic dimension, the pulse of the artist's body itself and what inhabits and fills it (Fig. 6).

The conflict of nature with urban space, the loss of ties, the cleavages between speed and slowness, the difference between ritual and commodity constitute structuring themes in Lia Chaia's visual and plastic reflection. One of her proposals consists in reformulating the idea of landscape (this is, incidentally, one of the topics that connect her to the cross-cutting discussion of this essay). Presenting a series of works titled "Paisagens Infinitas" [Infinite Landscapes] (2022), where we see watercolour cutouts on tracing paper placed in acrylic boxes, the artist brings abstraction codes into play. The boxes form coloured objects containing particles that recombine like a kaleidoscope. But these landscapes also use other materials and may result from tangled lines of electrical cables that fictionally create horizons and distances.

Now, the work of socialising nature, always with a critique of its destruction and the threats that loom over it, translates in this work of relevant originality and full contemporaneity, thanks to its plastic language, into a juxtaposition of body and green space. The installation "Fauna" (2006), using photography, wood, and audio where a human voice imitates animal sounds, presents various geometric figures suspended at different heights in the exhibition space. These figures cut out in wood serve as supports for photographs of plants and condensed patches of vegetation.

There is also an allusion to the art of gardening in its Chinese and French traditions. The metaphor of the jungle is here touched by a powerful form of irony, mixing the exactness that geometry establishes and the amalgam of an imagined, unexplored wild place. This irony is reinforced, of course, by the exhibition codes—the gallery and museum as spaces of observation, contrast, and dissemination.

Overall, Lia Chaia's aesthetic proposals are ways of interrupting the centripetal and tendentially totalitarian force towards which the city, as a synonym for a large and chaotic metropolis, converges. In this sense, her work participates in a desire to rethink the city by highlighting the disruptive side of urban imbalances or through a critical analysis of architecture. At the same time, Lia Chaia sets out to discover islands of well-being, to strip away urban camouflage, and to refute botanical illusions (Fig. 7).

One particularly emblematic work of this approach is the video "Minhocão" (2006). This is the name by which an area in the city of São Paulo is known, where an extensive viaduct was built, 3 kms long, exemplifying the degradation of urban space in that area. As if the difficulty in finding balances and compromises between concrete, transport routes, housing, and demagogically called green spaces were not enough, the "Minhocão" (an allusion to a hermaphroditic reptile whose habitat is humid terrain) is, for Lia and thousands of other inhabitants of the Brazilian metropolis, an affront with visual and social consequences. There are homeless people who survive under the viaduct, and the cars that drive over it pass close to the windows of nearby buildings. During weekends, on Sundays, car traffic is interrupted, and the enormous viaduct takes on a pseudo-recreational function, as if it were a park for entertainment purposes. Lia Chaia reacts aesthetically to this

**Fig. 7** Lia Chaia. Aleph: 2013, video, time 2 '50', 16:9 format. Video frame. other information from the video: performance: Fabíola Salles, photography: Fernanda Tanaka, editing: André Francioli, original audio: Ricardo Reis Chuí

situation with a video in which she symbolises what the gigantic, uncontrollable metropolis swallows and devours, as if it were a kind of surreal monster. In this video, we see the artist expelling pieces of her photographs, small pieces of paper, images of this tumultuous (and monstrous) urban figure. It seems a petty attitude, and in a sense, it may be. But it is, above all, a gesture over emptiness. A mix of repulsion and irony. Notes on a tragedy similar to so many others. This diegesis narrates, as frequently occurs in this artist's work, the drama of a loss of references due to the inaccessibility of nature. This, in turn, leads to another of her pieces, "Folíngua" (2003), a title resulting from the contraction of the words 'leaf' and 'tongue'. It is a photographic diptych self-portrait of the artist back to a tree with her mouth open. From her mouth, instead of a tongue, a green leaf emerges in each image.

This tense and questioning relationship with the anonymous and indifferent city is also embodied in a video, "Aleph" (2013) where, at a certain point, a transparent glass ball is observed sliding along a female arm and reflecting an inverted image of urban volume. More than an upside-down city (that is, a a city without logic or rationality), this piece transports the viewer to another, seemingly purified situation. At a certain moment, the ball (as if it were a magical crystal ball, like those in children's tales) no longer reflects buildings or disorderly high-rise construction but the sea, a poetic allusion to an infinite that awaits each and every one of us.

"Aleph" is also the title of a book by the Argentine writer Jorge Luis Borges (1899–1986), written in the late 1940s, which depicts a web of allusions to identities and dualities, realities that coexist, filled with erudite references, laborious philosophical conjectures constructed by means of an intriguing and seductive form of writing, where the author constantly questions himself. It is this dimension of parallel realities, some of them irreconcilable, that Lia Chaia explores in this video, which, even without mentioning Borges, takes him as an archetypal figure of inquiry into human destiny and happiness.

The "supermodernity" that Marc Augé (1992) speaks of is characterised by a permanent abundance of "excess" interpreted as an excess of time that implies congestion, that is, an "overload of events." It is in this sphere that the "non-places" the author discusses are inscribed—physical modifications in "urban concentrations," devices of "acceleration" of people and goods, a mix of anonymity, immense lack of character, or places of exasperation and discouragement, such as the refugee camps that traumatise our age, coexisting with indifference and orphanhood of a homeland. It is also at the core of this supermodernity that the image and contemporary artistic discourse on nature are inscribed. And thus, it breaks or interrupts a mental, psychological, political claustrophobia generated by the degradation of the conditions in which the masses live.

Lia Chaia's most recent exhibition, "Organoide" (2024), continues to explore the possibilities of the city as a support and as a body—now soft and relaxed, now contracted and convulsive. Also floating in the exhibition space are random gestures—hands that separate and come together according to an ambiguity that stimulates sensory and physical experience. Her gaze is always an incision in the urban fabric, a call to the body understood as an element of dance and movement, the choreography of meanings predisposed to read each other according to a personal aesthetic code that favours criticism as creation and pushes the boundaries of interpretation.[7] This is how Lia Chaia makes us say.

## 8 Conclusion

It is not a matter here, nor would it make any sense, to "take sides" for Nature or Culture but rather to assume that they mutually presuppose each other. Within this framework, it is necessary to reflect in a necessarily synthetic way on some philosophical and anthropological perspectives, considering the communicative power and ability of art. That is, its visual and discursive impact, the concrete aesthetic object, and in this specific case, nature as a theme and subject. The ability to engage in deciphering nature and its codes involves knowing how to look, considering a learning process constructed by means of contrasts and displacements that aim to preserve, as an aesthetic memory, objects and parts of a world that, unlike so many others, is still natural.

In a troubled era like this, where clichés and stereotypes abound about "the environment," many of them propagated by advertising and others by activism of different shades, all duly mediated and massified, a contemplative serenity is imperative. This is a space and a regime of communication that art can provide.

One example of the abundant clichés mentioned here is a multi-media advertising campaign conceived by a multinational company where consumers are challenged to celebrate "a contract with Nature".[8] According to the defined marketing strategy, the goal is to drive the consumption of the product—in this case, green energy—by creating the conviction among consumers that they are both "saving the planet and saving on their bills." In a society that constantly celebrates merchandise as one of the categories of spectacle, a situation well identified by texts from the Situationist movement in the second half of the twentieth century and successively revisited with different ideological angles, protecting the environment constitutes one of the predominant rhetorics but lacks genuine political solutions. It is in favour of stereotyped ecological good conscience and in rupture with demagogic extremisms that it makes sense, as is the case in the aesthetic relationship with the works invoked in this essay, to rethink the gaze. Be it the gaze that runs along the trunk of a tree or that other gaze (an increasingly disenchanted one) that slides like a tracking shot over any great city. Once placed within this visual equation, the works of Gabriela Albergaria and Lia Chaia exhibit a novelty where it is possible to establish the foundations of a sensitive and critical conscience and reorganise expectations about everything that surrounds us outside the screens.

---

[7] This is simply to highlight the semiotic issues that the original work of this artist raises in relation to Umberto Eco's book *Os Limites da Interpretação* (2024), Portuguese edition, Lisbon; original edition *I Limiti dell'Interpretazione*, Grupo Editorial Fabri, Bompiani, Sonzogno, Milan, 1990.

[8] See the video and news about this campaign at the following link: https://eco.sapo.pt/2024/06/06/campanha-da-iberdrola-desafia-consumidores-a-assinarem-contrato-com-a-natureza/.

## References

Argan CG (1999) História da Arte, do Iluminismo aos movimentos contemporâneos, S. Paulo, Companhia das Letras
Augé M (1992) (2007), Não Lugares—Introdução a uma Antropologia da Sobremodernidade, 90 Graus editora. Lisboa (edição original Seuil, Paris
Damisch H (1984) "Artes", in Artes, Tonal—Atonal, Enciclopédia Einaudi, volume 3, Lisboa, INCM, pp 11–65
Damisch H (1984) "Artista", in Artes, Tonal—Atonal, Enciclopédia Einaudi, volume 3, Lisboa, INCM, pp 66–90
Flan J (1996) Robert Smithson: the collected writings. University of Berkeley, California Press, Los Angeles, London
Freud S (2008) O mal estar na civilização, Lisboa, Relógio d´Água
Gombrich EH (2000) História de Arte, LTC, Lisboa (edição original Phaidon, Londres, 1950)
Heinich N (1998) Ce que l´Art fait à la sociologie. Minuit, Paris
Leach E. (1985) "Anthropos", in Anthropos-Homem, Enciclopédia Einaudi, volume 5, Lisboa, INCM, pp 11–66
Leach E (1985) "Natureza/Cultura" in Anthropos-Homem, Enciclopédia Einaudi, volume 5, Lisboa, INCM, pp 67–101
Perniola M (2005) A arte e a sua sombra, Lisboa, Assírio & Alvim
Sloterdijk P (2007) in Crítica do Contemporâneo, Porto, Fundação de Serralves, pp

**Eduardo Paz Barroso** is Full Professor of Communication Sciences in the Faculty of Human and Social Sciences of the University Fernando Pessoa. He is Researcher at LabCom, a research unit in communication and arts at the University of Beira Interior. Among being a lecturer, cultural manager, and programmer, he has been appointed to several public charges in Portugal in the cultural and artistic sector. Author of about a hundred works, such as books, book chapters, and articles in national and international journals.

# The Romantic Movement of Things: Ecofeminism and Slow Cinema in Manuela Serra's Work

**Caterina Cucinotta, Jesús Ramé López, and Ana Isabel Soares**

## 1 Introduction

Our aim in this essay is to identify and describe elements that help to understand the peculiarity of the construction of a particular gaze: the gaze of aesthetic contemplation at play in Manuela Serra's *O Movimento das Coisas* (*The Movement of Things*, 1985–2021). In doing so we hope to characterize the style unfolding in the artistic gestures that make up the film. For these purposes, we will emphasize the relevance of analyzing the filmic materiality, in an "ecumenical sense which encompasses artifactual objects and their functions, forms and uses, the spaces and times in which life takes place" (Broncano, 2023, p. 15). In this sense, an archaeological character will be deployed in the research to value a reflection on the survival of this film in time as well as the potency of a thorough understanding of its artistic foundations. Indeed, the aesthetic particularity of the proposal of Manuela Serra (b. 1948) has required several viewings of the work: here is a film worth successive glances, each bringing it greater depth and meaning.

We can divide *The Movement...* into two parts: while the first focuses on women working in the fields around the village of Lanheses, the second part depicts the life

C. Cucinotta (✉)
Facultad de Ciencias de La Comunicación, Departamento de Periodismo y Comunicación Corporativa, Universidad Rey Juan Carlos, Madrid, Spain
e-mail: caterina.cucinotta@urjc.es

J. R. López
Facultad de Ciencias de La Comunicación, Departamento de Comunicación Audiovisual y Publicidad, Universidad Rey Juan Carlos, Madrid, Spain
e-mail: jesus.rame@urjc.es

A. I. Soares
CIAC—Centro de Investigação em Artes e Comunicação, Universidade do Algarve, Faro, Portugal
e-mail: asoares@ualg.pt

 
W. Leal Filho et al. (eds.), *Composing Worlds: Humanities, Health and Wellbeing in the XXI Century Towards a More Sustainable World*, World Sustainability Series,
https://doi.org/10.1007/978-3-031-87108-5_7

in the village on a festive day, including a church service, the market and, finally, a family lunch. This chapter center on the first part of the film. There, Manuela Serra developed a visual work on the women of the village, presenting their social status, gestures, and experiences; in a word, the "movement" of the title, similarly offered as a supplement to the ways and rhythms of understanding (of the life portrayed as well as of the film). The expression "the movement of things" cuts across life, simultaneously associating the inexorable passing of time, a way of walking, or the direction of a gaze. All things in the world move; movement encompasses all things: in this sense, the breadth conferred by the title of the film seems to imply a philosophical position. What form does this positioning take? In Serra's film, the answer comes close to Emanuele Coccia's statement about the life of plants (and, before them, to an idea of nature): "It is in the face of the world and nature that man can truly think" (Coccia, 2019, p. 35). This is the direction of the analysis we propose, understood as thought resulting from a creator's confrontation with the world and nature.

## 2 The Romantic Movement and Slow Cinema

In 1750, the British poet Thomas Gray (1716–1771) composed a series of stanzas started in the Fall of 1742, presumably inspired by the death of his friend Richard West. At first, he simply and justly titled them "Stanzas Wrote in a Country Churchyard" and showed them to Horace Walpole, another companion he had met in Cambridge, and a fellow of West. Walpole circulated the handwritten verses among London *literatti*, now "An Elegy Wrote in a Country Churchyard". The applause they gathered led to their printing and immediate consumption in a larger scale. Ever since, the elegy is among the most often quoted verses in English, practically shadowing further work by Gray.

The present and continued celebrity of Thomas Gray's poem goes together with the author's fame as a precursor of Romanticism in England, even if William Wordsworth, more widely given that merit, wrote of one of Gray's sonnets that the poet was attempting to widen the gap between prose and poetry, a move contrary to the Romantic intents as they were put forward in Wordsworth's preface to *Lyrical Ballads*.[1] Yet, in his elegy Thomas Gray pays particular attention to an expression of human feelings which swerves away from the rationality propagated by the Enlightenment; he elevates the role of Nature in the communion with humans, and he states that such a communion—the key for the balance of life—is disappearing under the pretenses of fame and gain as they were boasted by the ascending social order. Above all, Gray underlines the value of the common man, the richness of humans in their simplicity and distance from social highlight. Lines such as "Full many a flower

[1] In the words of Wordsworth, Gray "was at the head of those who, by their reasonings, have attempted to widen the space of separation betwixt Prose and Metrical composition, and was more than any other man curiously elaborate in the structure of his own poetic diction" (1991, p. 252).

is born to blush unseen,/And waste its sweetness on the desert air"[2] point to the significance of the life of those without a social voice, those whose daily toil offers them little else besides domestic, family comfort; those, in sum, who silently die away from celebrity, anonymous people who dwell far from the uproar of power, in a small village which represents the opposite of what the poet could already witness, more than suspect, as industry and capitalism arose.

Portugal, 1979. Five years after the democratic revolution, cinema was a particularly attuned mode of recording the profound changes undergoing in a country stepping out of the lethargy of a long dictatorship, and one such change was into equality among men and women. Nonetheless, cameras were still predominantly directed by men. A few exceptions can be noted: Margarida Gil, working for the State television (RTP), had mostly directed news and documentary features—notably, *Clínica Comunal Popular de Cova da Piedade* (*Popular Communal Clinic of Cova da Moura*, 1975), centered on family planning; and *Para Todo o Serviço* (*For All Purposes*, 1976), which was about the formation of what would become a union of Portuguese women house servants. Solveig Nordlund, a Swedish immigrant in Portugal, was consolidating a film career begun when she was an assistant director to João César Monteiro, José Fonseca e Costa, Alfredo Tropa, and Alberto Seixas Santos (her opening fictional feature was *Nem Pássaro Nem Peixe* [*Neither Bird Nor Fish*, 1978]). The Luso-French director Monique Rutler also started to work as assistant director and editor, but she led the direction of the TV documentary features *O Aborto Não É Crime* (*Abortion Is No Crime*, 1975) and in 1979 she presents her first cinema work, a cross between fiction and documentary, *Velhos São os Trapos* (*Only Rags get Old*). Noémia Delgado directed the documentary *Máscaras* (*Masks*, 1974–76); and from 1976 to 1989, Margarida Cordeiro, the film mate of António Reis, codirected three long features for the big screen.

Arriving from the Institut des Arts de Diffusion in Brussels, Manuela Serra became assistant director to Rui Simões, a colleague in Belgium: firstly, in *Deus, Pátria, Autoridade* (*God, Homeland, Authority*, 1976), and afterwards in *Bom Povo Português* (*Good Portuguese People*, 1980). During the time that mediated between one and the other, in 1979 she started preparing what would be her initial and last feature film. Throughout the three years she took to produce the piece, its title was simply *Mulheres* (*Women*). The decision to name the film *The Movement of Things* uncovers the intention to widen the theme, to not restrict the work to one single, feminine view, even if it construes a feminist perspective.

Shooting began in the last decade of the revolution. However, the film premiered only in 1985, at Mannheim-Heidelberg International Film Festival in Germany, where it was awarded the FilmduKaten Prize. That same year it would receive the Agfa Prize in the Tróia International Film Festival in Portugal. Despite the acclaim of the critics, the circulation of the film was limited to Film Festivals and to few private and film club screenings. Public and ampler diffusion would only occur more than three decades afterwards, after Manuela Serra had abandoned the film world, and following the appearance of the documentary directed by Marta Ramos, Mário Fernandes, and José

[2] Gray, T. 1903, p. 101.

Oliveira, *35 Anos Depois, O Movimento das Coisas* (*35 Years After, The Movement of Things*, 2015). A version of the film, restored by the Portuguese Cinematheque, was screened on April 21st, 2021 at the Portuguese Cinematheque: in the meantime, it had become a cult movie.

According to Raquel Schefer, "filmed in the small village of Lanheses, Minho, [*The Movement…*] is part of a 1970s trend in Portuguese cinema which led the topics into the countryside; it is therefore close to the works of António Reis and Margarida Cordeiro, António Campos, Noémia Delgado, among others" (2016). A similar familiarity is noted by Francisco Valente writing about the capacity that cinema possesses of "filming a made-up fable with the people inhabiting it" and inscribing Manoel de Oliveira (and his *Acto da Primavera* [*Act of Springtime*], 1963), Reis and Cordeiro (with *Trás-os-Montes*, 1976), and Manuela Serra within the same impulse to "situate spectators on the dividing line between fiction and documentary" (2024, p. 45). Apparently different directions converge on the goals and accomplishments of these films: a search for the common being, for the simplicity of life and of events, the mere *moving of things* attracts those who turn the art of cinema into a mode of thinking about a reality that changes rapidly and radically. These directors seem to suggest that only outside of large urban centers does the possibility exist to reconcile humans with the world; that such a reconciliation can only be represented in cinema through a harmonized merging of documentary with fiction.

If a meeting point between ecology[3] and slow cinema must be expressed, aesthetic contemplation must be the strategic epicenter of this cinematographic proposal. It is the move deployed in *The Movement*…. From a filmic-philosophical point of view, Manuela Serra opted for "an aesthetic attitude or perspective, which consists of delighting in the appearance of things, in the way things appear to our senses, in their sensory qualities" (Tafalla, 2019, p. 21). This attitude implies that the contemplation of nature or of objects created by human beings does not pursue a specific end, that it is stretched out in time, and that it contrasts with "an accelerated way of life, guided by ideals of speed and productivity" (Tafalla, 2019, p. 22). The film picks up this character and connects it with an "aesthetics of everyday life" (Saito, 1998, pp. 135–149), which veers the viewer into direct contemplation, proposing we stop recording symbolic constructions upon what is observed, that is, imprinting different filmic rhythms. The tempo of *The Movement…* calls upon the viewer to listen and to look closely at nature and at human beings in their interaction, with a pedagogical swerve, since "learning to observe is learning to be silent" (Tafalla, 2019, p. 26). In this, Manuela Serra is following the path of Thomas Gray's elegiac narrator when he directly describes the abandoned village, the surroundings of which are now silenced by the absence of countryside life. Serra leads the contemplative gaze into a new aesthetic perspective of the world, which immediately imprints an ethical perspective towards the observed, an ethics of looking and of listening. This aspect of

[3] The first to refer and define the concept was Ernst Haeckel (1866) in his *Generelle Morphologie der Organismen* (*The General Morphology of Organisms*). "Under 'ecology' we understand the all-encompassing science of the relationships of the organism to its environment, where we can include, in the broader sense, all 'conditions of existence' (our translation, p. 286).

the gaze includes a political and ecological reflection of the filmic style itself, giving a filmic peculiarity to the resulting work. The tension between the particular and its relationship with the universal is the guideline supporting the aesthetic proposal of slow cinema: its gestures are represented in long shots to bring into contact the daily and collective life of the village, and to contrast with the emergence of a modern outside world of workshops, urban construction, and factories. Furthermore, the formal distance to reality does not propose a clearcut correspondence between good and beauty, but rather claims an imaginary that transcends stereotypes of commercial cinema (Tafalla, 2019, p. 41).

## 3 Ideas of Eco-Feminism: Romanticism and Contemplation

As Madalena Miranda pointed out in 2021, "the constellation of materialities of the film in a close-knit relation, expressed in the breakfast ritual, in the chest of dowry, in the blessing of the bread dough, reveals connections linking to nature elements as the earth, the river, and the mist, in which women's activities cleaning or tending organize a *matriarchal filmic cosmogony* working as alternative power structures and resistance forces against visible power relations." (Miranda, 2021)

Such visual elements denounce a relevant and rather affirmative aesthetic connection between ecofeminism and slow cinema, an expression which may be equal to *contemplative cinema*.[4] While ecofeminism underlines the value of feminine perspectives in the construction of a more sustainable—eventually fairer—world, that should entail the acknowledgement of traditional women practices in the communities. That is indeed the perspective of the film, aligned with what Leonor Areal (2011) identifies in it as the prioritizing of the "female characters' point of view" (p. 21) in the films directed by women in Portugal immediately after the Carnation Revolution. Manuela Serra explores the potentialities of tradition and offers images of tradition as the artistic dimension capable of elevating life, while connecting a concept of femininity with an inkling for contemplation that one can deem typical of cinema, as far as nature and as far as the wellbeing of individuals and of the planet are concerned.

As she catalogues Serra's film under the category of "documentary", Areal places it among those post-revolutionary works that describe "the rural way of living and its traditions" (p. 95). Traditional practices are indeed the center of Manuela Serra's gaze: from the outset, the camera weaves the imperceptible narrative of these quasi-ritualistic habits, following the day's flux of the toil of several individuals from sunrise to sunset.

Starting from the idea of film genre, formulations stem out of the archetypes arising from the construction of the myth (Altman). Hence, a taxonomy of film genres can refer to different perspectives. We will focus on the narrative components that define the filmic structure (Sánchez Noriega, 2006, p. 98): specifically, what Sánchez

[4] Cf. Thomas Elsaesser (2011), "Stop/Motion" in Eivind Røssaak (ed). *Between Stillness and Motion: Film, photography, algorithms*. Amsterdam: Amsterdam University Press, p. 117.

Noriega has defined as the canonical genres and among them the drama (Sánchez Noriega, 2006, p. 99), in which the conflict represented is based on expressing life in a direct way. In this sense, despite using documentary elements, the staging implies a romantic character.

Romanticism is indeed linked to ideals of moral and natural regeneration which deem science and rationalism as threatening of the existence on Earth and corrupting of humans. Not necessarily towards nationalistic or racist ideas of an artistic exploitation of rural contexts, as it happens within ecofascist development (cf. Biehl y Standenmaier, 2019, p. 153) identifiable in the famous genre of the Portuguese *rural drama* (Grilo, 2006, p. 16) during the "Estado Novo", epitomized by *As Pupilas do Senhor Reitor* (*The Pupils of the Dean*, José Leitão de Barros, 1935). The drama unfolding in *The Movement...* has a universal quality and it entails a close connection between feminism and the democratic and Romantic artistic movements.

We align with the aforementioned authors in including Serra's film within the scope of a certain Portuguese cinema that experimented with new languages in the early years of democracy while approaching an already peripherical nature. But we wish to take a more thorough look at the feature and to point *The Movement...* as a pivotal work in expanding concepts such as ecofeminism and slow cinema. Although these do not represent recent debates, they are rooted in time spans which still claim a deeper research and new perspectives. The fact that Serra's film was "underground" for so long and is now presented to new audiences—made up of people who often did not have direct access to the reality it portrays—seems to emphasize that claim.

The movement now known as ecofeminism started in France in the same year Portugal was liberated from the dictatorship. Most notably, it was expressed in the words and works of Françoise d'Eaubonne,[5] who proposed to understand the connections between the exploitation of nature and the oppression of women, and to terminate both. Ecofeminism acknowledges nature as the central dimension of human life and promotes the idea that the suppression and exploitation of nature is closely linked to women's subordination. It then seeks to foster a perspective in which the actual—physical as well as mental—closeness with the natural environment and the appreciation of feminine points of view form a part of the struggle for social and environmental justice. It seems logical to include Serra's film in this trend—even if it does not purport an activist attitude, nor a revolutionary one.[6]

Slow cinema began as a film movement opposing the frantic superficiality of the dominant film industry. It characterizes films fabricated upon slow rhythm, contemplative narratives, and an aesthetics of daily life. In the combination of these cinematographic experimentations with ecofeminism we propose to critically reflect upon nature while suggesting a strong and critical view of stereotypes and the dominant

[5] Cf. d'Eaubonne, 1974.

[6] Catarina Alves Costa, a documentarist herself, considers *The Movement...* outside of what she calls "revolutionary cinema", belonging instead to a group of cinematic works which "albeit in the path of an ethnographic and poetic swerve in Portuguese film, take this swerve as finished" (cf. Costa, 2012, pp. 290-291. Our translation.).

society. Indeed, both movements invite to assume a deeper consideration of the relationship between humans and the natural environment. Films proposed and understood as "slow cinema" frequently incorporate natural landscape and scenes from anonymous domestic and daily lives of simple people, to draw attention to the beauty and fragility of the natural world; ecofeminism, in turn, raises questions related to the care for the Earth and its connection with femininity and maternity.

In *The Movement of Things*, Manuela Serra clearly adopts an unconventional narrative and chooses not to align with commercial cinema. Instead, she prefers to create and evince sensorial experiences and landscape atmospheres rather than follow a linear plot. We are therefore strongly convinced that this narrative strategy can reflect an ecofeminist approach—not simply in representing the undeniable presence of women, but also in what we see as a deeper, more intuitive, and contemplative understanding of the links tying together human beings and the rest of nature.

## 4 Filmic Analysis

If we consider the definition of a genre as the triangulation between three aspects, one corner of the filmic triangle would correspond to the themes that frame narrative; another corner is expressive and defining of the tone of the film; finally, the third tip of this triangle would be the style, which shapes a film aesthetically. Thus arises this triangulated repertoire of themes, tone, and style in relation to the emergence of plot, myth, and geography (cf. Astre and Horarau, 1997, p. 42). In the following analysis of some sequences of *The Movement...*, we will pay special attention to these three aspects to define the dramatic peculiarity composing this work romantic, feminist, and slow film.

THEME

When referring to the themes developed in the film, we must emphasize the idea of agency, that "personal and collective capacity to act and transform something in the environment in a conscious and irreversible way" (Broncano, 2023, p. 19). The female characters seem to participate in this agency in the rural setting during all the sequences in which they appear; however, this occurs in an environment where a lack of external threatening agency from the non-rural world is at play, symbolized by the sequences of the construction of the bridge or of the factory that appears in the distance, as if it were Kafka's castle. This idea of agency appeared in films of the 1920s inaugurating a genre that focuses on the gaze over the changing landscape: that is the case of *The Man with the Movie Camera* (Dziga Vertov, 1924) or that of *Berlin—Die Sinfonie der Großstadt* (*Berlin, Symphony of a City*, Walter Ruttman, 1927). These examples refer to an attempt to contemplate what, in a changing world, is agency and what is not. However, the agency Serra points at anticipates themes that began to develop in the twenty-first century, starting with the definition of the concept of the "Anthropocene" as proposed by Paul Crutzen (cf. 2002, p. 23). Human beings affect the change in the planet's living conditions, but women play an invisible role that the film brings to the table with the choice of the people it shows. Hence, we can see in this film an ecofeminist proposal, as it hides a "critical thinking that allows us to question the present and set us on the path to a future worth living" (Puleo, 2019, p. 12), redefining the meaning of reality by forcing us to observe it closely.

The first shot takes place at sunrise, showing a lake and the sky with its changing light, until a slight camera movement to the left reveals a bell tower in the distance. (Gray's poem and the topic of the tolling bell seem to resound.) The same sequence echoes at the end of the film, now with a slight camera movement to the right and the steeple replaced by a shot of a factory spewing smoke from its large chimneys: the church, a symbol of the community convening and consolidating locus, is replaced by the factory, which separates people and eventually causes their disappearance. Everything in these two shots effectively speaks of the material and immaterial movement in a village, its inhabitants, and the social and cultural transformations they are about to experience.

TONE

The tone of the film derives from a tension between the particular and the universal. This suggests certain visual connections to *Os Verdes Anos* (*The Green Years*, Paulo Rocha, 1963), such as when Isabel is getting ready to leave the house and recalls the attitude of the uncle of the protagonist in Paulo Rocha's film. Yet, in the case of *The Movement…*, the cinematic gesture comes closer to a documental contemplation, a creative *Stimmung* which pervades the whole film. Such an atmosphere is best captured and experienced when viewing a theatre screening of the film instead of a home viewing: a small screen hides the nuances of the bigger picture.

This contemplative tone is also evidenced in the women's attitude throughout the film: they are almost always silent and leave the uttering of words to men. The woman acts, thinks with and through her hands, which gives the film a realistic quality translated as a manner of deploying different modes of relationship. Furthermore, this character of slow cinema results in a constant dialectic of shots, which instigates the viewer to assign a meaning to what she/he sees, thus turning the viewer into an investigator and setting into play a vivid and changing dramatic tonality.

As an example, in a short sequence (00:31:3–00:34:35) two men are cutting down trees, as the camera goes from them to another group: women and children who are mechanically and manually separating the wheat seeds from the chaff. Suddenly, a familiar noise takes the viewer to the image of the woman at the beginning of the film, who is working at the sewing machine with a piece of denim cloth, only to end with a panoramic view over groups of men working in the construction of a bridge in the aqueduct. This dialectic portraying the now devastating consequences of global warming were only clearly thought of in the 1970s and 1980s by people with open minds, on the one hand, and with close connections to local traditions, on the other. Interrogations about our future (the past the film represents but also its own suspected future) pervade the film, conferring it with an observational tone from beginning to end.

STYLE

The activist character of the film—as slow cinema—can be detected in its very aesthetic approach, since it chooses a decoupage of montage based on long sequence shots, which allows the actions of the characters to develop fully. This stylistic choice offers the viewer the possibility of a distance: but this comes as an estrangement which silences the voice of the observer and allows the images to speak, which implies an ethical sense of aesthetic contemplation, a gaze that, consequently, "we should not reduce other people's creations to a mere mirror where we observe ourselves, to mere excuses for narcissistic self-contemplation" (Tafalla, 2019, p. 142).

Cinema allows to demarcate, or to invite the viewer to focus her/his attention through filmic techniques. In *The Movement of Things*, zoom is a tool used to set off expectations the viewer must immediately systematize. There is indeed a frozen shot when the protagonist comes off the bus returning home: it serves the viewers contemplation, since it gives time and space for pondering and for suggesting the woman is also taking time to analyze what surrounds her at that moment. It is a subtle, swift gesture—and yet, it introduces an unfamiliar moment of terror before the whole setting around the characters.

By allowing the gaze to wander across the screen in contemplation of the gestures and interrelationships of people and things, successive viewings of the film invite us to discover new details that fill it with new artistic dimensions. Sound, for instance, is a fundamental component of style. In the case of *The Movement…*, it reinforces the narrative and the tone as these are experienced by the viewer. One of the aesthetic experiences that the spectator of this film goes through is to feel as if in her/his own skin a sort of listening contamination, represented by the sounds of the machines present within the shot or out of it. In this aspect it is ahead of current studies of soundscape or proposals such as the category of anthropophony—in the sense that this concept was developed by the musician and ecologist Bernie Krause (cf. 2012, p. 75).

Music fulfills a distancing function that allows the viewer a recognition of the fiction that leads to the drama. This is the case of the sequences in which a woman drives an ox cart, or when the woman goes off a series of unpopulated shots of the river, the town and the open air drying of laundry: all images that define the dramatic space in an exercise of estrangement through music. In this sense, we consider the most effective moment to be when the music rides over a shot of Isabel lying in bed, giving way to two shots of a man riding a bicycle through the rural space and then riding again over a shot of Isabel's family sitting around a table to eat. It constitutes a moment for questioning the nature of images.

Right after the first shot of the water, the seven minutes at the beginning of the film follow a young woman leaving her home to go to work. After a minute and a half of sequences exploring the nearly empty streets of the village, the screen shows a house and a light turned on within a window. The camera moves towards a house door through which a girl comes out and turns a radio on. The music heard coming from the device is of Portuguese guitar; the girl—Isabel—is getting coffee ready, doing her hair in front of a mirror, and then taking breakfast. "Canção" ("Song") by Carlos Paredes and the tic-tac of a clock accompany the silence of Isabel's gestures. Carlos Paredes is a straight remission to the initial sequence of *The Green Years*, a film in which the homonym composition by Paredes serves as a background to the interminable babble of the uncle going on about Lisbon and its women, while the tic-tac of the clock takes the viewer back to the anxiety caused by the passing of time in *O Mal-Amado* (*The Ill-Loved*, Fernando Matos Silva, 1973), a film in which this sound image seemed to anticipate the last days of the dictatorial regime. In *The Movement...*, human silence—or the absence of spoken words—is the relevant sound element. It is this element that makes the character of Isabel into the axis of a narrative following an all-feminine visual story. The clock seems to mark the tempo of Isabel's daily routine: getting ready for work, leaving the house with the sun already up and catching the bus at the stop. From that moment on, the camera refuses to follow her outside of Lanheses and into a modernized and unnatural life: it turns instead to the other women who remain in the village while they till the land, bake bread, or feed animals in their backyards.

The viewer only reconnects with the image of Isabel, in the textile factory where she works (outside of Lanheses) at 00:33:00. Then she is shown sitting in front of a noisy, speedy sewing machine. Despite the young woman's silence in both moments, there is an obvious difference between the first sequence at home and the second in the factory: time is distended, slow, and silent in the village, while it is rapid and blaring at the factory. Time changes with space—it changes according to the space inhabited and shown in the film. Throughout *The Movement...* the small village is understood and portrayed as an atmosphere of close connection with nature, to traditional practices and to a natural rhythm of living: women tend the fields and their animals so they can secure the food for the family (greens, milk, eggs appear and reappear on screen). These village practices demand the time of these women. Contrarily, leaving de Lanheses corresponds to a travel towards the future: men and motorcycles, and automobiles are shown, men building infrastructures (bridges connecting highways), men lumbering and putting down trees, the noise and the visual presence of chainsaws inflicting on the viewer the pressure of the unbalanced living of humans in nature. The future is speedy, noisy and does not regard natural times: if there is a valley to be crossed, a huge bridge shall be built to make transportations and movement more rapid. If several trees stop the swift dislocations of men, trees are chopped down. Distance is tricked by gasoline.

## 5 Final Words

This research has allowed us to link Manuela Serra's filmic proposal in *The Movement of Things* with the romantic drama in a documentary effort which composes, through the strategies of slow cinema, an ecofeminist artistic work. For this analysis, the idea of Romanticism comprises what early Romantic poets such as Thomas Gray or William Wordsworth expressed: a highlight on common individuals, and anonymous, domestic lives.

We focused our attention on Serra's creative gestures, which anticipate themes developed further on up the historical path of Portugal cinema, e.g., the renewal in the discourses of documentary cinema, sound contamination or hybridism, the exposure of social changes in the face of technological progress or the role of women in ecology and in the organization of society. All this makes *The Movement...* a film that anticipates current concerns and its revision a process that stimulates new readings, which underlines the relevance and contemporary character of the film.

We pointed out certain traits or elements in the film that establish it as a film work dwelling among a discourse of its own, not found so frequently in other Portuguese films: the narrative capacity to explain how the world—the visible, or perceptible landscape as much as the social order—changes, presenting human beings as the driving forces and the victims of that change. The episode of the bridge construction or the menacing industrial plant anticipate and illustrate the precipitation of the Anthropocene. In *The Movement...*, woman gains added significance as an element that ensures the stability of the social order; Serra suggests that an unacknowledged responsibility for the family and for the reliability of the domestic realm falls upon her. This role is made visible throughout the film, as if the duration of the shots within the house, mostly with no words spoken and privileging the motion of people and objects, did justice to the feminine action. The contemplative and the relational aspects are thus articulated in a dialectic between the particular and the general within the social sphere. Such an articulation guides the perspective of the viewer, leading him to observe the world differently and to discover social links and roles that were not so common to seen on cinema screens. Therefore, ethical aspects appear not only tangled into these newly presented themes, but also in the way of looking or the mode of the gaze. This adds moral value to the aesthetic ways of seeing the world, expressing in the creative gesture a large scope of political aspects contained in the concept of point of view. This questioning of the gaze of the spectator by installing it in a distanced place of contemplation allows the viewer to realize the extent to which film objects are political in their choices of perspectives or points of view: a metacinematographic reflection through style.

Manuela Serra has only directed this film. It is virtually incomprehensible how a film artist capable of such a strong rigor and such a clear and accurate perspective on space and time, has not had due recognition. Ecofeminism is merely one of the many aspects and approaches from which to consider The Movement... and to attempt to redeem the work and its director into visibility.

Through this film, we confront the past and the present, our ways of seeing and of their politics, in a sort of filmic renovation that makes us consider new paths for cinema and for what film objects can reveal about the human being.

## References


Areal L (2011) Cinema Português: um país imaginado. Após 1974. Lisboa: Edições 70

Astre G-A, Albert-Patrick H (1997) El Universo del Western. Fundamentos, Madrid

Biehl J, Standenmaier P (2019) Ecofascismo. Lecciones sobre la experiencia alemana. Virus, Barcelona

Broncano F (2023) La Escala de las Cosas. Humanismo y cultura material. Delirio, Salamanca
Coccia E (2019) A Vida das Plantas. Uma metafísica da mistura. Documenta/Sistema solar, Lisboa
Costa CSBA (2012) Camponeses do cinema: a representação da cultura popular no cinema português entre 1960 e 1970. [Ph. D thesis in Cultural and Social Anthropology]. FCSH—Universidade Nova de Lisboa, Lisboa
Crutzen P (2002) Geology of mankind. Nature 415
D'Eaubonne F (1974) Le Féminisme ou la Mort. Le Passager Clandestin 2020, Paris
Elsaesser T (2011) Stop/Motion. In: Eivind R (ed) Between stillness and motion: film, photography, algorithms. Amsterdam University Press, Amsterdam
Gray T (1751) Bradshaw J (ed) The poetical works of Thomas Gray. G. Bell and Sons, London 1903
Grilo JM (2006) O Cinema da Não-Ilusão. Histórias para o cinema português. Livros Horizonte, Lisboa
Haeckel E (1866) Generelle Morphologie der Organismen. Allgemeine grundzüge der organischen formen-wissenschaft, mechanisch begründet durch die von Charles Darwin reformirte descendenztheorie. G. Reimer, Berlin
Krause B (2008) Anatomy of the soundscape: evolving perspectives. J Audio Eng Society Audio Eng Soc 56(1/2)
Miranda M (2021) O Movimento das Coisas (Manuela Serra 1986). Un caso de ecofeminismo político en los elementos visuales. Paper presented at the conference Cine en el Ámbito Iberico: Género a foco. Madrid: URJC, 21-23 October, 2021
Rubin M (2003) Thrillers. Akal, Madrid
Saito Y (1998) Appreciating nature on its own terms. Environmental Ethics 20(20), Summer 1998:135–149
Sánchez Noriega JL (2006) Historia del Cine. Teoría y géneros cinematográficos, fotografía y televisión. Alianza, Madrid
Schefer R (2016) Uma transfiguração poética do "real"—O Movimento das Coisas, de Manuela Serra (1979/1985). Available at https://apaladewalsh.com/2016/11/uma-transfiguracao-poetica-do-real-o-movimento-das-coisas-de-manuela-serra-19791985/
Serra M (2000) Manuela Serra [entrevista] in Ilda Castro (org). Cineastas Portuguesas. Lisboa: Câmara Municipal de Lisboa, Ministério da Cultura, Instituto do Cinema, do Audiovisual e Multimédia, pp 180–200
Tafalla M (2019) Ecoanimal. Una estética plurisensorial ecologista de los sentidos. Plaza y Valdés, Madrid
Valente F (2024) Espelho Mágico: Uma história do cinema. Orfeu Negro, Lisboa
Wordsworth W (1798) Preface. In: Brett RL, Jones AR (eds) Lyrical ballads, 2nd ed. Routledge, London, 1991

# Innovations and Socio-environmental Approaches

# Green Knowledge in Green Roofs and Organizational Green Innovation

**José G. Vargas-Hernández, Carlos A. Rodriguez-Maillard, Francisco J. González-Ávila, and M. C. Omar C. Vargas-González**

**Abstract** This study aims to analyze the implications of green knowledge and technology in organizational green innovation, urban green innovation, and the implementation of green roofs. Green roofs can be an effective strategy for cities to enhance their thermal environment, save energy, and combat climate change. Furthermore, they serve as an efficient energy-saving solution. The analysis is predicated on the assumption that green technology is fundamental to the practices, operations, and activities in the realms of organizational and urban green innovation. The study endeavors to strike a balance between urban development and environmental concerns, considering the principle of sustainable development and its innovative green solutions. The methodology employed in the study draws upon analytical, reflective, and descriptive methods, supplemented by a review of theoretical and empirical literature. The analysis concludes that the sharing of green knowledge plays a pivotal role in creating and developing green technology, with positive implications for organizational green innovation, urban green innovation, and green roofs.

**Keywords** Organizational green innovation · Green knowledge · Green technology · Urban green areas

J. G. Vargas-Hernández (✉)
Research Professor Graduate and Research Division, Financial Literacy Research Center, Universidad Cristóbal Colón, Veracruz Veracruz, Mexico
e-mail: jvargas@ucc.mx

C. A. Rodriguez-Maillard
Universidad Cristóbal Colon, Veracruz, Mexico

F. J. González-Ávila
Tecnológico Nacional de México, ITSF, Fresnillo, Mexico

M. C. O. C. Vargas-González
Instituto Tecnológico de México, Ciudad Guzmán, Mexico
e-mail: omar.vg@cdguzman.tecnm.mx



W. Leal Filho et al. (eds.), *Composing Worlds: Humanities, Health and Wellbeing in the XXI Century Towards a More Sustainable World*, World Sustainability Series,
https://doi.org/10.1007/978-3-031-87108-5_8

## 1 Introduction

The COVID-19 pandemic, a major health crisis, has prompted numerous activities to become greener and more sustainable. Although the terms "sustainable" and "green" are often used interchangeably, they possess distinct meanings. "Sustainability" refers to the persistence and indefinite future of necessary and desirable characteristics of the human subsystem within the ecosystem (Hodge 1997). Meanwhile, rapid urbanization in megacities has led to a decrease in green spaces due to the rise in construction activities. Hence, it is crucial to ensure that urban development is sustainable, aligning the expansion of city facilities with social, economic, and environmental aspects (Gohari et al. 2023).

Organizational environmental learning and green knowledge are linked to green technology, a key stimulant for environmental protection, organizational green innovation, urban green innovation areas, and green roofs. Sustainable development pertains to decision-making considering economic and social impacts (Wiering et al. 2020), while meeting the needs of the present without compromising future generations' ability to meet their own needs (Brundtland 1987). Urban sustainability, including social and economic aspects, is one of the foremost features of cities (Dwift Consulting 2017). Green trends such as smart and sustainable cities propose using green and eco-friendly technology, offering a more sustainable environment with zero gas emissions, along with other opportunities and challenges.

Individuals living in dense communities tend to use public urban green innovation spaces and parks more frequently for relaxation and leisure. They may also travel to the countryside, experimenting with and adopting greener practices within organizations focused on environmental sustainability. This enables the introduction of green innovations (Anderson et al. 2010; Stubbs & Cocklin 2008). Organizations that adopt the concept of saving money through initiatives like recycling programs and thermostat monitoring can support non-governmental agencies involved in environmental preservation.

Frequent use of green spaces enhances individual satisfaction with public areas, improving social space and mental health (Hadavi 2017). People attach meanings, identities, and psychological experiences to several types of green spaces, as described in the Place Identity Integrated Model and Environmental Representation (Bernardini & Irvine 2007). Research on green organization identity has focused on the individual level (Chen & Chang 2013). A green organizational identity strengthens tasks related to the organization's environmental activities and enhances its ability to cope with green-oriented conditions. However, landscape connectivity varies between urban green innovation roofs and urban open space management.

Organizations face challenges in complying with green sustainability strategies, a crucial measure against the negative impacts of industrialization on the environment. Organizational business sustainability encompasses green growth, green branding, and green sustainable reporting. Branding sustainability enhances an organization's appeal to environmentally conscious customers.

Green sustainable development, coupled with economic growth, social justice, progress, and environmental security, are pertinent research issues in areas such as green entrepreneurship in organic farming (Gupta & Vegelin 2016; Ihnatenko & Novak 2018; Kucher 2019; Mohd & Norhidayah 2016; Savickiene & Miceikiene 2018; Shevchenko & Petrenko 2020; Skydan et al. 2020). Achieving a balance between urban development and environmental concerns while combating climate change and global warming requires the consideration of sustainable development principles and innovative solutions (Gohari et al. 2023).

Organizations undertake green initiatives to develop sustainable competitive advantages (Galdeano-Gómez et al. 2008; López-Gamero et al. 2008; Wysocki 2021). These sustainable components offer organizations a competitive edge (Bansal 2005; Barnet 2007; Chang & Fong 2010; Namkung & Jang 2013). Organizational business systems form the foundation for any sustainable business, formulating and implementing green initiatives to achieve sustainability and competitiveness. The ecological partnership between organizations and society facilitates sustainable competitiveness.

Organizational ecological and sustainable procedures follow a strategic orientation (Fernandes & Solimun 2017). Although sustainability was previously considered an additional cost for organizations, it is now perceived as a strategy and tool to derive value and foster innovation (Van Holt et al. 2020). A shared organizational vision promotes the development of green behaviors, contributing meaningfully to the organization's overall goals (Aragon-Correa et al. 2008; Renwick et al. 2008).

There is a link between contemporary socio-ecological challenges and the search for solutions that respect people, nature and the planet is crucial in addressing the complex problems we face today. One approach that can be particularly useful is the integration of green technology and innovation into urban and organizational environments.

## 2 Background

Numerous studies, including one by Castro & Orellana (2019), explore the concept of green roofs. They examine how the introduction of green roofs can enhance the sustainability and creativity of cities. According to these researchers, green roofs can help establish urban green innovation zones, aimed at promoting sustainable urban development through the integration of green infrastructure and innovation. The paper includes a literature review discussing the benefits of green roofs, such as their incorporation into urban green innovation spaces and their ability to mitigate the impacts of urban heat islands, improve stormwater management, and boost biodiversity.

Green roofs have demonstrated their energy efficiency, environmental friendliness, and long-term economic viability. The implementation of green roofs has become imperative as an effective and sustainable approach that aids in achieving urban sustainability (Alqahtany 2023). The design of a green roof system is a complex process involving numerous factors, including social, cultural, climatic, and environmental considerations (Williams et al. 2010).

There are diverse types of green roofs. These include the extensive green roof system, the semi-intensive living roof planted with horizontal cover plants, and the intensive green roof, which includes growth medium, plantation, and irrigation and must be established over a concrete floor base (Karachaliou et al. 2016). The innovation in creating these roofs has become significant in recent years. Klemm et al. (2019) investigate the role of green roofs as a part of urban green infrastructure and their potential to contribute to sustainable and innovative cities. The authors discuss the challenges and obstacles in implementing green roofs, such as the lack of knowledge and expertise and the need for supportive policies and regulations. They conclude that green roofs are a crucial part of urban green infrastructure for innovative cities, and their full potential should be realized through supportive policies and regulations.

Additionally, Hemsley-Brown & Laing (2019) focus their research on potential hindrances to the adoption of green roofs as a method for sustainable urban development. They identify several potential barriers, such as a lack of understanding and knowledge about green roofs, high initial installation costs, the absence of encouraging policies and regulations, and a lack of incentives for property owners to install green roofs, all derived from a literature review. The study also examines social and cultural variables, including aesthetic preferences and perceptions of the value of green space, which could influence the adoption of green roofs. The authors propose that to overcome these obstacles, a multifaceted strategy involving collaboration between various stakeholders, including policymakers, researchers, practitioners, and the public, is needed.

Palomino & Durán (2020) provide a comprehensive analysis of the study on the impact of green roofs on the urban heat island effect. The authors discuss the urban heat island effect—a phenomenon where urban areas experience higher temperatures than their rural surroundings due to the built environment and human activity—and its causes. The study underscores the potential of green roofs to reduce the urban heat island effect and improve thermal comfort, offering insights for policymakers, practitioners, and scholars interested in the design and implementation of green roofs in urban settings.

## 3 Main Focus

Wu & Jim (2021) provide an extensive review of the literature on the roles green roofs play in sustainable urban development. They highlight the benefits of green roofs, including their capacity to reduce the urban heat island effect, enhance biodiversity, improve stormwater management, and provide social benefits, all in line with the broader goals of sustainable urbanization. The authors posit that green roofs are an integral part of urban green infrastructure—a network of green spaces and technologies offering numerous social and environmental benefits—and that they hold immense potential to support sustainable urban growth. In the same vein, Jamei et al. (2023) suggest that installing green roofs reduces heat exchange between indoor and outdoor environments, thereby lowering building energy consumption. This demonstrates the importance of understanding and considering the impacts of green roofs, which stem from a combination of knowledge, technology, and green innovation.

Green roofs present a promising alternative to traditional rooftop materials, realigning urbanized spaces with the natural environment. A green roof typically consists of a roofing membrane, drainage layer, geotextile filter, growth substrate, and vegetation suitable for the regional climate. These components together aid in stormwater retention, urban heat island mitigation, and reduction in building energy consumption (Gunn et al. 2022). Understanding the value of green roofs in addressing environmental issues in cities, including mitigation of the urban heat island effect, is crucial for efforts to decrease air pollution and enhance urban species biodiversity.

## *3.1 Green Knowledge and Technology*

Today's business ecosystem presents uncertainties and volatility in the development of green organizational and business models, challenging corporate risk management and decision-making. Technological knowledge about the economic, social, and environmental benefits of constructing green roofs promotes their efficient construction and maintenance. Organizational green knowledge shares issues adopting a multi-level framework, contributing to green technology. For environmental protection and to stimulate green innovation, organizational management must formulate green knowledge and green technology. Knowledge management of environmental issues serves as a resource for learning, problem-solving, and creating competitiveness (Liao et al. 2004). Green knowledge management is a potent tool for organizational improvement in a transformative environment (Raudeliūnienė et al. 2018). Environmental knowledge learning and green technology form the dynamic complexity of organizational green innovation (Li et al. 2021; Redman 2014).

Knowledge sharing forms the foundation of knowledge management (Zhang & Sundaresan 2010), and in the context of sustainable development goals like eliminating pollution, environmental protection, and Sustainable Development Goals (SDGs), sharing knowledge with other organizational members is beneficial (Bock et al. 2005). Green Knowledge Sharing (GKS) refers to the behavior of organizational members who disseminate information, data, and knowledge about green issues to others, thereby creating new knowledge and learning opportunities (Chen & Chang 2013; Norton et al. 2015). Green property psychological ownership and green knowledge sharing are influenced by green organizational identity. Knowledge sharing of individuals under time pressure is reduced (Chang et al. 2020). Organizational knowledge sharing focuses on fostering an atmosphere of environmental protection to create green knowledge sharing related to green issues.

Green technology is a socio-environmental healing technology that can minimize environmental damage and degradation while conserving natural resources.

Sustainable green technologies used in processes and applications leave no footprint (Aithal & Shubhrajyotsna 2016). Urban green innovation technology can be transferred by public government and private businesses aiming for inclusive green growth and ensuring social integration by eradicating poverty among vulnerable groups and keeping humanity's footprint within ecological boundaries. Building an inclusive and green organization of available and retaining talent is a formidable task (Brands & Fernandez-Mateo 2017). Green roof technology is employed to address social and environmental concerns and enhance climate resilience. Green plant-based green chemistry principles are used for green synthesis, nanomaterials, and nanoparticles (Maas & Hox 2005). Green and eco-friendly nanotechnology based on green chemistry techniques seeks to decrease the risks associated with industrial applications.

Green eco-friendly nanotechnology solutions used in sustainable development goals reduce the technification threat. Environmental degradation and climate change pose threats to global sustainable development and the Sustainable Development Goals. The idea of achieving sustainable development goals through the use and development of proven green eco-friendly nanotechnology processes furthers the proliferation and growth of many other systems and devices. Green nanotechnology is based on miniature-sized communication and computational devices, nano-sensor devices, high-density memory chips, etc. Ensuring ecological communication among stakeholders discloses the relevance of environmental concerns (Abimbola et al. 2010). Sustainable green nanotechnology principles and processes in the primary sector of the economy relate to the extraction of raw materials from nature.

Long-term effects of human food are required by green manufactured nanomaterials designed according to principles of green nanotechnology, complemented with current regulations aimed at fostering sustainable development of green technology. Increased instances of illnesses have been linked to the consumption of manufactured and processed inorganic food, leading to concerns about food safety. Green design and development of health and environment using nanoscale materials is a concern aimed at improving sustainable solutions. Green synthesis methods can mitigate the adverse effects of nanomaterials on the environment and user health. The complementarity of mixed methods helps detect contradictions, improve the results of one method using the results of the other, and enhance the scope and breadth of different components of research (Greene et al. 1989).

Green nanotechnology in the service industry is intangible and affects directly and indirectly. Green nanotechnology fosters innovation in educational technology through high-quality, low-cost, and ubiquitous online education. ICCT technologies working with green nanotechnology provide intelligent services in artificial intelligence, the internet of things, cloud and quantum computing, 3D printing, etc. Green nanotechnology supports ICCT technologies in the development of super-intelligent machines and human beings. Green nanotechnology enhances the durability, speed, and reach of digital entertainment instruments (Aithal & Shubhrajyotsna 2016).

## 3.2 *Organizational Green Innovation*

The principles, values, and norms of economic concern, societal parity, and ecological accountability underpin organizational green business innovations (Alawattage & Fernando 2017). Green organizational innovations can be generated either internally or externally, driven by green context-oriented factors. Based on green organizational identity, organizations have attributed more significance to environmental issues. This identity enhances organizational green innovation and competitiveness and ensures organizational competitive advantages (Geraie 2015). Theories of organizational identity and psychological ownership are utilized to construct a conceptual model.

Organizational green process innovation is eco-friendly, developing goods manufacturing processes in accordance with their life cycles (Kam-Sing 2012). Organizational green innovation and actions are conceived to mitigate adverse environmental effects. The green innovation behaviors carried out under a strategy establish a relationship between organizational green innovation strategy and actions. Motivational mechanisms stimulate innovation to facilitate organizational green innovation. Prosocial motivation, identified as a mechanism linking contextual factors and innovative outcomes, motivates organizational stakeholders to pursue collective efforts towards green innovative outcomes with their teammates, leading with values of organizational green innovation (Norton et al. 2015; Liu et al. 2016).

Potential risks and challenges of implementing innovative resources and inputs supported by green efficacy are addressed by organizational green innovation (Liu et al. 2015). Green organizational innovation is associated with individual tasks of organizational environmental activities to strengthen the organization. Green organizational identity enhances organizational creativity and innovation and ensures sustainable competitive advantages (Geraie 2015). This leads to social acceptance, enabling the organization to develop sustainable organizational green competitiveness capability.

The green competitive advantage describes the conditions necessary for a firm to hold a position in sustainable and ecological management and innovation, which cannot be imitated by others (Lin & Chen 2017). The resource-based view of organizational research focuses on the characterization of organizations (Pereira & Vence 2012) and their resources and capabilities (Fong & Chang 2012; Marchi 2012). This view includes factors such as organizational redundancy, innovation capabilities, network embedding, strategy orientation, and green innovation strategy as drivers of green innovation practices. To enhance environmental sustainability, organizations rely on intangible resources (Singh et al. 2020). A green pioneer organization in a resource-intensive sector must have an organizational model (Elkington 1994) with a social and environmental impact.

Human resources green management deals with valuable organizational resources and assets considered based on environmental sustainability. Evidence indicates that the workforce strongly engages with environmental sustainability development and becomes more committed and satisfied organizationally. Human resource green management is engaged in organizational environmental sustainability management (Shoeb & Nisar 2015; Wirtenberg et al. 2007). Green dynamic capabilities lie within organizational processes and are consistently reliable in organizations (Chen 2008).

## 3.3 Urban Green Innovation Spaces and Green Roofs

The rapid growth of the urban population, coupled with the low rate of economic development due to the pandemic, is leading to significant civilization challenges. These include the scarcity of urban green innovation spaces, pollution, and health issues. Smart growth, fostered by sustainable and inclusive growth, is based on the adoption of greener, more efficient technologies and job creation to support social cohesion (Gunn & Mintrom 2016). To remain inclusive, organizations need to embrace elements like equity and social justice for goal creation, strategy formulation, and public commitment development (Johnson 2017).

Organizational green innovation emphasizes the cultivation of organizational green culture, environmental pollution prevention, and waste reduction to enhance sustainability. Urban green innovation spaces are expanding in green metropolises, driven by the preservation and expansion of sustainable environmental development (Kalandides 2021).

Urban green innovation spaces should be delineated by construction and protection limits. People's interaction with these spaces can be public or private (Vijayaraghavan 2016). Public and private urban green innovation spaces can coexist and include green roofs and multifunctional open spaces framed within effective landscape management. These spaces, despite having different perceptions (Coolen & Meesters 2012), should ideally coexist (Mesimaki et al. 2017).

People's perceptions of private and green urban spaces vary, with a focus on social identity and environmental sustainability. Green behaviors, defined as employee actions that contribute to environmental sustainability, extend beyond job tasks and are considered part of organizational citizenship (Ones & Dilchert 2012). The loss of green urban areas due to residential and commercial expansion, exacerbated by urban sprawl, leads to a decrease in habitats, biodiversity, recreational areas, natural resources, and elements (Burszta-Adamiak & Fiałkiewicza 2019). Urban green spaces with high social and environmental vulnerability, especially urbanized areas with high land temperature, should be prioritized for the implementation of urban green innovation areas and green roofs.

Traditional streets can transform into green urban connection areas, serving as an open space system and parkland. An urban green innovation belt can prevent urban sprawl by merging with nearby settlements. The greenbelt concept conserves open space accessibility along roadways, connecting to a regional trail. However, the greenbelt has not successfully prevented urban expansion and sprawl, but it has facilitated access to open spaces (Maruani & Amit-Cohen 2007). Investments to develop accessible green spaces designed to meet economic, social, and leisure needs, contribute to community building, and alleviate construction pressures. Urban green innovation areas are open to enhance urban green innovation space networks and extend green infrastructure functions, including green roofs (Fung 2018).

Green infrastructure comprises a network of green open spaces that provide economic, social, and environmental benefits to city, town, and village residents. It creates an appealing environment, increases urban resilience benefits by managing stormwater, mitigates the urban heat island effect, and purifies the air. Green roofs, regarded as passive open green spaces, offer environmental benefits and provide open spaces for greater urban livability (Mesimaki et al. 2017). New city infrastructures and facilities should incorporate eco-roofs. Multi-functional urban green innovation infrastructure should allocate funding for green development initiatives.

Scholars have used quantitative and qualitative analysis to understand people's needs and perceptions of green roofs. The environmental effectiveness of green roofs can enhance public health and meet the social needs of vulnerable populations (Vijayaraghavan 2016). Green roofs, solar roofs, and living roofs, considered multi-benefit assets, can bring significant economic benefits beyond stormwater management, energy efficiency, and air quality improvement. They can also reduce the urban heat island effect. The loss of green spaces intensifies the heat island effect and stormwater runoff. Hence, the development of urban green innovation spaces and roofs should comply with regulations on stormwater control. An incentive program for the installation of eco-roofs can provide better stormwater management.

Historically, green roofs have evolved from the Hanging Gardens of Babylon, providing aesthetic value, and fostering human interaction, and even contributing to insulation. Le Corbusier integrated green roofs into modern architecture (Berardi et al. 2014). Countries like Germany, France, Switzerland, the USA, Canada, and Japan have researched and implemented green roofs. For instance, in Portland in 2005, authorities required city-owned buildings to incorporate green roofs. Similarly, Toronto passed a bylaw requiring all newly developed buildings to include green roofs. In Tokyo, all new buildings must have green roofs (Vijayaraghavan 2016).

Urban green innovation spaces should facilitate green infrastructure solutions, such as building green roofs, which positively influence the urban environment. Green infrastructure's operational stage involves maintenance solutions, focusing more on effective stormwater management and ecological landscape conservation than social and economic benefits (Mell 2010). Urban green innovation roofs, integrated into open space management, offer multifunctional opportunities to develop green infrastructure and benefit urban communities and neighborhoods. Both private and public green roofs, as open spaces, should serve as multifunctional green infrastructure and meet residents' urban open space needs (Meerow & Newell 2017).

A well-developed urban green innovation infrastructure requires a multifaceted approach, with increasing awareness of green roofs to alleviate global and local environmental impacts. Vegetated green roofs naturally sequester carbon dioxide through photosynthesis (Rowe 2011). Studies confirm the capacity of green roofs to retain rainwater, reduce air pollutants, and improve the microclimate, positively influencing the urban heat island effect (Burszta-Adamiak et al. 2019; Shafi que et al. 2018). There are several types of intensive and extensive green roofs, each incorporating different soil depth, weight, irrigation, plant species, and maintenance.

Urban green innovation projects include urban gardening, permaculture, guerrilla gardening, smart city, and building architecture. Productive urban green innovation combines urban agriculture, allotment gardens, and a culture of self-sufficiency, all ecologically motivated (Chen et al. 2020).

Greenhouse gas emission mitigation techniques to improve agriculture's environmental impact need to consider animal welfare, the largest anthropogenic contributor (Kucher et al. 2018; Llonch et al. 2017; Reisinger & Clark 2018). Unused spaces can transform into productive greenery and develop into natural and green innovation areas that preserve and increase ecological value. They can also serve as joint creation and learning spaces (Green 2020; Nachhaltiges 2020).

Energy optimization in building rehabilitation should consider creating new alternative green spaces transformed into green innovation areas for recreational purposes. Green roofs offer leisure spaces, enhance the aesthetic value of buildings (Sutton 2014), reduce the urban carbon footprint (Ugai 2016), and manage the direct water footprint (Fialkiewicz et al. 2018). Identifying urban places that need urban green innovation areas and suitable buildings for green roofs is critical for community and neighborhood implementation. Administrative and technical support is essential in designing, constructing, developing, and maintaining urban green innovation areas and green roofs. This task enhances the allocation of urban green innovation areas based on the structural design of buildings and surrounding environmental characteristics in the community. Green roof design and construction should consider the selection of plant species and the development of small.

Urban green innovation areas and green roofs are sustainable resources that contribute to the attractiveness of the city. The integration of open green areas in the city is creating a natural structure where green roofs in private properties contribute to the natural structure of the cities (Corporation 2016). Urban green innovation spaces and green roofs contribute to the increase in the efficiency and effectiveness of energy consumption in urban areas (Lee et al. 2018). The green infrastructure maintenance depends on the available space and configuration and the different purposes of green spaces and should consider the social benefits to the local community.

In the green economic innovation model, urban green innovation areas and green roofs are crucial parts of the open green spaces. The economic performance of companies in urban green innovation areas and green roofs is based on sustainability, environmental friendliness, and energy efficiency. Urban green innovation spaces and green roofs are part of the green economic innovation model for green economic growth that should develop social benefits and the welfare of the community.

The use of urban green innovation areas and green roofs as natural areas and as multi-functional open green spaces in cities, towns, and villages is highly recommended. Green roofs and urban green innovation spaces are significant for sustainable environmental management and sustainable urban development. The strategies and policies for green growth must consider the balance between economic development and environmental sustainability to make the cities more livable and resilient.

### 3.4 Urban Green Innovation Spaces and Green Roofs

The rapid growth of the urban population and the low rate of economic development due to the pandemic are leading to serious civilization problems, among others, the lack of urban greener innovation spaces, pollution, and lack of health. Smart growth is supported by sustainable growth based on the adoption of greener, more efficient technologies, inclusive growth, and fostering employment and social cohesion (Gunn & Mintrom 2016). Organizations must take an approach to stay in inclusive green organizations, embracing elements such as equity and social justice needed to create goals, formulate and implement strategies, and develop public commitment (Johnson 2017). Organizational green innovation refers to the innovation that emphasizes the implementation of organizational green culture, environmental prevention of pollution and waste reduction to enhance organizational sustainability. Urban green innovation areas are increasing in green metropolis driven by the preservation and expansion of sustainable environmental development (Kalandides 2021).

Urban green innovation spaces must be defined by the limits of construction and protection. The interactions of people and urban green innovation spaces are divided into public and private (Vijayaraghavan 2016). Public and private urban green innovation spaces co-exist and can be incorporated into green roofs as open green spaces aimed to use multi-functionality framed ineffective landscape management. These spaces have differences in perception (Coolen & Meesters 2012) although they should co-exist (Mesimaki et al. 2017).

People have different perceptions of private and green urban spaces with social identity and environmental sustainability. Green behaviors are defined as the actions and behaviors in which employees engage, linked with, and contribute to environmental sustainability (Ones & Dilchert 2012). This concept includes voluntary green behaviors beyond the job tasks, referred to as organizational citizenship, also considered as task-related environmental behavior (Bissing-Olson et al. 2013; Boiral & Paillé 2012). The disappearance of green urban areas due to residential and commercial purposes, enhanced by the urban sprawl, leads to the loss of habitats, biodiversity, and recreation areas and a decrease in natural resources and elements (Bursztta-Adamiak & Fiałkiewicza 2019). Urban green spaces with high social and environmental vulnerability are the most urbanized areas with high land temperature and might be prioritized for the implementation of urban green innovation areas and green roofs.

Traditional streets can be transformed into green urban connecting areas considered as an open space system and parklands. An urban green innovation belt prevents urban expansion and urban sprawl by merging with nearby settlements. Greenbelt conserves accessibility of open spaces along roadways and connects to a regional trail. Nevertheless, the greenbelt did not successfully prevent urban expansion and sprawl but facilitated access to open spaces (Maruani & Amit-Cohen 2007). Intensification of investments to develop accessible green spaces conceived to cover economic, social, and leisure needs and activities contributes to creating a sense of community and responding to the pressures of construction. Urban green innovation areas are open to enhancing urban green innovation space networks and extending the green infrastructure with the functions of green roofs (Fung 2018). Green infrastructure is the urban physical environment in a network of green open spaces that brings economic, social, and environmental benefits to residents of communities in cities, towns, and villages.

Green infrastructure is a way to create an attractive environment and increase the urban resilience benefits by better-managing stormwater, mitigating the urban heat island effect, and purifying air. Green roofs are considered passive open green spaces that have environmental benefits and provide open spaces for more urban livability (Mesimaki et al. 2017). Eco roofs are required for new city infrastructures and facilities. Multi-functional urban green innovation infrastructure should allocate funding for green initiatives of development.

Scholars have applied quantitative and qualitative analysis to address the needs and perceptions of people on green roofs. The environmental effectiveness of green roofs should improve public health and connect the social needs of vulnerable people (Vijayaraghavan 2016). Green roofs and solar and living roofs are considered multi-benefit assets bringing significant economic benefits beyond stormwater management, energy efficiency for cooling, improving air quality, and reducing urban heat islands. The loss of green spaces intensifies the heat island effect and storm water runoffs. The development of urban green innovation spaces and roofs is required to comply with regulations on stormwater control. A motivation program for the installation of eco-roofs provides incentives for better management of the stormwater.

Green roofs have historically evolved from the hanging gardens in Babylon, embracing aesthetic values facilitating human interactions and increasing insulation. Le Corbusier applied green roofs to modern architecture. (Berardi et al. 2014). In Germany, followed by France and Switzerland, the research and implementation of green roofs. In the USA, in Portland in 2005, the authority requires city-owned buildings to build green roofs; in Canada, in Toronto, the authority passed a bylaw requiring all the newly developed buildings; meanwhile, in Japan, in the city of Tokyo, the newly constructed buildings must have green roofs (Vijayaraghavan 2016).

Urban green innovation areas should provide spaces for green infrastructure solutions such as the building of green roofs, which have a beneficial influence on the urban environment. The stage of operation of green infrastructure is the maintenance solutions. The green infrastructure is a single function derived from the ability to effectively manage stormwater and focus on ecological landscape conservation rather than social and economic benefits (Mell 2010). Urban green innovation roofs integrated into open space management have multifunctional opportunities to develop green infrastructure and benefits in urban communities and neighborhoods. Private and public green roofs as open spaces a multi-functional green infrastructure and should accommodate the urban open space needs of residents (Meerow & Newell 2017).

Well-developed urban green innovation infrastructure needs to be supported by a multi-faceted approach. The increasing awareness of green roofs to alleviate global and local environmental impacts. Vegetated green roofs reduce carbon dioxide in the atmosphere naturally sequestered through the photosynthesis processes (Rowe

2011). It has been confirmed the capacity of green roofs to retain rainwater, reduce air pollutants, and improve microclimate has a positive influence on the urban heat island effect (Burszta-Adamiak et al. 2019; Shafi que et al. 2018). There are diverse types of intensive and extensive green roofs, which include vegetation with different soil depths and weights, irrigation, plant species, and maintenance. Urban green innovation projects include urban gardening, permaculture, guerrilla gardening, smart city and building architecture. Productive urban green innovation combines urban agriculture, allotment gardens, and a self-sufficiency culture that is ecologically motivated (Chen et al. 2020).

Mitigation of greenhouse gas emission techniques to improve the environmental impact of agriculture needs to consider animal welfare as the largest anthropogenic contributor (Kucher et al. 2018; Llonch et al. 2017; Reisinger & Clark 2018). Wasteland spaces can be converted into a place of productive greenery and developed into natural and green innovation areas that preserve and increase the ecological value and are also used as a place for joint creation and learning (Green 2020; Nachhaltiges 2020).

Rehabilitation of buildings using energy optimization should consider the creation of new alternative green spaces transformed into green innovation areas for free recreational purposes. Green roofs provide leisure spaces and enhance the aesthetical values of buildings (Sutton 2014) reduce carbon footprint in urban areas (Ugai 2016) and direct water footprint (Fialkiewicz et al. 2018). Matching urban places needing urban green innovation areas and green roofs in suitable buildings to determine opportunities to implement in communities and neighborhoods. Administrative and technical support is necessary for designing, constructing, developing, and maintaining urban green innovation areas and green roofs. This task recommendation enhances the allocation of urban green innovation areas based on the structural design of buildings and the surrounding environmental characteristics in the community. The design and construction of green roofs need to consider the plant species and the development of small fauna.

The construction and application of green roofs can be achieved in large urban areas located near other urban constructed areas and should not be a single investment. Stovin et al. (2012) calculate that the surface of roofs accounts for almost 50% of the sealed urban areas with the potential for urban green innovation roofs. Green roofs decrease the costs of discharging snowmelt and stormwater. The green stormwater initiative's infrastructure reduces greenhouse gas emissions and improves air quality. The development initiatives of green roofs are linked with stormwater management to meet the required flood management control. Local building codes must allow green roofs exempted from the building floor area and related to open spaces. The incentives for zoning to encourage the implementation of the urban green innovation roof should include the development of parking landscapes as one measure adopted in a stormwater management plan that may provide solutions (Murray 2017).

The development of green roofs value investments as an additional incentive for potential customers. Meeting the ecological needs of customers (Jain & Kaur 2004). There is a trend for organizations to shift operations toward the direction of environment-friendly processes despite the challenges of customer cynicism regarding organizational green actions (Kumar & Christodoulopoulou 2014). Co-financing and local legal regulations for the construction of green roofs are the most often used incentives. Green development finance has been assigned to Multilateral Development Banks, alongside public–private partnerships (PPPs) and new forms of finance, blended finance, bond instruments, green bonds, social impact bonds and development impact bonds. Urban green innovation roofs are the result of the implementation of motivational tools and other incentives for the different stakeholders.

The existence of types of barriers encourages local authorities to improve the use of incentives for the creation and functioning of urban green innovation areas and urban green innovation roofs and the most common use of incentives are the direct ones and legal regulations for the construction of regulations (Mentens et al. 2006). Incentives must be translated from sufficient interest into motivation for the excessively strict requirements for the construction of green roofs, such as the minimum surface area, to become eligible for co-financing and tax allowance. Market demand-related incentives, such as promotional instruments, are tools to determine demand for buildings with greenery in comparison to traditional ones. The implementation of financial and non-financial incentives is a crucial factor for green roofs as a relevant element of the urban environment (Brudermann & Sangkakool 2017). Financial incentives include subsidies or donations granted to reimburse the costs of investments that are aimed at supporting potential investors.

The development of rural and urban green innovation tourism close to nature and based on settlements is becoming more relevant for the impact on the green economy and ecology, with the self-employment of residents to provide environmental accommodation and food to the visitors (Kolomiets & Tomashuk 2021; Mazur & Tomashuk 2019; Tomashuk & Baldynyuk 2021). The green economy development and the pursuit of environmental protection is bringing profits to the organizations (Chen et al. 2006).

## *3.5 Case Studies*

Green roof applications have long been popular in Germany and Norway (Köhler and Poll 2010). Recently, reports have emerged highlighting the different advantages of green roofs in tropical areas, characterized by hot and humid climatic conditions. These advantages include mitigating urban heat islands and reducing energy consumption for cooling. Studies are now beginning to enumerate the impacts generated by green roofs, with details provided below (Table 1).

Green roof studies have primarily focused on reducing energy consumption. In developed countries, buildings' air conditioning consumes a lot of electricity. These studies show that green roofs are an effective method for saving energy in hot climates, improving the energy efficiency of indoor spaces, enhancing indoor air quality, reducing outdoor air temperature, and mitigating the urban heat island effect.

**Table 1** Main articles on green roofs innovations and urban spaces

| Article | Author(s) | Main topic |
|---|---|---|
| Evaluation of cool roof and vegetations in mitigating urban heat island in a tropical city, Singapore | (Li & Norford 2016) | "The result reveals that the city-wide deployment of cool roofs can significantly reduce near-surface air temperature and surface skin temperature during the daytime, especially at noon, although the effect is less pronounced during nighttime in Singapore" |
| Green Roof: Its Awareness Among Professionals and Potential in Malaysian Market | (Rahman et al. 2013) | "It discusses the growing awareness of green roof technology among professionals in Malaysia's built environment industry. A total of thirty local green roof projects have been identified, indicating a growing tendency to deploy green roof technology in Malaysia across various professions, including construction, landscape architecture, urban planning, project management, and academia" |
| Using Green Roofs for Social Housing to Improve Energy Consumption in New Cities. (An Applied Study of Social Housing in Egypt's New Cairo City) | (Goda et al. 2023) | "The study shows the improvement in energy consumption of cooling and heating energy in social housing in new urban cities in Egypt, reduces high temperatures inside and outside buildings, and decreases the urban heat island effect, the study suggests a method that is based on modifying the traditional roof of residential construction in Egypt by converting it to a green roof. An integrated analysis model including a strategy shows the environmental and climatic benefits of using green roofs to provide the external outdoor temperature of social housing buildings. The result revealed that green roofs, which are used by 50% of green roofs in traditional buildings, were shown to be efficient in reducing annual energy consumption by 12% and decreasing the average temperature inside buildings by 2.44 °C, whereas the average temperature outside was reduced by 3 °C." |
| Green Roofs as an Approach to Enhance Urban Sustainability: A Study of Public Perception in Riyadh, Saudi Arabia | (Alqahtany 2023) | Most of the population of Riyadh understands the benefits related to green roofs, and they have shown a positive response to a few factors. Almost 94% of people agreed with the fact that green roofs enhance the aesthetics of the building, and the same number of people agreed that they play a role in controlling the air quality. Similarly, 93% of people that green roofs provide additional recreational space in the building. On the other hand, 91% of the respondents identified the climate of the area as the biggest challenge in implementing green roofs in the buildings |

(continued)

**Table 1** (continued)

| Article | Author(s) | Main topic |
|---|---|---|
| The use of green roofs in urban green innovation spaces: A case study in Barcelona | (Nogue & Salas 2018) | The study offers a comprehensive perspective on the motivations, advantages, and challenges of implementing green roofs in urban green innovation spaces. It investigates the reasons for installing green roofs in these areas, the benefits, and challenges associated with it, and the role stakeholders play in the planning and maintenance of these green roofs |
| How to design green roofs to enhance urban green innovation space? Evidence from Beijing | (Gong et al. 2020) | The study provides a more in-depth analysis of the design features and effectiveness of green roofs. It further examines the design elements of green roofs and their efficiency in enhancing urban green innovation spaces. The study also offers suggestions on designing and deploying green roofs to support sustainable urban development and innovation, providing valuable insights for policymakers and practitioners |
| The application of green roofs in urban green innovation spaces: A case study in Jinan City, China | (Li et al. 2021) | The case study focuses on the application of green roofs in urban green innovation spaces in Jinan City, China, providing a comprehensive analysis of their design and implementation in these spaces. It assesses the effectiveness of green roofs in yielding various environmental and social benefits, such as diminishing the urban heat island effect, improving air quality, and providing community activity spaces |
| Green roof innovation in Singapore: Policies, performance, and potential | (Tan et al. 2019) | A case study on green roof innovation in Singapore presents a broader perspective on the policies and regulations that have facilitated the widespread implementation of green roofs throughout the city. The study examines the laws and policies that have streamlined the installation of green roofs in Singapore, evaluating their efficacy in providing a host of environmental and social benefits, while also identifying any existing obstacles to their utilization. Furthermore, the study offers recommendations on how to design and implement green roofs not only in Singapore but also in other urban contexts, providing valuable insights for policymakers and practitioners |

*Source* Own elaboration based on references included

They also present technological developments and their applications in urban green innovation spaces.

This topic has become increasingly relevant, as it presents the challenges that countries must face regarding climate change and the increasing heat island effect. It also explores the obstacles green roofs might confront, as countries revamp their laws in response to the new economic, political, and social circumstances they face.

## 4 Conclusions

This study analyzes how organizational environmental and green knowledge learning is linked to green technology for environmental protection, stimulating organizational green innovation, urban green innovation spaces, and green roofs. Green roof technology can be considered an effective solution for sustainable urban development in megacities due to its numerous economic, environmental, and social benefits. However, the importance of these benefits varies for diverse types of buildings, warranting further investigation.

Organizations transform to mitigate and neutralize environmental impact and to adapt to environmental sustainability. Green transformation is a multifaceted process involving interconnected and overlapping processes, posing several managerial challenges. Environmental community involvement promotes and improves the natural environment, aiming for the sustainable growth of society and inspiring voluntary participation in social activities. The green government should ensure access to affordable, sustainable housing in urban green innovation spaces and open buildings to create an urban social cohesion process consolidated in urban social movements. The design of an effective green roof system is a complex procedure that depends on several sustainability factors, including social, cultural, climatic, and environmental considerations.

Organizational networks are vital for connecting each other in the green movement space. Despite efforts to increase diversity and inclusion, environmental movements, non-government organizations, and environmental foundations lack racial diversity. Residents express their intention to participate in discussions and debates to develop a green urban framework. It is recommended that local authorities initiate strong campaigns in favor of green technologies, particularly for new or under-construction buildings. They must educate people about the benefits green roofs can provide and facilitate residents and homeowners in designing, executing, and installing such systems.

Incorporating green components into technology can create sustainable green technologies capable of preventing environmental degradation and promoting a cleaner environment. Green roof technology addresses social and environmental concerns and enhances climate resilience. Green nanotechnology offers both opportunities and challenges in the industry to foster growth by supporting nanotechnology usage. Green roof technology has been recommended as an effective solution for increasing per capita green spaces and improving sustainability means of urban development due to its diverse advantages.

Gaining a competitive advantage through implementing a sustainable organizational strategy is crucial for survival. The organizational green shared vision should include sustainable business goals. Long-term organizational green innovation strategy must encourage followers to implement green operations. Organizational socio-ecological strategies are formulated and implemented to achieve green competitiveness. Despite significant benefits, other factors, including high construction costs and maintenance issues, need to be considered for future global applications.

Most green roof studies have focused on reducing energy consumption. In developed countries, air conditioning in buildings consumes a large amount of electricity; these studies show that green roofs are an appropriate method of saving energy in hot climates, to improve the energy efficiency of indoor areas, the air quality of indoor buildings, and the air temperature of outdoor buildings and decrease the urban heat island effect, present technological developments and how to use in urban green innovations spaces.

Finally, this study emphasizes the balance between urban development and environmental concerns, considering sustainable development principles and innovative green solutions. This approach addresses ecological challenges and promotes energy efficiency and biodiversity protection. By incorporating these innovative and sustainable practices, we can create solutions that respect and benefit all aspects of our world, helping to compose the many possible worlds we envision. This topic takes greater relevance these days, as it presents the challenge that countries must face to ensure that climate change and the heat island effect are increasing, as well as the obstacles that green roofs can confront, while the countries redesigning the laws in this new economic, political, and social moment they are facing.

## Questions

1. Why and how technological developments can minimize the occurrence of environmental problems through the adoption of green innovation and green roofs in urban areas?
2. What are the main technological factors that this study aims to identify in the use of urban green innovation spaces?
3. What are the main factors that this study aims to identify in the use of technology in green roofs?
4. What are some of the technological factors that drive the adoption of green roofs in horizontal property by means of a technology acceptance of the place integrated identity model.
5. Under what assumptions are sustained the technological dynamics of the urban green innovation?

6. How can be described the perceived use of a technology in urban green innovation areas when addressing concerns such as attitude toward use, intention to use, perceived ease of use, and perceived usefulness?
7. Which methodology you may suggest being used to determine the perceived use of technology in green roofs?
8. For an efficient use of technology in urban green innovation areas and green roofs, which criteria suggest helping to minimize the chaos, the occurrence of environmental breakdowns, such as air pollution, increased temperature, urban heat islands, etc.
9. How could technology help to address such socio-ecological problems through the implementation of simple strategies?

## References


Abimbola T, Lim M, Hillestad T, Xie C, Haugland SA (2010) Innovative corporate social responsibility: the founder's role in creating a trustworthy corporate brand through "green innovation." J Prod Brand Manag 19(6):440–451

Aithal P, Shubhrajyotsna A (2016) Nanotechnology innovations and commercialization—opportunities, challenges & reasons for delay. Proceedings of national conference on changing perspectives of management, IT, and social sciences in contemporary environment 14, 1–12

Alawattage C, Fernando S (2017) Postcoloniality in corporate social and environmental accountability. Acc Organ Soc 60:1–20. https://doi.org/10.1016/j.aos.2017.07.002

Alqahtany A (2022) Green roofs as an approach to enhance urban sustainability: a study of public perception in Riyadh Saudi Arabia. Buildings 12:2202. https://doi.org/10.3390/buildings12122202

Anderson R, Amodeo M, Harzfeld J (2010) Changing business cultures from within. In: 2010 State of the World Report. The World watch Institute 96e102

Aragon-Correa JA, Hurtado-Torres NE, Sharma S, García-Morales VJ (2008) Environmental strategy and performance in small firms: A resource-based perspective. J Environ Manage 86(1):88–103. https://doi.org/10.1016/j.jenvman.2006.12.032

Bansal P (2005) Evolving sustainably: a longitudinal study of corporate sustainable development. Strateg Manag J 26(3):197–218

Barnet ML (2007) Stakeholder influence capacity and the variability of financial returns to corporate social responsibility. Acad Manag Rev 33(3):794–816

Berardi U, GhaffarianHoseini AH, GhaffarianHoseini A (2014) State-of-the-art analysis of the environmental benefits of green roofs. Appl Energy 115:411–428. https://doi.org/10.1016/j.apenergy.2013.10.047

Bernardini C, Irvine KN (2007) The 'nature' of urban sustainability: private or public green spaces. Sustain Dev Plan III 2:661–674. https://doi.org/10.2495/SDP070642

Bissing-Olson MJ, Iyer A, Fielding KS, Zacher H (2013) Relationships between daily affect and pro-environmental behavior at work: the moderating role of pro-environmental attitude. J Organ Behav 175:156–175

Bock GW, Zmud RW, Kim YG, Lee JN (2005) Behavioral intention formation in knowledge sharing: examining the roles of extrinsic motivators, social-psychological forces, and organizational climate. MIS Q 29:87–112

Boiral O, Paillé P (2012) Organizational citizenship behaviour for the environment: measurement and validation. J Bus Ethics 109:431–445

Brands RA, Fernandez-Mateo I (2017) Leaning out: how negative recruitment experiences shape women's decisions to compete for executive roles. Adm Sci Q 62(3):405–442
Brudermann T, Sangkakool T (2017) Green roofs in temperate climate cities in Europe - an analysis of key decision factors. Urban for Urban Green 21:224–234
Brundtland G (1987) Our common future: the world commission on environment and development. Oxford University Press
Burszta-Adamiak E, Stańczyk J, Łomotowski J (2019) Hydrological performance of green roofs in the context of the meteorological factors during the 5-year monitoring period. Water Environ J 33:144–154
Burszta-Adamiak E, Fiałkiewicza W (2019) Review of green roof incentives as motivators for the expansion of green infrastructure in European cities. Sci Rev – Eng Environ Sci 28(4):641–652. http://iks.pn.sggw.plhttps://doi.org/10.22630/PNIKS.2019.28.4.5
Castro AJL, Orellana F (2019) Green roofs as a strategy for urban green innovation spaces. Sustain Cities Soc 44:697–705. https://doi.org/10.1016/j.scs.2018.10.034
Chang NJ, Fong CM (2010) Green product quality, green corporate image, green customer satisfaction, and green customer loyalty. Afr J Bus Manage 4(13):2336–2344
Chang TW, Chen YS, Yeh YL, Li HX (2020) Sustainable consumption models for customers: investigating the significant antecedents of green purchase behavior from the perspective of information asymmetry. J Environ Plan Manag 64(9):1668–1688. https://doi.org/10.1080/09640568.2020.1837087
Chen YS (2008) The driver of green innovation and green image–green core competence. J Bus Ethics 81(3):531–543. https://doi.org/10.1007/s10551-007-9522-1
Chen YS, Chang CH (2013) The determinants of green product development performance: green dynamic capabilities, green transformational leadership, and green creativity. J Bus Ethics 116:107–119
Chen YS, Lai SB, Wen CT (2006) The influence of green innovation performance on corporate advantage in Taiwan. J Bus Ethics 67:331–339
Chen YS, Lin SH, Lin CY, Hung ST, Chang CW, Huang CW (2020) Improving green product development performance from green vision and organizational culture perspectives. Corp Soc Responsib Environ Manag 27:222–231
Dwift Consulting (2017) 12 mal Berliner Leben, 12 mal Berlin Erleben. Konzept für einen stadtverträglichen Berlin-Tourismus 2018+; VisitBerlin https://about.visitberlin.de/sites/default/files/201802/Tourismuskonzept_Berlin_Studie_2017.pdf
Consumption in New Cities. (An Applied Study of Social Housing in Egypt's New Cairo City). Futur Cities Environ 9(1):1–11. https://doi.org/10.5334/fce.165
Coolen H, Meesters J (2012) Private and public green spaces: meaningful but different settings. J Housing Built Environ 27(1):49–67
Elkington J (1994) Towards the sustainable corporation: Win-win-win business strategies for sustainable development. Cal Manag Review 36(3):90e100
Fernandes AAR, Solimun (2017) The mediating effect of strategic orientation and innovations on the effect of environmental uncertainties on the performance of business in the Indonesian aviation industry. Int J Law Manag 59(6):1269–1278
Fialkiewicz W, Burszta-Adamiak E, Kolonko-Wiercik A, Manzardo A, Loss A, Mikovits C, Scipioni A (2018) Simplified direct water footprint model to support urban water management. Water 10(5):630. https://doi.org/10.3390/w10050630
Fong CM, Chang NJ (2012) The impact of green learning orientation on proactive environmental innovation capability and firm performance. Afr J Bus Manage 6(32):727–735
Fung KL (2018) Expanding the green network on rooftops: a study of integrating green roofs as a part of urban green innovation space planning (Order No. 10932095). Available from ProQuest One Academic. (2124999351). Retrieved from http://wdg.biblio.udg.mx:2048/login?url=https://www.proquest.com/dissertations-theses/expanding-green-network-on-rooftops-study/docview/2124999351/se-2?accountid=28915

Galdeano-Gómez E, Céspedes-Lorente J, Martínez-del-Río J (2008) Environmental performance and spillover effects on productivity: evidence from horticultural firms. J Environ Manage 88(4):1552–1561. https://doi.org/10.1016/j.jenvman.2007.07.028

Geraie MS, Rad FM (2015) Mediator role of the organizational identity green in relationship between total quality management and perceived innovation with sustainable competitive advantage. Int J Biol, Pharm Allied Sci 4:266–276

Goda Z, Foda M, Elsayyad N (2023) Using green roofs for social housing to improve energy

Gohari A, Gohari A, Bin A (2023) Importance of green roof criteria for residential and governmental buildings: a multi-criteria decision analysis. Environ Sci Pollut Res 30:3707–3725. https://doi.org/10.1007/s11356-022-22472-4

Gong Y, Li X, Li Z (2020) How to design green roofs to enhance urban green innovation space? Evidence from Beijing. Habitat Int 107:102–113. https://doi.org/10.1016/j.habitatint.2020.102113

Greene J, Caracelli V, Graham W (1989) Toward a conceptual framework for mixed-method evaluation designs. Educ Eval Policy Anal 11:255–274

Gunn A, Mintrom M (2016) Higher education policy change in europe: academic research funding and the impact agenda. Eur Educ 48(4):241–257. https://doi.org/10.1080/10564934.2016.1237703

Gunn P, Burak H, Van Geel P, Baldwin C (2022) Development of resistor-capacitor and finite difference models to evaluate green roof thermal performance. Build Environ 226:1–15. https://doi.org/10.1016/j.buildenv.2022.109700

Gupta J, Vegelin C (2016) Sustainable development goals and inclusive development. Int Environ Agreem: Polit, Law, Econ 16(3):433–448. https://doi.org/10.1007/s10784-016-9323-z

Hadavi S (2017) Direct and indirect effects of the physical aspects of the environment on mental well-being. Environ Behav 49:1071–1104

Hemsley-Brown T, Laing A (2019) Green roofs and sustainable urban development: an exploratory review of potential barriers to adoption. Sustain Cities Soc 50:101648. https://doi.org/10.1016/j.scs.2019.101648

Hodge T (1997) Toward a conceptual framework for assessing progress toward sustainability. Soc Indic Res 40:5–98

Ihnatenko M, Novak N (2018) Development of regional programs for the development of agrarian enterprises with organic production based on the European and international experience. Balt J Econ Stud 4(4):126–133. https://doi.org/10.30525/2256-0742/2018-4-4-126-133

Jain SK, Kaur G (2004) Green marketing: an Indian perspective. Decision 31(2):161–209

Jamei E, Chau H, Seyedmahmoudian M, Mekhilef S, Alzahraa F (2023) Green roof and energy—role of climate and design elements in hot and temperate climates. Heliyon 9:1–16. https://doi.org/10.1016/j.heliyon.2023.e15917

Johnson SK (2017) What 11 CEOs have learned about championing diversity. Harvard Business Review, August 17th, 2017

Kalandides A, Grésillon B (2021) The ambiguities of "Sustainable" Berlin. Sustainability 2021(13):1666. https://doi.org/10.3390/su13041666

Kam-Sing Wong S (2012) The influence of green product competitiveness on the success of green product innovation: Empirical evidence from the Chinese electrical and electronics industry. Eur J Innov Manag 15:468–490

Karachaliou P, Santamouris M, Pangalou H (2016) Experimental and numerical analysis of the energy performance of a large scale intensive green roof system installed on an office building in Athens. Energy Build 114:256–264. https://doi.org/10.1016/j.enbuild.2015.04.055

Klemm W, Zinngrebe Y, Heidenreich S (2019) Green roofs as part of urban green infrastructure for innovative cities. Environ Innov Soc Trans 31:50–56. https://doi.org/10.1016/j.eist.2018.11.003

Köhler M, Poll PH (2010) Long-term performance of selected old Berlin greenroofs in comparison to younger extensive green roofs in Berlin. Ecol Eng 36(5):722–729. https://doi.org/10.1016/j.ecoleng.2009.12.019

Kolomiets TV, Tomashuk IV (2021) Entrepreneurship and development of rural areas in Ukraine. Colloquium-J 9(96):29–42
Kucher L, Heldak M, Orlenko A (2018) Project management in organic agricultural production. Agric Resour Econ 4(3):104–128. https://doi.org/10.22004/ag.econ.281753
Kucher A (2019) Sustainable soil management in the formation of competitiveness of agricultural enterprises. Acad Publ House «Talent». https://doi.org/10.13140/RG.2.2.19554.07366
Kumar V, Christodoulopoulou A (2014) Sustainability and branding: an integrated perspective. Ind Mark Manage 43(1):6–15. https://doi.org/10.1016/j.indmarman.2013.06.008
Li J, Yin Z, Li Y, Sun S, Li M, Li Y (2021) The application of green roofs in urban green innovation spaces: A case study in Jinan City China. Environ Sci Pollut Res 28(3):3525–3537. https://doi.org/10.1007/s11356-020-11544-2
Li X, Norford L (2016) Evaluation of cool roof and vegetations in mitigating urban heat island in a tropical city, Singapore. Environ Sci Pollut Res 30:3707–3725 Urban Climate 16:59–74. https://doi.org/10.1016/j.uclim
Liao SH, Chang JC, Cheng SC, Kuo CM (2004) Employee relationship and knowledge sharing: a case study of a Taiwanese finance and securities firm. Knowl Manag Res Pract 2:24–34
Lin YH, Chen YS (2017) Determinants of green competitive advantage: the roles of green knowledge sharing, green dynamic capabilities, and green service innovation. Qual Quant 51(4):1663–1685. https://doi.org/10.1007/s11135-016-0358-6
Liu J, Chen J, Tao Y (2015) Innovation performance in new product development teams in China's technology ventures: the role of behavioral integration dimensions and collective efficacy. J Prod Innov Manag 32:29–44
Liu D, Jiang K, Shalley CE, Keem S, Zhou J (2016) Motivational mechanisms of employee creativity: a meta-analytic examination and theoretical extension of the creativity literature. Organ Behav Hum Decis Process 137:236–263
Llonch P, Haskell MJ, Dewhurst RJ, Turner SP (2017) Current available strategies to mitigate greenhouse gas emissions in livestock systems: an animal welfare perspective. Animal 11:274–284
López-Gamero MD, Claver-Cortés E, Molina-Azorín JF (2008) Complementary resources and capabilities for an ethical and environmental management: a qual/quan study. J Bus Ethics 82(3):701–732. https://doi.org/10.1007/s10551-007-9587-x
Maas CJ, Hox JJ (2005) Sufficient sample sizes for multilevel modeling. Methodology 1:86–92
Marchi VD (2012) Environmental innovation and R&D cooperation: Empirical evidence from Spanish manufacturing firms. Res Policy 41(3):614–623
Maruani T, Amit-Cohen I (2007) Open space planning models: a review of approaches and methods. Landsc Urban Plan 81(1–2):1–13
Mazur KV, Tomashuk IV (2019) Governance, and regulation as an indispensable condition for developing the potential of rural areas. Balt J Econ Stud 5(5):67–78
Meerow S, Newell JP (2017) Detroit, spatial planning for multifunctional green infrastructure: growing resilience. Landsc Urban Plan 72:62–75
Mell Ian C (2010) Green infrastructure: concepts, perceptions, and its use in spatial planning. Doctoral Thesis, School of Architecture, Planning and Landscape Newcastle University
Mentens J, Raes D, Hermy M (2006) Green roofs as a tool for solving the rainwater runoff problem in the urbanized 21st century? Landscand Urban Plan 77(3):217–226
Mesimaki M, Hauru K, Kotze DJ, Lehvavirta S (2017) Neo-spaces for urban livibility? Urbanities' versatile mental image of green roofs in the Helsinki metropolitan area, Finland. Land Use Policy 61:587–600
Mohd HA, Norhidayah S (2016) Sustainable food production: insights of Malaysian halal small and medium sized enterprises. Int J Prod Econ 181(part B):303–314. https://doi.org/10.1016/j.ijpe.2016.06.003
Murray N (2017) Urban disaster risk governance. A systemic review PPI-Centre. UCL Institute of Education

Nachhaltiges Berlin (2020) Stadt der grünen Trends-Wie Berlin den Weg der urbanen Nachhaltigkeit geht. Available online: https://about.visitberlin.de/nachhaltiges-berlin

Namkung Y, Jang S (2013) Effects of restaurant green practices on brand equity formation: do green practices really matter? Int J Hosp Manag 33(2):85–95

Nogué S, Sala M (2018) The use of green roofs in urban green innovation spaces: a case study in Barcelona. Urban For & Urban Green 33:19–29. https://doi.org/10.1016/j.ufug.2018.02.008

Norton TA, Parker SL, Zacher H, Ashkanasy NM (2015) Employee green behavior: a theoretical framework, multilevel review, and future research agenda. Organ Environ 28:103–125

Ones DS, Dilchert S (2012) Employee green behaviors. In Jackson DSSE (ed) Managing human resource for environmental sustainability, Jossey-Bass

Palomino JC, Durán P (2020) The influence of green roofs on the urban heat island effect: a review of the current state of knowledge. Urban Climate 34:100696. https://doi.org/10.1016/j.uclim.2020.100696

Pereira Á, Vence X (2012) Key business factors for eco-innovation: an overview of recent firm-level empirical studies. Cuadernos De Gestion 12:73–103

Rahman S, Ahmad H, Rosley M (2013) Green roof: its awareness among professionals and potential in Malaysian market. Procedia Soc Behav Sci 85:443–453. https://doi.org/10.1016/j.sbspro.2013.08.373

Raudeliūnienė J, Davidavičienė V, Jakubavičius A (2018) Knowledge management process model. Entrep Sustain Issues 5:542–554

Redman C (2014) Should sustainability and resilience be combined or remain distinct pursuits? Ecol Soc 19(2):190–202

Reisinger A, Clark H (2018) How much do direct livestock emissions contribute to global warming? Glob Change Biol 24:1749–1761

Renwick DWS, Redman T, Maguire S (2008) Green human resource management: a review and research agenda. Int J Manag Rev 10(1):1–18. https://doi.org/10.1111/j.1468-2370.2007.00226.x

Rowe DB (2011) Green roofs as a means of pollution abatement. Environ Pollut 159:2100–2110. https://doi.org/10.1016/j.envpol.2010.10.029

Savickiene J, Miceikiene A (2018) Sustainable economic development assessment model for family farms. Agric Econ—Czech 64:527–535. https://doi.org/10.17221/310/2017-AGRICECON

Shafi que M, Kim R, Rafi q M (2018) Green roof benefits, opportunities, and challenges–a review. Renew Sustain Energy Rev 90:757–773

Shevchenko A, Petrenko O (2020) Current state of micro and small agribusiness in Ukraine. Agric Resour Econ 6(1):146–160. https://doi.org/10.51599/are.2020.06.01.10

Shoeb A, Nisar T (2015) Green human resource management: policies and practices. Cogent Bus & Manag 2(1):1030817–1031103. https://doi.org/10.1080/23311975.2015.1030817

Singh SK, Del Giudice M, Chierici R, Graziano D (2020) Green innovation and environmental performance: the role of green transformational leadership and green human resource management. Technol Forecast Soc Chang 150:119762

Skydan O, Nykolyuk O, Pyvovar P, Martynchuk I (2020) Methodological approach to the evaluation of agricultural business system flexibility. Manag Theory Stud Rural Bus Infrastruct Dev 41(4):444–462. https://doi.org/10.15544/mts.2019.36

Stovin V, Vesuviano G, Kasmin H (2012) The hydrological performance of a green roof test bed under UK climatic conditions. J Hydrol 414:148–161

Stubbs W, Cocklin C (2008) Conceptualizing a sustainability business model. Organ Environ 21(2):103e127

Sutton R (2014) Aesthetics for green roofs and green walls. J Living Arch 1(2):1–20

Tan PY, Wong NH, Tan TK, Wong KW (2019) Green roof innovation in Singapore: policies, performance, and potential. J Environ Manage 233:128–135. https://doi.org/10.1016/j.jenvman.2018.12.060

Tomashuk IV, Baldynyuk VM (2021) Identification of problems and prospects of rural infrastructure development of Ukraine. Econ Sci Colloq-J 13(100):58–70. https://doi.org/10.24412/2520-6990-2021-13100-58-70

Ugai T (2016) Evaluation of sustainable roof from various aspects and benefits of agriculture roofing in urban core. Procedia Soc Behav Sci 216:850–860

Van Holt T, Statler M, Atz U, Whelan T, van Loggerenberg M, Cebulla J (2020) The cultural consensus of sustainability-driven innovation: strategies for success. Bus Strateg Environ 29(2):3399–3409. https://doi.org/10.1002/bse.2584

Vijayaraghavan K (2016) Green roofs: a critical review on the role of components, benefits, limitations, and trends. Renew Sustain Energy Rev 57:740–752

Wang Y, Hu H, Dai W, Burns K (2021) Evaluation of industrial green development and industrial green competitiveness: evidence from Chinese urban agglomerations. Ecol Ind 124:107371. https://doi.org/10.1016/j.ecolind.2021.107371

Wiering M, Liefferink D, Boezeman D, Kaufmann M, Crabbé A, Kurstjens N (2020) The wicked problem the water framework directive cannot solve. The governance approach in dealing with pollution of nutrients in surface water in the Netherlands, Flanders, Lower Saxony, Denmark, and Ireland. Water 12(5):1240. https://doi.org/10.3390/w12051240

Williams N, Rayner J, Raynor K (2010) Green roofs for a wide brown land: opportunities and barriers for rooftop greening in Australia. Urban For & Urban Green 9(3):245–251. https://doi.org/10.1016/j.ufug.2010.01.005

Wirtenberg J, Harmon KD, Russell WG, Fairfield KD (2007) HR's role in building a sustainable enterprise. Hum Resour Plan 30:10–20

Wu W, Jim CY (2021) Urban green infrastructure and sustainable urbanization: a comprehensive review of the roles of green roofs. Landsc Urban Plan 212:104097. https://doi.org/10.1016/j.landurbplan.2021.104097

Wysocki J (2021) Innovative green initiatives in the manufacturing SME sector in Poland. Sustainability 13(4):2386. https://doi.org/10.3390/su13042386

Zhang Z, Sundaresan S (2010) Knowledge markets in firms: knowledge sharing with trust and signaling. Knowlege Manag Res Pract 2010(8):322–339

**José G. Vargas-Hernández, M.B.A, Ph.D.** is Research professor at Tecnológico MM Unidad Zapopan, formerly at University Center for Economic and Managerial Sciences, University of Guadalajara. Member of the National System of Researchers of Mexico. Professor Vargas-Hernández has a Ph. D. in Public Administration and a Ph.D. in Organizational Economics. He has been visiting scholar at Carleton University Canada, University of California Berkeley and Laurentian University, Canada. He holds a Ph.D. in Economic, Keele University; Ph.D. in Public Administration, Columbia University; studies in Organizational Behavior at Lancaster University and has a Master of Business Administration; published nine books and more than 300 papers in international journals and reviews (some translated to English, French, German, Portuguese, Farsi, Chinese, etc.) and more than 300 essays in national journals and reviews. He has obtained several international Awards and recognitions. He has also experience in consultancy. His main research is in organizational economics and strategic management. He teaches for several doctoral programs.

**Carlos A. Rodriguez-Maillard, Ph.D.** has more than 25 years of experience in diverse industries such as logistics, shared services centers, construction products and banking. Working in countries like Brazil, Colombia and México. I am a process and transformation leader at Grupo DEACERO, a leading steel producer in Mexico and America. My mission is to improve the business strategy across the organization, using my skills in Digital Transformation, problem-solving, change management and business management. I am also a professor at EGADE Business School (Tecnológico de Monterrey). I share my knowledge in Organizational Performance Management, User-Centered Design, and Project Management classes with MBA and MBM students. I have a Ph.D. in business administration, an MBA, a master's in finance and technology administration and multiple articles and book chapters published in academic and business journals. I am fluent in Spanish, English, and Portuguese and enjoy learning new things and facing new challenges.

**M. C. Omar C. Vargas-González** is Professor, research assistant and former Head of Computer Systems Department at Tecnológico Nacional de México Campus Ciudad Guzmán, professor at Telematic Engineering at Centro Universitario del Sur Universidad de Guadalajara with a master's degree in Computer Systems. Has been trained in Innovation and Multidisciplinary Entrepreneurship at Arizona State University (2018) and a Generation of Ecosystems of Innovation, Entrepreneurship and Sustainability for Jalisco course by Harvard University T.H. Chan School of Health. At present conduct research on diverse fields such as Entrepreneurship, Economy, Statistics, Mathematics and Information and Computer Sciences. Has collaborated in the publication of many scientific articles and conducted diverse Innovation and Technological Development projects.

# Social Engagement in Innovation for Human-Environmental Futures: A Brief Reflexive Sight for Nature-Based Solutions projects

**Cláudia Pereira, Emília Araújo, Elisabete Teixeira, and Paula Urze**

**Abstract** Contemporary political and scientific discourses strongly emphasise issues related to the construction, communication, and contestation of the future. Nature-Based Solutions (NBS) represent high-impact social innovations aimed at reshaping practices and enhancing the resilience of infrastructures and populations through natural solutions. This chapter revisits the primary objectives of NBS and discusses the significance of social engagement in NBS creation, implementation, and maintenance to elevate both the level and quality of social participation in future endeavors. Therefore, it addresses key aspects of the social engagement process, including stakeholder definition, methodologies in use, and the relationship between participation and innovation within the context of NBS design. A narrative analysis of part of the existing literature focused on engagement and NBS, allow us to conclude that social engagement in NBS projects contributes to the design of democratic futures and can be further enhanced through interdisciplinary and transdisciplinary efforts, as well as the inclusion of memory and history-related information throughout the process. Deeper intertwining between social participation approaches and open innovation frameworks from the outset can lead to more sustainable outcomes for thinking about how to build in practice resilient societies.

C. Pereira · E. Teixeira
ISISE, ARISE, Department of Civil Engineering, University of Minho, Guimarães, Portugal
e-mail: elisabeteteixeira@civil.uminho.pt

E. Araújo (✉)
Institute of Social Sciences, University of Minho, CECS, Braga, Portugal
e-mail: emiliararaujo@gmail.com; era@ics.uminho.pt

P. Urze
NOVA School of Science and Technology, CIUHCT, Lisbon, Portugal
e-mail: pcu@fct.unl.pt



W. Leal Filho et al. (eds.), *Composing Worlds: Humanities, Health and Wellbeing in the XXI Century Towards a More Sustainable World*, World Sustainability Series,
https://doi.org/10.1007/978-3-031-87108-5_9

## 1 Introduction

There has been a proliferation of diverse and innovative projects to strengthen societal resilience to potential climate-related disasters encompassing NBS (Nature-Based Solutions). NBS are defined by the EU Commission as "Solutions that are inspired and supported by nature, which are cost-effective, simultaneously provide environmental, social, and economic benefits and help build resilience. Such solutions bring more, and more diverse, nature and natural features and processes into cities, landscapes, and seascapes, through locally adapted, resource-efficient and systemic interventions".[1] Nevertheless, these normally entail physical modifications to spaces, alterations in practices, and adoption of new technological tools, aiming to improve the quality of life (Xie & Bulkeley 2020).

Examples of such projects addressing critical infrastructures include the construction of wetlands in flood-prone areas, the installation of solar panels for harnessing solar energy, the removal and replacement of trees to increase shaded areas in cities, the utilisation of wind energy in mountainous or coastal regions, and the implementation of large-scale digital artificial intelligence systems for disaster preparedness and monitoring. These projects share a common trait: their significant scientific and technological scope, which holds the potential to have a more positive effect on local communities and their future prospects (Mahmoud et al 2024).

The core question to debate is that NBS imply changes that impact local populations, what brings about more complex issues concerning the nature of participative projects. In recent years, the discourse on public participation in science and innovation has been wide, emphasising the need for social, scientific and political responsiveness. In addition, efforts have been focused on enhancing communication and collaboration methods to overcome the limitations of participatory strategies centred solely on information and consultation.

However, the future is subject to contestation and dispute (Araújo 2019), and despite the scientific, technological, and political interest in these changes, researchers grapple with fundamental questions: Do people genuinely want, and need these changes as envisioned by scientific and technical solutions? Do they understand the effects and implications of these changes? To what extent are experts and decision-makers willing to conceptualise and implement changes in these projects to genuinely contribute to sustainable development? How can populations, experts, and policy makers be empowered to participate actively and take responsibility in NBS projects, avoiding use of participation only as a electoral flagship?

Answering those questions requires both an interdisciplinary approach and conceptualising localities as complex and interdependent ecosystems where various stakeholders play a pivotal role in evaluating and deciding on the adoption of NBS (Ferreira et al. 2022). Therefore, we hypothesise that enhancing participatory sovereignty, by identifying the history of places, memories of past ecosystems,

[1] European Commission, European Research Executive Agency, Nature-based solutions—EU-funded NBS research projects tackle the climate and biodiversity crisis, Publications Office of the European Union, 2023, https://data.europa.eu/doi/10.2848/879543.

needs, possibilities, and aspirations for the future, is crucial to contour some of the persistent gaps in participation and engagement, strengthening efforts to improve open innovation methodologies (Chesbrough 2003).

In this sense, the key purpose of this paper is to demonstrate how NBS projects can represent a significant step forward building sustainable futures by improving the theoretical basis that underpins their acceptability, which implies delving into the cultural, historical, and sociological aspects of the populations affected by NBS projects. The text is based on a narrative literature review that highlights how NBS projects are considering social engagement with a focus on interdisciplinarity, past experiences, memories, and future aspirations. Finally, it discusses the interlinks between these concepts, methodologies, and their alignment toward the "future we want", as defined by the United Nations.

## 2 Method

This paper is based on a narrative literature review. Firstly, we considered core reflections made by precedent authors about the concepts of future (Adam 2011), time perspectives and care (Bellacasa 2017; sustainable futures, and responsiveness in science (Martins 1998). Secondly, we took into consideration existing literature reviews on social engagement in NBS projects, that focus on innovation, memory, and the future. To guide our research, we have used the following keywords and year range, encompassing articles published between January 2019 and January 2024:

- ("nature-based solutions" OR "green infrastructure" OR "blue infrastructure" OR "ecosystem-based approaches) AND ("stakeholder engagement" OR "social engagement" OR "public involvement" OR "participatory decision-making" OR "citizen participation" OR "citizen co-design) AND ("memory" OR "future" OR "identity") AND ("innovation");

The databases used in this narrative analysis were Scopus, Web of Science, b-on and Google Scholar. Search operators were used to increase the efficiency of the search, such as Boolean operators[2] and Field Codes (for example the Boolean operators "AND" and "OR"). Among these texts, we find particularly important literature reviews as they refer to information published before 2019. In order to present a reflexive view on the topic and clarify the connections between several concepts, our analysis also included other studies, to provide a more comprehensive understanding of basic questions worth addressing about the conceptualisation of the future. Despite the efforts to reference all the studies encountered, we were confronted with some difficulties, especially due to the diversity of NBS and the subsequent differences concerning social engagement. This constitutes a limitation in the study to be overcome in future research.

[2] In this research, key words and publication years were not taken into account, only the theme and appropriation of the articles were taken into account, for inclusion.

## 3 Participation, Engagement and "the Future We Want"

The study of the future encompasses a broad range of topics within social studies of science and technology, as well as social studies of time and temporalities. Discussions surrounding the nature of the future, its resonance with the past and present, and its portrayal as a fictional, imaginary, utopic, or dystopic scenario are prevalent in academic discourse. One pressing issue gaining prominence is the role of social engagement in empowering people's agency towards shaping the future on the basis of socially driven values with regard to social issues as poverty, inequality and oppression (Martins 1998, Ramos 2009, Adam 2011, Appadurai 2013, Bellacasa 2017, Armiero 2021).

For Appadurai (2013), development entails the right to aspire—to consider alternatives and possibilities beyond the scope of present political and technoscientific offers. This ability to aspire allows communities to signal what matters to them, what they wish to preserve, and how they envision reshaping their futures, thus rejecting the determinism that stills characterizing many political and technoscientific projects. Appadurai's concept aligns with the notion put forward by the United Nations regarding "the future we want," which emphasizes the inclusion of diverse perspectives and the search for consensus amidst the singularities and differences of cultures and natural worlds.

The UN[3] emphasises the correlation between participation, engagement, and the realization of this desired future; thus, participation is construed broadly to encompass access to information, consultation, and direct involvement in drafting, implementation, and monitoring of laws and policies. Furthermore, the UN (2020) underscores that engagement must be inclusive, transformative, and proactive, alerting that it is necessary to recognise stakeholders' views impartially, giving voice to vulnerable groups, counting on sociodemographic characteristics such as language, income, or social network. In this line, transformative engagement fosters dialogue and collaboration, while proactive engagement ensures that individuals have access to information, understand how to participate, and comprehend what is expected from them (UN 2020).

### *3.1 The Basis of the Concept of Stakeholder*

Often, decision-makers and even scientists and technicians, overlook or refuse to involve the affected populations in projects, believing that they do not contribute to technical solutions, out of fear of contradicting their political or technical interests (Brill et al. 2022). This denial, devaluation, or detachment regarding participation, coupled with the ease with which it can be disguised and manipulated, results in many

[3] Stakeholder Engagement for the 2030 Agenda—ESCAP and IAP2: https://www.unescap.org/sites/default/files/Stakeholder%20Engagement%20Indicator%20Framework%20Brochure_180518_0.pdf.

processes considered "participatory" either not involving any participation mainly involving groups with direct political or financial interests in the project. Additionally, even teams and companies doing participation using innovative technological tools also fall in the same fault.

The approach developed by the UN (2020)[4] considers stakeholders in a broad perspective assuming that only considering decision makers with the power to determine or influence the projects in courses is strongly limitative. Thereby, stakeholders are considered actors representing not only those who may influence political interests, but especially those who may be potentially affected, beneficiaries, users or consumers, as quoted below (UN 2020, p. 2–3):

- Women
- Children & youth
- Indigenous peoples
- Non-governmental organizations
- Local authorities' workers & trade unions
- Business & industry
- Scientific & technological community
- Farmers
- Local communities
- Educational & academic entities
- Faith groups
- Foundations & private philanthropic organizations
- Migrants & their families
- Older persons
- Parliamentary networks & associations
- Persons with disabilities
- Volunteer groups

Experts, *think tanks* and national and international organisations relevant to the field can also be included. The crucial lesson in defining stakeholders lies in understanding their relationship with a project. Stakeholders should be selected based on their potential to offer critical opinions or act as allies and enthusiasts from the project's inception. Diversity is essential, encompassing factors such as ethnicity, religion, gender, age, and physical condition, ensuring representative involvement from various groups. However, while ethical protocols often address diversity, it may be overlooked or distorted. Thus, there is no definitive list of stakeholders, but principles must be respected, including anonymity, confidentiality, information structuring to represent the diverse groups involved. The Table 1 summarises the stakeholders' level of involvement in the project on the basis on existing literature (UN 2020, Ferreira et al. 2022, Brill et al 2022).

## 4 Main Conclusions About Social Engagement and Public Participation in NBS Projects

Existing literature on NBS primarily emphasises the importance of stakeholder engagement as a fundamental element in contextualising the development of NBS within urban settings. d'Hont & Jill H. Slinger (2022) explore collaborative activities in Dutch coastal management, highlighting the involvement of local stakeholders in

[4] https://hlpf.un.org/sites/default/files/migrated/documents/27114MGoS_TOR18_Dec_2020.pdf.

**Table 1** Stakeholder level

| Stakeholder level | Description |
|---|---|
| Information | Stakeholders who receive balanced and objective information to understand the project's nature and objectives. They do not provide feedback directly |
| Feedback | Stakeholders who offer feedback and opinions on specific decisions or activities. Their input is considered but may not have final decision-making power |
| Engagement | Stakeholders are actively involved in decision-making and implementation, ensuring their concerns are understood and considered in the project |
| Co- partnership | Stakeholders who partner with the project team, play an active role in decision-making and share responsibilities for project aspects |
| Capacity building | Stakeholders who receive training, resources, or skills for being involved in the project creation and maintenance |

knowledge sharing. Their analysis underscores the significance of local expertise and stakeholder willingness to participate in developing solutions, emphasising the need to integrate approaches from social-ecological systems. Additionally, they stress the importance of policy analysis in understanding the role of each actor in the process and how their interests can impact the project. Derlukiewicz et al. (2023) present best practices for smart mobility in European cities, advocating for community engagement as a means to promote environmental protection grounded in a shared sense of ownership.

However, the authors also acknowledge challenges stemming from urbanisation, capital movements, human displacements, and gentrification, which can erode community ties and historical knowledge. Willems et al. (2020) analyse green infrastructure projects in European cities, highlighting the complexities of participation, particularly during project maintenance phases. They emphasise the need for more effective instruments to democratise public service delivery and suggest longitudinal studies to understand participation dynamics.

Brown (2021) introduces the Green Stormwater Infrastructure Social Spatial Adoption (G-SSA) model, which simulates the adoption behaviour of private property owners in implementing green stormwater infrastructure, while stressing the role of engagement with people. Zieglere et al. (2022) proposes three cross-cutting issues for future research, practice, and policy: social innovation for NBS, participatory governance, and technology addressing biodiversity loss. The authors identify overarching themes of social innovation including NBS, technology, and participatory governance. These studies underscore the importance of stakeholder engagement, policy analysis, and social innovation for the development of NBS solutions that are socially and environmentally sustainable, integrating historical knowledge and reflecting diverse community needs and perspectives. In summary, it is important to retain a few ideas regarding the existing literature on the need and the interest of engagement in NBS projects:

- Involvement of local participatory structures in the course of the project

- Relevance of local expertise and willingness of stakeholders to participate
- Integration of social-ecological systems theory in designing and implementing local solutions
- Need for policy analysis to understand the role and interests of each actor in the process of NBS development
- Call for community engagement and participation in sustainable development to foster a shared sense of ownership and responsibility for the environment
- Recognition of challenges posed by urbanisation, capital movements, and immigration fluxes in decision-making for future challenges
- Importance of citizen participation in green infrastructure projects
- Challenges in achieving comprehensive participation, particularly in project maintenance phases (both stakeholders and citizens)

## 5 Main Methodologies Used for Stakeholder Engagement and Participation in NBS´s Projects

As Mitic-Radulovic et al. (2021) note, most of the methodologies and existing tools for social engagement in projects involving interventions in space follow action research assumptions. These methodologies entail diagnosis, co-design, solution implementation, reflexive monitoring and community-based participatory action research (MacDonald 2012). Sanz-Mas et al. (2024) describe a research project aiming at evaluating the effectiveness of transforming 11 primary schools in Barcelona into climate shelters. The authors employed different research techniques, including ethnography, photovoice, and focus groups to assess the project and ascertain the degree of social engagement.. In line with these recommendations, Derlukiewicz et al. (2023) argue that NBS needs to be planned in such a way that contemplates a set of requisites that count for the identification of the locals and generation of a sense of ownership. Teladia & Windt (2022) are very clear about the need to use participatory methods to give voice and autonomy to citizens, and not only to decision-makers, or other stakeholders with the power of influencing decisions. The authors refer to Arnstein's (1969) eight types of participation and non-participation: from manipulation and therapy that "are at the lower levels of the ladder and are levels of "non-participation" to delegated power and citizen control, when "the citizens are the key decision makers, either by holding a majority of decision-making seats or by having full managerial power". Similarly, Court et al. use that research to further propose cross disciplinary work, community involvement and the generation of narratives that are deeply analysed and taken into consideration. The authors point out the need to distinguish between listening and including people (of different social groups), and allow them to co-create and participate in a final well-accepted and desirable solution.

Zeng et al (2023) defend the use of principles of NBS to change agricultural practices in a manner that agricultural technologies do not conflict with or terminate conventional practices and knowledge stating that there is a need to employ methods

**Table 2** Some of the main methods used in NBS projects for social engagement

| Main methods used in NBS projects for social engagement | |
|---|---|
| • Ethnography<br>• Photovoice, design fiction, narratives<br>• Focus groups, surveys, interviews<br>• Living labs<br>• Draw and use of written and oral narratives<br>• Co-creation activities<br>• Several communication methods for fostering trustworthy relations; | • Participative techniques (assemblies, workshops, consensus conference, world café);<br>• Platforms like web interfaces or newsletters for knowledge sharing<br>• Co-design and prototyping (design thinking, 3D mapping);<br>• Case studies in web platforms |

that instill a sense of ownership among stakeholders and citizens. In line with many other authors that mention the need to involve people in co-design (Basnou 2020), Bhatta et al. (2024) analyse the evolution of the concept of living labs for the case of nature-based solutions implementation. These authors state that living labs facilitate collaborative innovation addressing environmental and socio-economic challenge, as they favour contexs for real life experimentation and collaboration.

Below we outline the main methods used in NBS projects for social engagement (Table 2):

As noted before, the operationalization of the idea around the "future we want" demands deep and further reflection on the use of each of these methods, and their implications on the type of information provided.

## 6 Interdisciplinarity and Trans Disciplinarity for Social Engagement in NBS Projects

Basta et al (2021) argue that the COVID-19 pandemic underscored the urgent need to address the disruptions caused by human activities on natural systems, highlighting the importance of enhancing resilience and preparedness against future disturbances. The authors claim that it is crucial to prioritise sustainability in the post-pandemic era, embedding NBS into strategies and policies for sustainable land-use planning, resource efficiency, environmental management, and supporting a green economy and stressing the need to involve people. Therefore, local-level NBS initiatives benefit from collaboration between diverse stakeholders, including scientists, engineers, policymakers, community members, and "native" knowledge holders. This collaboration contributes to ensure that NBS projects are contextually relevant, socially inclusive, and environmentally sustainable. As mentioned, existing studies emphasise the need to employ multilevel perspectives in NBS projects (Mitic-Radulovic & Lalovic 2021). Furthermore, literature review underscores the relevance of interdisciplinary, as well as transdisciplinary approaches and the necessity to move beyond the concept of participation in initial phases of co-creation, and engage citizens in maintenance of NBS projects. d'Hont & Slinger (2022) highlight the importance of transdisciplinary learning has for many scholars working in the field. Scott et al. (2022), propose a workflow towards innovation which takes into consideration the need for dialogue between sectors, and most important collaboration between scientific areas, as "a prerequisite for effective mainstreaming" (p. 214).

Silva (2022) draws on literature that highlights the role of interdisciplinarity, as a way to build knowledge convergence, claiming that biogeochemistry is a solid approach to accelerate innovation. Moreover, the author put forwards suggestions for climate change mitigation that encompasses working with social sciences. In the author's view, apart from adopting measures, it is necessary to create opportunities to test and generate baselines of understanding of what actions can be developed, and at what costs. Authors state that:

> "Contributions from other environmental sciences (e.g., ecology, geology, soil sciences, etc.)—as well as from the social sciences and humanities (e.g., economics, law, philosophy, sociology, etc.)—are expected to increase the positive impact of biogeochemistry. Beyond academia, collaborative research in and with historically excluded communities is critical to ensuring that carbon capture projects are socially and ecologically sustainable" (p. 34).

Stroud et al (2022) speak about ecosystem service provision (p. 84) signaling the need to gather information from social sciences to understand ecosystems in their totality. The authors say that "Research should build upon these findings, undertaking full life-cycle assessments of vegetation, nature-based solutions, and study the interaction of multiple interventions to fully capture how ecosystem service supply is impacted throughout the management lifespan of areas or green infrastructure" (p. 85). Advocating for a "clear and detailed economic analysis of the benefits or costs of ecosystem services provided by plants in urban areas", the authors (p. 96) understand that the "inclusion of nature-based solutions should have sound economic grounding, and without a cost–benefit analysis it is difficult to provide a strong evidence base for the financial benefits often attributed to nature-based solutions (Cousins 2021; Pretty & Barton 2020)".

As established, important literature focus on inter and transdisciplinarity in projects involving sustainability plans, innovation in NBS, or transformation using green infrastructure projects. Several papers highlight the importance of interdisciplinarity and the need for a relationship between knowledge to better enhance the design and implementation of projects. The main reason lies primarily in the desired social involvement in the projects and of stakeholders—the most mentioned agents in all papers. Drawing on expertise from the social sciences—sociology-history along with knowledge from other fields, interdisciplinary emerges as a measure or suggestion. However, it is observed that the projects did not include a clear description of how this interdisciplinarity objectively occurs in the field, nor how it is promoted. It is also not stated how it effectively contributes to the design.

## 7 The Relevance of the Past and Memory in NBS Projects

Incorporating local ecological memory and cultural heritage into NBS planning and implementation helps preserve traditional water management practices, community identities, and ecosystem resilience. This integration fosters a deeper connection to local landscapes and strengthens the sustainability of NBS projects over time. As for the typology of data gathered, and notwithstanding the references to historical and sociological information, characterising the places involved, the literature focuses heavily on information that is explicitly helpful in understanding the social dynamics of spaces and how they can become open to the changes resulting from project implementation. We find little reflection on the importance of transdisciplinary historical and longitudinal knowledge for the design of NBS and their maintenance. In this case, it is important to refer to knowledge that materialises the concept of an ecosystem—considering natural elements (including fauna and flora), sociological traits, and the socio-historical and socio-emotional relationships created around these same social and technical infrastructures. Knowledge that can represent added value in the design of substantially significant innovation for the resident population, but also for the region and the country, as they are identity marks of places and spaces that can be enhanced in the form of public collective projects, or shaped to meet the needs of populations, becoming economically viable.

This review allows us to infer a set of essential aspects concerning NBS and engagement:

- Limited focus on the relevance of transdisciplinary historical and longitudinal knowledge for NBS design and maintenance
- Lack of reflection on how historical and socio-emotional relationships shape spaces and their potential openness to NBS implementation
- Emphasis on gathering information explicitly helpful in understanding social dynamics and changes in project implementation, rather than historical context
- Recognition of the potential added value of historical and socio-cultural knowledge in designing innovative NBS that resonate with local identities
- Need for incorporating socio-historical and socio-emotional aspects into NBS projects for enhanced community engagement and long-term sustainability

## 8 The Relevance of Future in NBS Projects

Considering future-present challenges such as climate change, urbanisation, and population growth is crucial for designing resilient NBS solutions at the local level. Forward-thinking approaches anticipate future water-related risks and opportunities, ensuring that NBS projects remain adaptive, inclusive, and effective in the face of evolving environmental and societal conditions.

As Ferreira et al (2020) assert, "the design of urban environments must embrace the individual ideals of citizens and stakeholders, which can only be achieved if effective methods of communication, involvement, and feedback are ensured" (p. 1). science and technology policies, on the sustainable development of cities, and the aspirations and expectations of their populations. These policies must reflect communities' desires and visions for their past, present, and future, from an empowered perspective where citizens feel reflected in urban innovations. As Ferreira et al. (2020)point out, "the design of urban environments must embrace the individual ideals of citizens and stakeholders, which can only be achieved if effective methods of communication, involvement, and feedback are ensured" (p. 1). Many of the papers reviewed discussed the future, often in the context of prediction and the need for policy to provide knowledge that enables cities to be more proactive in addressing risks and disasters. However, there appears to be a scarcity of data on how social involvement can facilitate the integration of these aspirations into the design of future projects. This limitation is significant not only academically but also from political and sociological standpoints, as it undermines the role of citizen involvement in shaping their communities. Most of the literature stresses the need for effective science and technology policies and local governance to reflect citizens' ideals and aspirations for sustainable urban development.

- In the literature, future-oriented discussions often focus only on predicting and mitigating climate change effects and preparing for potential disasters, and not on populations and communities.
- There's a scarcity data and discussion about how participatory methodologies can incorporate temporalities and the diverse aspirations of populations in the design of future projects
- Urbanization brings profound sociological changes as new populations may lack connections to the places they reside, intensifying symbolic struggles for the ownership of "future choices.
- NBS aim to preserve environments and social aspects of places, but their effectiveness relies on accumulated knowledge and meaningful interaction among residents
- In the era of AI and smart cities, involving populations in future planning becomes crucial for promoting human capital and fostering socio-emotional sustainability
- The integration of sociological and anthropological concepts of the future, emphasising co-production and continuous dialogue between communities and nature is notably absent

## 9 The Relevance of Open Innovation in NBS Projects

Local-level NBS projects can drive innovation by integrating traditional ecological knowledge with modern technologies and design approaches. Innovative solutions tailored to local conditions enhance the effectiveness, efficiency, and resilience of NBS initiatives. Ziegler et al. (2022), drawing on a review paper, proposes that "three cross-cutting issues that can be a focus for future research, practice, and supportive policy: social innovation for nature-based solutions, social innovation for participatory governance, and social innovation for technology that tackles biodiversity loss.“ (p. 8). Addressing the role of Social Innovation (SI) in participation, the authors call for a broader understanding of innovation processes for biodiversity conservation and sustainability, highlighting the importance of recognising SI's role in shaping transformative pathways.

Court et al. (2022) and Basta et al. (2021) also point out the role of social engagement for innovation through living labs—dynamic platforms for innovation, learning, and collaboration to address complex environmental challenges. These allow for practice-based exploration and experimentation to derive solutions, particularly in social-ecological innovation. Co-creation is emphasised, involving collaborative work across phases like co-initiation, co-design, co-implementation, co-evaluation, and co-maintenance or co-management, ensuring shared responsibility and integrated management. In conclusion, even if innovation appears to be a very relevant word for characterising NBS projects, its relationship with social engagement for building the "future we want" seems to be vague, demanding, in our view, a broader understanding of innovation processes for biodiversity conservation and sustainability with the recognition of the role of SI in shaping transformative pathways for NBS, with multi-actor involvement. In this regard, an important method to consider during NBS design and implementation is open innovation (Chesbrough 2003), going beyond the expertise of any single organisation or sector, and integrating collaboration and knowledge sharing across boundaries, recognising that insights and expertise can come from various sources, including citizens, community groups, academia, industry, and government agencies, providing platforms for citizens to contribute their knowledge, observations, and ideas. This would foster a sense of ownership and empowerment, leading to contextually relevant and culturally sensitive NBS solutions that address the unique needs and challenges of local environments. In practical terms, innovation is a galvanising concept that involves a diverse range of stakeholders to foster innovation development. The notion of OI represents a further advancement in the discourse on innovation, specifically by highlighting the significance of engagement and the active involvement of diverse stakeholders in the creation of novel solutions (Godin and Vinck 2017). This approach allows for greater flexibility, creativity, and access to a broader pool of ideas and talents, ultimately leading to enhanced innovation outcomes. Open innovation (Chesbrough 2003) can be utilised to pilot and assess NBS interventions in practical environments. Through collaboration with local communities, governments, and other stakeholders, organizations can call for feedback, assess efficacy, and enhance NBS solutions to effectively address the needs of both populations and the natural environment.

The involvement of communities in innovations that affect their daily lives and future is a multidimensional issue because there are various understandings of who the stakeholders are and what participation entails throughout the process. This idea becomes even more relevant when dealing with innovations aimed at mitigating risks from potentially harmful events for communities, which may not be immediately perceived as such. Social sciences have developed a wide range of methodologies for involving communities in the processes of identifying needs and designing intervention alternatives. Most of these methodologies stem from the notion that it is necessary to overcome the temptation to impose innovations on communities and individuals, prevent these innovations from being "extractive" or "disruptive to the futures of the involved populations," and encourage effective co-creation through the participation of communities in the design and implementation of solutions. As put forward by Mazzucato (2018) innovation is not solely the result of individual entrepreneurship or private sector investment but often relies on collective action involving various stakeholders, including governments, research institutions, businesses, and civil society.

Through collaborative efforts, these entities combine their expertise, resources, and perspectives to co-create innovative solutions. Such developments are extensions and improvements of the assumptions and rules embedded in sociological and anthropological methods, as they entail acknowledging the need to share research findings with the communities themselves who are the subjects of scientific studies. Issues related to construction, communication, and the questioning of the future are at the forefront of current discussions. In practice, the actions of decision-makers, influenced by large-scale and immediate political and economic interests, often skip many stages of engagement at various levels, even at the local level. This results in high levels of insecurity and estrangement among populations, leading to increased individualisation a loss of interest and a sense of justice within the community. Prejudices persist regarding participation and engagement, while historical contextualisation and the experiential and lay knowledge in decision-making processes are disregarded. Engagement at various stages of processes, including maintenance, requires anticipation and efforts to consolidate collaboration networks that make solutions sustainable while also being socially acceptable because they are meaningful to the populations using them or positively affected by them.

There is a wide variety of NBS each with its own peculiarities, but it is evident that there is a need to continue deepening interdisciplinarity and the relevance of the past and memory in thinking about present innovation. Interdisciplinarity implies that there are no preconceived recipes for who to involve or how, as this depends on the type of issue under analysis and the level of demand associated with a particular NBS. Social engagement is very time-consuming (Felt 2015; Kuitert & Buuren 2022). In this sense, despite the existing bulk of literature on NBS, and the focus is given to social related barriers (Rahman 2023), it is essential to foster more disciplinary integrated and precise specification of the concept and its relationship both with practices of innovation, a collaboration between institutions and individuals and social participation, particularly in the case of large projects (Collier et al 2023).

## 10 Final Considerations

Examples of NBS projects include ecosystem restoration, sustainable water and soil management, green infrastructures, and other nature-inspired approaches to mitigate extreme climate events such as floods, fires, heatwaves, or man-made hazards (Cook-Patton et al. 2021; Van Zatten 2023). Their design and implementation require extensive dialogue among cultural, environmental, socio-technical, and political dimensions. including the assumption of an effective collaboration between scientists, project developers, and decision-makers throughout the design implementing process and alignment with the sustainable infrastructure development goals (Van Zanten et al. 2023). As Adam (2011) acknowledges as regards the future-present, there is a need for innovation to consider the complexity of temporal relationships established in the dependencies between human-environmental ecosystems, emphasising the relevance of time as a legacy to "preserve" considering the complex human-nature ecosystems. So far, the literature review points out pivotal lessons for NBS promoters to consider a long-term perspective, recognising the impacts of NBS on communities' futures. Even so, in face of a strong technological push, new avenues must be paved to gather existing information and better improve practices for social engagement with "the future we want".

**Acknowledgements** This work was partly financed by FCT / MCTES through national funds (PIDDAC) under the R&D Unit Institute for Sustainability and Innovation in Structural Engineering (ISISE), under reference UIDB / 04029/2020 (doi.org/https://doi.org/10.54499/UIDB/04029/2020), the Associate Laboratory Advanced Production and Intelligent Systems ARISE under reference LA/P/0112/2020. This work was supported by *FCT/* Portugal, through the CIUHCT– Interuniversity Research Centre for the History of Science and Technology, NOVA University (UIDB/00286/2020). This work was supported by *FCT*, Portugal, through the Research Center on Communication Studies, CECS—under the reference UIDB/00264/2020 (CECS). This work is also financed by national funds through FCT—Foundation for Science and Technology, under grant agreement [Ref 101121210] attributed to the 1st author. This work is funded by NBSINFRA CITYNATURE-BASED—Solutions Integration to Local Urban Infrastructure Protection for a Climate Resilient Society", with Ref. 101121210 Funded by the European Union. Views and opinions expressed are however those of the author(s) only and do not necessarily reflect those of the European Union or REA. Neither the European Union nor the granting authority can be held responsible for them.

## References

Adam B (2011) Towards a Twenty-First-Century sociological engagements with the future. Insights 4(11):1–18

Appadurai A (2013) The future as cultural fact: essays on the global condition. Rassegna Italiana di Sociologia 14(4):649–650

Araújo E (2019) Science and technology studies and engineering innovation: the concepts of future and diversity. In: Innovation, engineering and entrepreneurship (pp 849–855). Springer International Publishing

Armiero M (2021) Wasteocene: Stories from the global dump. Cambridge University Press

Arnstein SR (1969) A ladder of citizen participation. J Am Inst Plann 35(4):216–224

Basta C, Kunseler E, Wamsler C, Van Der Jagt A, Baró F, Balenciaga I, Wickenberg B (2021) Inclusiveness, equity, consistency, and flexibility as guiding criteria for enabling transdisciplinary collaboration: lessons from a European project on nature-based solutions and urban innovation. Front Clim 3:63007

de La Bellacasa MP (2017) Matters of care: speculative ethics in more than human worlds. U of Minnesota Press

Bhatta A, Vreugdenhil H, Slinger J (2023) Characterizing nature-based living labs from their seeds in the past. Environ Dev 100959

Brill G, Carlin D., McNeeley S, Griswold D (2022) Stakeholder engagement guide for nature-based solutions. United Nations CEO Water Mandate and Pacific Institute, Oakland, CA. Available from https://ceowatermandate.org/nbs/wp-content/uploads/sites/41/2022/11/CEOWater_SEG_F2.pdf

Brown S, Ferreira C, Houck M, Liner B (2021) Conceptual ex ante simulation for Green Stormwater Infrastructure Adoption on private property using agent-based modeling. Water Environ Res 93(11):2648–2669

Chesbrough HW (2003) Open innovation: the new imperative for creating and profiting from technology. Harvard Business Press

Collier MJ, Frantzeskaki N, Connop S, Dick G, Dumitru A, Dziubała A, Xidous D (2023) An integrated process for planning, delivery, and stewardship of urban nature-based solutions: the connecting nature framework. Nat-Based Solut 3:100060

Cook-Patton SC, Drever CR, Griscom BW, Hamrick K, Hardman H, Kroeger T, Ellis PW (2021) Protect, manage and then restore lands for climate mitigation. Nat Clim Chang 11(12):1027–1034

Court A, Kelly A, Hardman M (2022) Exploring the need for innovation in greening urban environments: reflecting on radical practice in Greater Manchester UK. Cogent Soc Sci 8(1):2109261

d'Hont FM, Slinger JH (2022) Including local knowledge in coastal policy innovation: comparing three Dutch case studies. Local Environ 27(7):897–914

Derlukiewicz N, Szałata Ł, Mańkowska D (2023) Scaling Up innovation-exemplary initiatives for sustainable development in european smart cities with a focus on smart mobility. Eur J Sustain Dev 12(3):229–229

Felt U (2015) The temporal choreographies of participation: Thinking innovation and society from a time-sensitive perspective. In: Remaking participation (pp 178–198). Routledge

Ferreira V, Barreira AP, Loures L, Antunes D, Panagopoulos T (2020) Stakeholders' engagement on nature-based solutions: a systematic literature review. Sustainability 12(2):640

Fox N, Campbell-Arvai V, Lindquist M, Van Berkel D, Serrano-Vergel R (2022) Gamifying decision support systems to promote inclusive and engaged urban resilience planning. Urban Plan 7(2):239–252

Godin B, Vinck D (eds) (2017) Critical studies of innovation: alternative approaches to the pro-innovation bias. Edward Elgar Publishing

Kuitert L, van Buuren A (2022) Delivering blue-green infrastructure: Innovation pathways for integrating multiple values. Front Sustain Cities 4:885951

MacDonald C (2012) Understanding participatory action research: a qualitative research methodology option. Can J Action Res 13(2):34–50

Mahmoud I, Morello E, Bisello A, Kolokotsa D (2024) Augmented nature-based solutions: a possible taxonomy of technologies "in" and "for" urban greening strategies. In: Smart and sustainable planning for cities and regions (pp 125–139). Springer

Martins H (1998) Technology, modernity, politics. The politics of Postmodernity 150–181

Mazzucato M (2018) The value of everything: making and taking in the global economy. Hachette UK

Mitić-Radulović A, Lalović K (2021) Multi-level perspective on sustainability transition towards nature-based solutions and co-creation in urban planning of Belgrade Serbia. Sustainability 13(14):7576

Rahman HT, Manuel P, Sherren K, Rapaport E, van Proosdij D (2023) Characterizing social barriers to nature-based coastal adaptation approaches. Nature-Based Solutions 100099
Ramos Torre R (2009) Metáforas del tiempo en la vida cotidiana: una aproximación sociológica
Rosner L (2013) The technological fix: How people use technology to create and solve problems. Routledge
Sachs JD (2015) The age of sustainable development. Columbia University Press
Sanz-Mas M, Ubalde-López M, Borràs S, Brugueras S, Continente X, Daher C ... López MJ (2024) Adapting schools to climate change with green, blue, and grey measures in barcelona: study protocol of a mixed-method evaluation. J Urban Health 1–14
Scott A, Holtby R, East H, Lannin A (2022) Mainstreaming the Environment: exploring pathways and narratives to improve policy and decision-making. People Nat 4(1):201–217
Silva LC (2022) Expanding the scope of biogeochemical research to accelerate atmospheric carbon capture. Biogeochemistry 161(1):19–40
Solan M, Bennett EM, Mumby PJ, Leyland J, Godbold JA (2020) Benthic-based contributions to climate change mitigation and adaptation. Philos Trans R Soc B 375(1794):20190107
Stroud S, Peacock J, Hassall C (2022) Vegetation-based ecosystem service delivery in urban landscapes: a systematic review. Basic Appl Ecol 61:82–101
Takavakoglou V, Pana E, Skalkos D (2022) Constructed wetlands as nature-based solutions in the post-COVID agri-food supply chain: challenges and opportunities. Sustainability 14(6):3145
Teladia A, van der Windt H (2022) A new framework for analyzing local participation in community energy initiatives. In: IOP conference series: earth and environmental science (Vol 1085, No 1, p 012034). IoP Publishing
Voland N, Saad MM, Eicker U (2022) Public policy and incentives for socially responsible new business models in market-driven real estate to build green projects. Sustainability 14(12):7071
Willems JJ, Molenveld A, Voorberg W, Brinkman G (2020) Diverging ambitions and instruments for citizen participation across different stages in green infrastructure projects. Urban Plan 5(1):22–32
Xie L, Bulkeley H (2020) Nature-based solutions for urban biodiversity governance. Environ Sci Policy 110:77–87
Van Zanten BT, Gutierrez Goizueta G, Brander LM, Gonzalez Reguero B, Griffin R, Macleod KK ... Jongman B (2023) Assessing the benefits and costs of nature-based solutions for climate resilience: a guideline for project developers. World Bank. http://hdl.handle.net/10986/39811
Zeng S, Li J, Wanger TC (2023) Agroecology, technology, and stakeholder awareness: Implementing the UN Food Systems Summit call for action. iScience, 26(9)
Ziegler R, Balzac-Arroyo J, Hölsgens R, Holzgreve S, Lyon F, Spangenberg JH, Thapa PP (2022) Social innovation for biodiversity: a literature review and research challenges. Ecol Econ 193:107336

**Cláudia Pereira** is researcher and scholarship holder in the European Horizon project, NBSINFRA—Citynature-Based Solutions Integration to Local Urban Infrastructure Protection for A Climate Resilient Society (https://nbsinfra.eu/). She holds a master's degree in science communication from the University of Minho and a bachelor's degree in biology from the University of Coimbra. She has participated in various colloquiums and conferences on participation and science communication and conducted her master's dissertation integrated with a final internship at I3s, under the title "Science communication through art in Portugal: a study on scientific institutions."

**Emília Araújo** is an Associate Professor with aggregation at the University of Minho, in the Department of Sociology, at the Institute of Social Sciences. She teaches qualitative research methods. She is an integrated researcher at CECS and collaborator at SOCIUS/UL. Her research focuses on the study of time, the future, culture, and science, with a focus on interdisciplinary work, participation, and methods for social engagement in public policies for governance of the future. She is author and co-author of diverse national and international publications. Recently, she has been interested in analyzing the changes brought about by AI for participative governance, with a focus on public policies at the local level. She has coordinated Erasmus+ projects and, in 2023-24, participated as a co-editor and co-author of two books dealing with social studies of science, time/future, and participative approaches in local decision-making.

**Elisabete Teixeira** is an Invited Assistant researcher at the University of Minho, she is Environmental Engineering with a PhD in Civil Engineering. At this moment, she is the Coordinator of a European Horizon project, NBSINFRA—Citynature-Based Solutions Integration to Local Urban Infrastructure Protection For A Climate Resilient Society (https://nbsinfra.eu/). Her field of competencies involves the sustainable performance of eco-efficient low-cement concrete, life cycle assessment, sustainability assessment, risk analysis, protection of critical infrastructures and nature-based solutions. She authored 27 ISI Journal papers (8 as first author with 50 average citation and 2.462 FWCI), with a Scopus h-index of 13. Until now, she was the supervisor of 5 master dissertations, she is the supervisor of 4 PhD theses (on-going), and supervisor of two research fellow. She is a team member and Lecture in the NoRISK, an International Master on Risk Assessment and Management of Civil Infrastructures (recently aproved for funding, and with the first edition 2024/2025).

**Paula Urze** is associate professor with aggregation at NOVA School of Science and Technology. She is associated researcher at CIUHCT—Interuniversity Centre for the History of Science and Technology, with research interests in innovation, science, technology, work, and professions. She has led various national projects, participated in several international projects, and is author and co-author of diverse national and international publications. She has served in scientific societies in the field of sociology of organizations and work. Her recent research is particularly focused on two key topics: the projectification of work and the social studies of artificial intelligence. She is co-principal investigator of the project "Projectifying work: network organisation models in contemporary capitalist societies" (2022. 04212.PTDC).

# Technologies and Mental Health How Does Social Media Redefine Loneliness in the 21st Century?

**Rui Estrada and Marina Prieto Afonso Lencastre**

"But a person gets used to everything, even to what repulses them. That's the truth."

**Abstract** In the 21st century, the intersection of rapidly advancing technologies with social media, mental health and a pervasive sense of loneliness has become a focus of scientific research. This chapter explores the intricate web of connections between these elements, drawing inspiration from Franz Kafka's seminal work, "Metamorphosis", and the thought-provoking film, "Sick of Myself" to unravel the complexities of social recognition through the lens of horror. The digital age has ushered in an unprecedented era of connectivity, marked by the proliferation of new technologies and the ubiquity of social media platforms. While these advances promise greater communication and global connectivity, they also bring with them a paradoxical reality—the rise of loneliness and its profound impact on mental health. Moreover, the pursuit of social recognition through horror, as explored in both Kafka's literature and contemporary film, highlights society's twisted fascination with the macabre. The digital realm, with its potential for anonymity and detachment, becomes a breeding

This chapter was developed in the context of the network project "Compor Mundos. Humanidades, Bem-estar e Saúde", based at the Fundação Fernando Pessoa https://compormundos.fundacaofernandopessoa.pt/ and initially follows from a communication presented by the two authors at the International Colloquium Solitude, solitude, aloneness: social problem or personal choice?, organized by CITCEM: Faculty of Arts, University of Porto, 22 and 23 November 2023.

Javier Cercas, 2022, Terra Alta II, p. 64.

R. Estrada · M. P. Afonso Lencastre (✉)
University Fernando Pessoa, Porto, Portugal
e-mail: mlencast@ufp.edu.pt

R. Estrada
e-mail: restrada@ufp.edu.pt

 
W. Leal Filho et al. (eds.), *Composing Worlds: Humanities, Health and Wellbeing in the XXI Century Towards a More Sustainable World*, World Sustainability Series,
https://doi.org/10.1007/978-3-031-87108-5_10

ground for disturbing and often horrifying interactions. The fear of social exclusion, a central theme to Kafka's narratives, finds a modern manifestation in the fear of being socially shunned in the digital realm. The chapter navigates the complex landscape of new technologies, social media, mental health, and loneliness through the prism of Kafka's existential literature and the contemporary lens of "Sick of Myself". The research sheds light on the transformative impact of the digital age on individual identities and highlights the need for a nuanced understanding of the interplay between technology, social dynamics, and mental well-being in the 21st century.

## 1 Metamorphosis and Seclusion

Gregor Samsa wakes up one morning transformed into a monstrous insect. This marks the beginning of Franz Kafka's famous work titled *Metamorphosis*. An industrious son working as a travelling salesman, Gregor supported his family (parents and sister) and even intended to finance his sister Greta's education at the conservatory, despite the sacrifices this would entail (2002, pp 22–23). Everything comes to an end on that morning: the monstrous Gregor, speaking in an unintelligible animal voice while remaining secluded in his room and eventually dying, "thought back of his family with emotion and love." (2002, p 44).

For the purposes of this chapter, Gregor's horrific metamorphosis is something that should not be shown or seen by himself, and especially not by his family. Several passages in the work illustrate this persistent concealment: when his parents and the office manager see him for the first time, the latter quickly flees the house, the mother faints, and the father does everything he can to bring him back into the room: "He was merely fixed on the idea that Gregor should be got back into his room as quickly as possible." (2002, p 17). A month after the metamorphosis, Greta, who is supposed to bring him food every day, sees him for the first time, fully exposed, looking out of the window. She hastily leaves the room, prompting Gregor to observe, "It made him realise that she still found his appearance unbearable and would continue to do so, she probably even had to overcome the urge to flee when she saw the little bit of him that protruded from under the couch." (2002, p 25). He then decides: "(…) in order to spare her even this sight, he spent four hours carrying the bedsheet over to the couch on his back and arranged it so that he was completely covered and his sister would not be able to see him even if she bent down." (2002, p 25).

Perhaps the most dramatic moment occurs when the mother decides to see him again. Gregor was hanging on the wall of the room and Greta tried in vain to prevent her mother from looking: "(…) she stepped to one side, saw the enormous brown patch against the flowers of the wallpaper, and before she even realised it was Gregor that she saw screamed: "Oh God, oh God!" Arms outstretched, she fell onto the couch as if she had given up everything and stayed there immobile." (2002, p 29).

Gregor had no stage beyond the room that kept him secluded. The few times he was seen outside this room, in the room next to the bedroom and by the members of his family, were met with horror. His father's reaction was equally violent, leading him back into captivity, to which Gregor finally resigned himself. His hideous appearance even led his sister, desperate in the face of the situation, to depersonalise him: "You've got to get rid of the idea that that's Gregor. We've only harmed ourselves by believing it for so long. How can that be Gregor? If it were Gregor he would have seen long ago that it's not possible for human beings to live with an animal like that and he would have gone of his own free will." (2002, p 43). Indeed, Gregor's reclusiveness in every sense is crucial to the dramatic outcome of this narrative: he transforms into an insect, does not leave the house, especially the room, is seen by a limited number of people, including family members, and dies alone.

We use this famous work by Kafka as a starting point for the discussion we want to develop about loneliness and its consequences in a world where technology has, for better or worse, banished the inescapability of the seclusion to which Gregor was subjected since the metamorphosis. Indeed, mediatisation and the widespread use of social media mean that people can now interact with others at a distance and present themselves to the world, even when confined to any space, either through writing or, above all, through images. In this context, recognition, the inevitable gaze of others on us, has become the norm.

But what if, despite being constantly in the spotlight, no one notices us?

It is precisely this loneliness of non-recognition that we want to discuss. First, we will address the painful paradox of social media: many of us are alone, despite having technical skills that can take us anywhere. In the second part of our text and drawing from the recent film "Sick of Myself" by Kristoffer Borgli (2022), we will analyse a possible and dramatic response to the need for recognition and reflect on the place of loneliness in a century that aims to erase the distance between people with a click.

## 2 Stage and Loneliness

In a recent study titled "Loneliness and Social Media: Why do we feel so alone in the most crowded places in the world?" by Rui Miguel Costa, Filipa Pimenta, and Alexandra Ferreira-Valente,[1] the authors reflect on the apparent paradox of feeling alone even when technology allows for easy networking and endless interactions with others. The authors state: "(...) it has been demonstrated that disappointment due to the lack of attention from others on social media (and, to a lesser extent, negative social comparisons) are among the factors that explain why people more dependent on social media are more likely to feel lonely." (p 11) In other words, loneliness comes not from the inability to be seen, but rather from the neglect of the possibility. Once the technical challenge has been overcome, the cruel social question remains: why does no one notice me? And there are so many who could. It is this

[1] Cf. https://oobservatoriosocial.fundacaolacaixa.pt/-/a-solidao-e-as-redes-sociais-por-que-nos-sentimos-tao-sos-nos-locais-do-mundo-mais-apinhados-de-gente

disparity between the enormity of the stage I am on and the scarcity of comments, interactions, likes, or others, that leads to loneliness. Truly brutal, considering that nothing foreshadowed it.

In most cases, a viral moment of glory, whatever that means, is preceded by another form of loneliness: the stage lights shine briefly, only to fade tragically. Jaron Lanier, the author of the book "Ten Arguments for Deleting Your Social Media Accounts Right Now," commented thereon in an interview with the newspaper Público: "In the book, I don't tell people to give them up [social media]. What I do is present the arguments for why they should give them up, in the hope that they can at least give up some and refuse to turn their lives into a gigantic search for attention. (…) If we want to be popular, if we're looking for attention, what social media does is give us a bit of that. We can go viral for a while, but then it disappears."

Of course, this "gigantic search for attention" competes with thousands of others; in other words, the stage is huge, and everyone wants the same from everyone: recognition. On the other hand, in this immense virtual *agora*, we also witness the success of others, more or less familiar or unknown, which causes discomfort and inflated behaviours: "It appears that although more socially anxious and lonely individuals may wish for online social connection, they may exhibit inhibited behaviour, engage in maladaptive cognitive patterns such as rumination, or negatively compare themselves to others while using social media, all of which may prevent them from experiencing the social benefits of using social media." (O'Day and Heimberg 2021, p 8).

Without ignoring that studies on loneliness and other emotional states on social media are controversial,[2] our focus in this presentation is to highlight this sensation that social media has created: the feeling of being alone amidst many.[3]And that brings us to the last point before moving on to the film "Sick of Myself". If social media recognition is the result of fierce competition, what are we willing to do to get it? Anything, in fact: unlike the unfortunate Gregor Samsa, today no one remains secluded in their room. In other words, the quest for attention at all costs leads to forms of social recognition through horror and other means that distort the idea that recognition has a positive semantic value.[4] Inflated too by algorithms, these unusual

[2] - Cf. Jessica Mueller-Coyne, Claire Voss, Katherine Turner, 2022, "The literature examining loneliness and social media use has produced conflicting and unclear results. Hunt et al. (2018) found that lonely individuals feel that going online makes it easier for them to make friends, while Whaite et al. (2018) found that greater amounts of social media use is linked to increased feelings of real-life isolation.".

[3] The French anthropologist Marc Augé argues: "Today, we can say that the non-place is the context for all possible places. We are in the world with entirely artificial references, even in our home, the most personal space possible: sitting in front of the TV, looking at the cellphone, the tablet, and with headphones on at the same time… We are in a permanent non-place. These devices continually put us in a non-place. We carry this non-place on top of us, with us…" (https://brasil.elpais.com/brasil/2019/01/31/tecnologia/1548961654_584973.html).

[4] Consider, for example, the definition of "reconhecimento" (recognition) in the Dictionary of the Portuguese Language by the Academy of Sciences of Lisbon: "3. attribution of importance, of merit; acknowledgment of value synonyms: valorisation. His efforts earned him recognition from his boss. 4. feeling of gratitude for a benefit received; attitude of gratitude. His recognition towards the friend

behaviours end up being the last resort to satisfy the demand for attention. "Speak ill of me but speak" is the *vox populi* that takes on another dimension in the realm of social media. It is the price of a useless but compulsive minute in the virtual arena.

Let's go back to Jaron Lanier: "If we look at people as a whole, we see that we are capable of paying attention to very positive things: we are capable of love, cooperation, and optimism. But negative emotions of fear, paranoia, aggression, these things arise more quickly. Before algorithms came along, the fact that they were faster didn't mean they were more powerful: there used to be a balance between positive and negative emotions. But the algorithm responds to these faster mechanisms that trigger and exercise negative emotions, and with that, we become more tribal, more paranoid, we start seeing each other as enemies, thinking the world is against us; we become more aggressive, in short."

The next section analyses the film "Sick of Myself", which also describes a metamorphosis through horror, but this time the *urbi et orbi* is announced.

## 3 Sick of Myself: Public Metamorphosis

The film "Sick of Myself", directed by Kristoffer Borgli, was released in 2022 and tells the story of Signe, a young girl seeking social attention and resorting to extreme methods to achieve it. Signe's boyfriend, Thomas, is self-centred and gains recognition in the art scene for his uniqueness in creating art from stolen objects. Signe has an ordinary job waiting tables in a café and Thomas is primarily concerned with his own needs, leading to a disconnected relationship. Signe, jealous and even envious of Thomas's growing recognition in the art and media circles, tries to get his attention by staging a simulated dog attack and faking a severe allergic reaction at a dinner party. After discovering a Russian anti-anxiety drug online that has a transformative effect on the body, especially the skin, Signe takes large doses and experiences intense symptoms that prompt a medical consultation. She is diagnosed with a rare disease. On the same day, Thomas is interviewed by a major magazine and fails to pick Signe up from the hospital. Signe then overdoses and is hospitalised, forcing Thomas to spend more time with her. As she continues to take the drug, Signe gradually transforms into a monstrous figure. Her inexplicable appearance attracts immense social attention through the media, social networks, and friends. She is interviewed and even becomes a model of inclusiveness.

Signe, who outdoes her boyfriend in social attention, remains seriously ill. Her divorced parents live separate lives, leaving her alone. Signe's mother advises her to join a holistic therapy group, but the group is suspicious of her honesty. The film ends with the discovery of the drug use, which leads to Signe being abandoned by

---

who helped him when others failed him was great. 5. reward, retribution, prize, or award. He was decorated in recognition of the high services rendered to the homeland. 6. observation, inspection, exploration. A group of scouts went ahead to "reconhecer" (in English, inspect) the campsite area." (https://dicionario.acad-ciencias.pt/pesquisa/?word=reconhecimento).

everyone in a deteriorating state of health. The final scene shows her attending a self-help group with tears streaming down her face.

The film "Sick of Myself" vividly depicts a serious problem inherent in the use of social media: the dependence on virtual forms of attention, and the gradual abandonment of meaningful, real-life relationships. In the anonymity and transience of urban life, which is increasingly focused on quick and flashy success, Signe becomes both a victim and an agent of cultural narcissism. This narcissism replaces the challenges of emotional life with notions of personal fulfilment dependent on visits and likes. The film raises various questions about the overuse of social media, loneliness, and mental health. In Signe's case, there are obvious signs of narcissistic personality disorder from the outset, as seen in the scenes with the invented dog attack and the fabricated allergy. In addition, her emotional relationships with her parents and boyfriend, together with the cultural organisation of the social attention system, condemn her to a lonely and self-destructive personal trajectory. This trajectory could potentially be reversed if Signe's tears in the final scene correspond, for the first time, to a genuine experience of loneliness rather than a cunning strategy to return to the spotlight.

The central question is: does Signe really feel alone, or is her experience more one of humiliation and anger, the core themes of narcissism? In other words, is the film about loneliness or about one of the most explored traits by social media—an exaggerated sense of self? What is loneliness, and where is its place in the 21st century? This question explores the complex interplay between individual psychology, interpersonal relationships, and the cultural dynamics of attention in the digital age.

Valuable insights into the complex relationship between social media, loneliness and mental health can be gleaned from certain themes in the film. The themes of narcissism and social comparison are evident in the behaviour of both Signe and Thomas, reflecting the egocentric and superficial nature of many contemporary social interactions. Attempting to present an idealised version of oneself and comparing it to the equally idealised lives of others can lead to feelings of inadequacy, aggression, and low self-esteem. The theme of validation and self-esteem is evident in the obsession of both individuals with garnering attention by any means necessary (such as stealing objects or faking an illness). This reflects a constant desire for recognition through the pursuit of likes, comments, and followers on social media. This process relies on, and can lead to, a fragile sense of self-worth, as it is determined by external validation rather than a genuine sense of self-esteem built on managing the challenges and joys of life.

The desperation for the attention of others highlights how addiction to social media and social attention is similar to drug use. This serves as a metaphor for the extreme behaviours people sometimes go to in order to stand out in the overcrowded social media landscape. However, the metaphor has a distinctly real aspect: we now know that social media applications such as Facebook, Instagram, TikTok, and others stimulate brain regions and trigger neurotransmitters such as dopamine, similar to those activated during experiences of pleasure and pain, respectively (Westbrook et al. 2021). The speed and transience of online experiences increase the need and become compulsive as individuals seek new pleasure experiences, depleting the brain's capacity to react. Psychological fatigue, anxiety, and depression can ensue, dragging individuals into greater isolation and a sense of loneliness.

Despite Signe's recent social attention, she remains emotionally isolated, living in a lie and a broken relationship with her boyfriend and distant parents. The fleeting nature of online popularity stimulates emotional peaks that may provide temporary satisfaction, but often fade quickly, leaving people yearning for more and perpetuating a cycle of external validation. The paradox of social media lies in the large number of online relationships that can mask feelings of loneliness and emotional disconnection offline. The authenticity of relationships is compromised, and the film highlights the dangers of sacrificing one's true self for online popularity.

To conclude, and going back to Kafka's Metamorphosis, we could comment as follows: Gregor and Signe have a similar narcissistic problem in some aspects, and very different ones in others. The 19th and 21st centuries serve as different cultural stages for the social and clinical expression of psychic change. Kafka's 19th century narrative unfolds within a restricted patriarchal family circle, emphasising hierarchy and guilt. Signe's 21st century context expands into a global environmental and social sphere, emphasising exposure and shame. In both cases, however, there is a metamorphosis that implies dissociation and depersonalisation.

As is always the case with mental health, the significance lies in how one deals with the experience. It would not be surprising if Kafka at some point personally experienced the repulsive insect-like state into which he felt transformed, especially in the presence of his father. He turned this experience into a work of art. Signe, on the other hand, underwent a different kind of transformation, seemingly more conscious but equally radical, with moments of delirious abnormality, seeking attention in extreme ways to escape the non-existence in which she felt trapped. Unlike Gregor, whose parents and sister were actively present, Signe grew up with separated parents, each of whom followed their own path, lost in the global and liquid modernity of contemporary non-places. Signe's metamorphosis is dictated by triumph, Gregor's by defeat, the two poles of narcissism. In both cases, however, there is a common search for love and recognition.

In the Anthropocene, at a time when the environment is in crisis, biodiversity is threatened and many people are living in precarious situations, it is urgent to reconnect with significant persons and local living creatures, to find a certain frugality of life, namely of social media, that allows us to refocus, love, and care for our concrete surroundings. In this way we can be recognised as persons and feel real and alive, both in the space and in the time we are given to live.

## References

Augé M (2023) Com a tecnologia já carregamos o 'não lugar' em cima, connosco. https://brasil.elpais.com/brasil/2019/01/31/tecnologia/1548961654_584973.html. Accessed 01 Sept 2023

Borgli K (2022) Farta de Mim Mesma. IMDbPro

Cercas J (2022) Terra alta II. Independência. Porto Editora, Porto

Costa RM, Pimenta F, e Ferreira-Valente A (2023) A solidão e as redes sociais: por que nos sentimos tão sós nos locais do mundo mais apinhados de gente? https://oobservatoriosocial.fundacaolacaixa.pt/pt/-/loneliness-and-social-media-why-do-we-feel-lonely-in-the-most-crowded-spaces-in-the-world1/13. Accessed 11 Sept 2023

Coyne J, Voss C, Turner K (2022) The impact of loneliness on the six dimensions of online disinhibition. https://www.sciencedirect.com/science/article/pii/S2451958822000033. Accessed 11 Sept 2023

Kafka F (2017) Metamorphosis. Gyan Publishing House, New York

Lanier JJ (2023) De cada vez que o Facebook ganha um cêntimo, alguém foi prejudicado. https://www.publico.pt/2023/10/22/sociedade/entrevista/facebook-ganha-centimo-alguem-prejudicado2065986&trackingId=%7Bkpdx%7DAAAAmEN9vKDzXAoKWk5YZEtlUk5wZRIQbGg2eG0zMTltNmYxMmpneBoMRVhOU1lKWlBaMEhCIiUxODIzYWg4MGJnLTAwMDAzMXYyZmFxcGxuY2lkamJ2NWduZGxnKhJydW5Kc1BLTkxBUjJVTlI3NjhSEnYtXQDwF3JldG5qMmxyZ1oMNjYuMjQ5LjcyLjQwYgNvYmJogb_PogZwAXgI. Accessed 15 Sept 2023

McCain J, Campbell WK (2018) Narcissism and social media use: a meta-analytical review. Psychol Pop Media Cult 7(3):308–327

O'Day EB, Heimberg RG (2021) Social media use, social anxiety, and loneliness: a systematic review. https://www.sciencedirect.com/science/article/pii/S245195882100018X. Accessed 16 Sept 2023

Westbrook AG, Bosch R, Maata JI, Hofmans L, Cools R (2021) Striatal dopamine synthesis capacity reflects smartphone social activity, iScience 24:102497

**Rui Estrada** full Professor at the Fernando Pessoa University, Integrated Researcher at CITCEM, Coordinator of the network "Composing Worlds: Humanities, Well-being, and Health" and researcher of Permanent Observatory Violence and Crime, University Fernando Pessoa, 4249–004 Porto, Portugal.

**Marina Prieto Afonso Lencastre** retired Full Professor at the University of Porto, Full Professor at the University of Fernando Pessoa, Psychotherapist and Supervisor Specialist at the Portuguese Society of Clinical Psychology, Coordinator of the network "Composing Worlds: Humanities, Well-being and Health".

# Humanities, Environment, and Health in the Anthropocene: How does climate change force us to think and act differently?

**Marina Prieto Afonso Lencastre, Rui Estrada, and Juan del Río**

**Abstract** The Anthropocene, marked by unprecedented human impact on the Earth's geology and ecosystems, has brought humanity to a pivotal moment where conventional ways of thinking are proving inadequate to address the challenges posed by climate change. This chapter explores the intersection of climate change and new ways of thinking such as non-binary thinking, relational ontology, analogism, and transversality. These are articulated with the sustainable practices championed by the *Alternativas* project. As climate change accelerates, the need for a paradigm shift becomes imperative. Conventional binary thinking has often perpetuated exploiting the environment, and reinforced the human/nature divide. In response to this, new ways of thinking are emerging that challenge traditional dualisms and embrace ideas that recognise the interconnectedness of all life forms and ecosystems. Relational ontology serves as a foundational concept in this paradigm shift, emphasising the interdependence of entities and rejecting the reductionist view that separates humanity from nature. As a way of thinking, analogism emphasises the importance of recognising patterns and relationships betwen different systems, fostering a holistic understanding that is crucial to addressing the multiple challenges posed by climate change. Transversality, a concept rooted in interdisciplinary collaboration, emerges as a key strategy for navigating the Anthropocene, facilitating the development of

This article was written in the context of the network project "Composing Worlds: Humanities, Well-being, and Health." https://compormundos.fundacaofernandopessoa.pt/ and initially follows from a communication presented by the two authors at the Tempos Cruzados Interdisciplinary Colloquium. Dialogues between History, Anthropology and Archaeology, organised by SPAE—Faculty of Arts of the University of Porto, on 24 October 2023.

M. P. A. Lencastre (✉)
Fernando Pessoa University, Porto, Portugal
e-mail: mlencast@ufp.edu.pt

R. Estrada
University Fernando Pessoa, Porto, Portugal
e-mail: restrada@ufp.edu.pt

J. del Río
Alter Natives, Porto, Portugal
e-mail: Info@alternativasdocumental.info



W. Leal Filho et al. (eds.), *Composing Worlds: Humanities, Health and Wellbeing in the XXI Century Towards a More Sustainable World*, World Sustainability Series,
https://doi.org/10.1007/978-3-031-87108-5_11

sustainable practices that transcend disciplinary boundaries. *Alternativas* is an exemplary sustainable practices project that seeks to explore and implement alternative solutions to pressing environmental issues. The project demonstrates how transversal non-binary approaches can pave the way for innovative solutions to climate change by fostering collaboration and embracing a relational ontology that recognises the intrinsic value of all living beings.

## 1 Introduction

It is now clear that a significant part of the challenges of the 21st century, particularly in the area of health and well-being, are related to how humans interact with local and global ecological contexts, cultures, and technologies. In the Anthropocene (Crutzen and Stoermer 2000), human presence in nature and the natural responses of the planet Gaia (Lovelock 2007) are inextricably intertwined, forming new local and global environments for life. How do we think about these places and how do we live in them? This is a fundamental question for a new eco-centered anthropology developed by the Composing Worlds network.

The current climate scenario and new biophysical conditions are leading to significant changes in biotopes and evolving forms of life. Environmental changes in human niches not only trigger migrations but also generate new social movements and cultural products, questions about the future, new scientific investigations, new narratives, and unprecedented visual conceptions. Currently, science and public awareness coexist through controversies and collaborations in knowledge production. Information, media, and digital realities add complexity to psychosocial and cultural meanings, often extrapolating beyond the immediate bodily experience of life situations. The metamorphosis of the world, as Ulrich Beck (2016) calls it, expresses this radical change of the planet as we knew it until the 20th century, and a new estrangement that keeps us in a perplexed position in the face of the great transformations of our habitats and of the world in general. These are epochal changes in views and ways of life, brought about by changes in climate regimes, global social and political life, and technologies.

Inspired by the significant material and symbolic changes taking place in the world today, our starting question is: How can the humanities contribute to a better understanding of the current environmental and social situation, and thereby enable action? This is an exploratory and, at this stage, descriptive investigation developed in the context of social epistemology understood as "a body of knowledge about the social dynamics of intellectual activity of human collectives." (Oddone 2007, p. 114).

## 2 Mutations in the paradigms of thought in the Anthropocene

As the Earth has entered the Anthropocene, a geological era shaped by human activity, probably since the Industrial Revolution, but certainly since the presence of radioactive records, resulting from human nuclear activity in the geological strata studied by scientists, it would seem that the current paradigms of thought have also entered a new epistemological era. Although the term Anthropocene is still debated among experts, it allows us to have a synthetic view of the current situation of the planet. Popularised in 2000 by Paul J. Crutzen, Nobel Prize for Chemistry, and the biologist Eugene F. Stoermer, the term is widely used in the social sciences and humanities. This neologism suggests that modern thinking was wrong in predicting that humans would completely dominate "nature" and unlock its secrets. Not only has this project failed to materialise, but the delicate balances of the Holocene have been upset, and we now face unprecedented material challenges. This is a central theme in the work of Bruno Latour (1991, 2017). At the macro level, with climate change, and at the micro level, with the release of microorganisms as a result of the invasion of habitats previously reserved for other animals, this new geological era reminds us that our relationship with the Earth is complex and interconnected. In many fields, from the exact sciences to the humanities and the arts, the importance of delicate interrelationships with ecosystems is now recognised, urging us to forge new alliances with the Earth and its inhabitants, considering ourselves as members of ecological communities, just like our non-human co-inhabitants, whose alterity remains irreducible. It is essential that we move towards a more integrated and eco-centric view of the world.

The modern representation of the world as a machine, typical of the previous centuries and culminating in the idea that physics would be the quintessential science—an idea confirmed in the 20th century by the enormous power of nuclear physics—is giving way to a more organic representation of the world, the famous Gaia of the ancient writings, recommended by Latour (2017) as the best metaphor for understanding contemporary environmental changes. In fact, modernity elevated humans to superior beings, different from other living beings, and gave them the right, even the moral duty, to exploit natural resources that were considered inexhaustible. This radical distinction between humans and non-humans buried with it the divinities and powers associated with the natural elements. But the naturalistic thinking in f the West is not uniform. Various groups, such as Romantics, the Vitalists, among others, have tried to escape Cartesian physicalism. Today, an ontological shift requires new images and metaphors, and new ways of thinking about our being in the world. The environmental crisis is a crisis of imagination and thought, actively seeking better ways of imagining nature and defining how it relates to humanity.

This chapter explores these ideas, taking as its starting point an observation made by Ph. Descola during a conference at the Vuitton Foundation in Paris (Descola 2018). According to Descola, naturalism is coming to an end and may be replaced by what he calls "a metastable form of analogy," which attempts to gather the diverse contemporary realities that challenge our understanding. How do we invent words for a new world of life? How do we weave new relationships that allow us to think and act on an unstable planet full of other and multiple beings? For Descola (2018), a new analogy will be non-anthropocentric. Humans will not be the source of the anthropisation of the world, but this source will rather be the living environments with their human and non-human elements. Examples of these attempts are the extra-moderns, those who are not pre-modern but live in modernity and try to escape it by inventing new ways of living. An example of extra-modernity is the legal recognition of the status of subjects of natural elements such as rivers, mountains, animal species, linking them to the aesthetic, ethical and spiritual dimensions of a new world vision. Another example is the *Alternativas* project, which will be described later in this chapter.

It is against this background that we have posed the main question of this chapter: How do climate change and the environmental crisis force us to think and act differently? In the first part of our chapter, we will mainly discuss forms of thought mutation. In the second part, we will go down to the practical level and see how changing ways of living are intertwined with new ways of thinking.

## 3 Methodology

This is essentially an exploratory and theoretical essay. We start off with the responses of 12 experts from different fields to an interview on the intersections between the humanities, health, and well-being in the 21st century. This interview was constructed and answered in the context of the "Compor Mundos. Humanidades, Bem-estar e Saúde" network (Campelo et al. 2022).[1] We analysed these texts and identified four main ideas that were either implicit, or suggested by the experts, as objects for reflection (Lencastre and Estrada 2022). These ideas are:

i. Non-binary Thinking
ii. Relational Ontology
iii. Analogism
iv. Transversality

More generally, we asked where and how we can find these ideas in contemporary thought?

### 3.1 *Non-binary Thinking*

The idea of overcoming binary and dichotomous thinking has been present in literature since the systemic theory developed in the 1960s by von Bertalanffy (1968),

[1] The methodology used in the project is described in detail in Lencastre and Estrada (2022). *Compor Mundos. Humanidades, Bem-estar e Saúde*, Vol. 2, Porto: UFP.

which was later taken up by various fields of knowledge, from the natural sciences to the social and psychological sciences.

The contrasting concepts of nature/culture; subject/object; body/mind/spirit; human/animal/plant/geological; self/other; sacred/profane, which were clearly separated in the naturalistic tradition of thought, allowed the simplification of natural objects to be examined by scientific methods that reduce the complex to the simple and thus act on a parts of it. From the last third of the 20th century, the social, political, and environmental issues, as well as the encounters with other cultures, especially in Europe, slowly imposed different ways of thinking and interpreting. A more complex, subtle, and inclusive view of the world emerged, allowing for the recognition of diversity and thinking in spectrums, which seems closer to reality than thinking in opposites. Different degrees of quality and nuance result from cognitive empathy, with others, but also with objects traditionally interpreted as compact physical wholes acting as mechanical monads—this is the case with our approach to animals (de Waal 2001), plants (Wohleben 2015), but also to geological and chemical elements. Questioning and contextualising assumptions allows us to embrace dynamism, but also ambiguity and uncertainty, without the need for simple and definitive answers. The complexity of social, political, scientific, and ethical issues requires dialogue and inclusive participation, cultivating a cosmopolitan and open mindset that accepts self-criticism and relationships.

An example in genomic science of overcoming binary thinking is that of Barbara Mclintock, who in the 1950s hypothesised that genetic regulation could explain how multicellular organisms could diversify the characteristics of each cell, even if their genomes were the same. This idea completely changed the concept of the genome, which until then had been interpreted as a simple set of static, separate instructions. McClintock's work on the regulatory and controlling elements of genes was so complex and innovative that other scientists were reluctant to accept it. Despite this, McClintock moved forward and continued her research, but due to the reaction of the scientific community at the time and McClintock's perception that she was deviating from the mainstream, she stopped publishing her findings. She later commented that her laboratory experience had shown how challenging it is to make other scientists aware of their tacit assumptions. Due to the entrenched ideas of many scientists, some advances could not be shared, and she had to wait for a more general conceptual shift in the community and communicate them at the right time. She understood the role of evolutionary agents before other scientists did. She was awarded the Nobel Prize for Physiology in 1983.

## *3.2 Relational Ontology*

Non-binary thinking leads us to a relational paradigm (relational ontology), emphasising the importance of relationships and interactions between entities, rather than focusing solely on individual characteristics. Reality is seen as a complex network of relationships in which everything is interconnected and interdependent.

Relational ontology is widely used in ecology. For example, the food web shows how living beings in an ecosystem are interconnected through food chains and nutrient cycles. Changes in one population can affect other populations and the overall balance of the ecosystem. Originating in the 1960s, systems theory takes a relational ontological approach to understanding complex systems (Bertalanffy 1968). It analyses how the parts of a system relate to each other and how these relationships affect the behaviour and function of the system. In sociology, social network analysis and actor-network theory (Latour 1991) study the interactions between individuals, social groups, and institutions. They help to understand a highly topical phenomenon that is exacerbated by social networks: how social relationships shape the spread of information, the diffusion of ideas, and the formation of epistemic communities.

Processual ontology is another related field that, following authors such as Whitehead, Bergson, Dewey, William James, and more recently B. Latour or I. Stengers, shows that reality is composed of fluid processes in constant change, rather than fixed and isolated entities. This situation has been described in sociology by Zygmunt Bauman (2000) through the notion of liquid modernity as opposed to solid modernity. Things are seen in terms of their relationships and constant change, rather than as static objects, thus presenting a more organic and non-mechanical view of reality.

Relational ontology can also be applied to the fields of knowledge theory and epistemology. Knowledge is seen as constructed through interactions and communication between individuals and communities, and learning and understanding are influenced by social and cultural relationships. In 2022, an interview with physicist, mathematician, and historian of science Evelyn Fox Keller discusses the challenges she has faced in expanding the boundaries of science studies with her pioneering work on language, gender, and science. Her views changed the way we think about the history of science (Rocha et al. 2022). She drew on evidence from genetics, neuroscience, developmental psychology and other fields to show the growing inadequacy of both reductionist atomistic models and conventional attempts to construct wholes from parts. In addition, she emphasised the importance of exploring themes of interactivity and constructing parts from wholes. There was also evidence of the need to position individual traits more effectively within social dynamics or, more broadly, to consider causality from the top down. As Evelyn Fox Keller stated, "If I were to identify a single unifying theme running throughout my entire intellectual trajectory, from my work in mathematical biology to my work in gender and science (where it is probably especially evident), this unifying theme would be a persistent pursuit of a relational epistemology." (in Rocha et al. 2022, p. 735).

### 3.3 *Analogism*

Analogical thinking is a cognitive process that draws connections and identifies similarities between disparate concepts or ideas. Analogism uses similarity, contiguity, metaphor, or metonymy to make one object or idea understandable in terms of another. Rhetorical expressions such as metaphor and metonymy transcend linguistic barriers by transforming abstract concepts into visual or descriptive representations. For example, the Eipo people of New Guinea express emotions by comparing sadness to the idea of "a broken bridge under the feet" and joy to the sensation of "the sun shining on the chest" (Eibl-Eibesfeldt 1989). These associations are based on common individual bodily experiences, making them universally understandable. In addition, descriptive concepts of emotional states, such as the use of the term "heat" to describe excitement, highlight the expressive power of language in different cultures. For example, the Eipo describe intense emotion as "heat in the liver", while Thais and Westerners associate heat with sexual arousal (Eibl-Eibesfeldt, Op. cit.). The relationship between emotional concepts and physiological experiences is evident in language, as in the case of anger, which is associated with sensations of heat and boiling, resulting in expressions such as "boiling with rage" or "red with rage" (Lakoff 1987). These metaphors not only convey meaning, but can also be imaginatively extended to other art forms such as cinema, painting, and music, and acquire new semantic and syntactic significance.

According to Saraiva (1996), linguistic concepts play a crucial role in shaping perception by influencing the way society associates perception with certain linguistic classes. The author illustrates this by citing experiments that demonstrate how labelling ambiguous perceptive figures can affect memorisation by triggering a process of semantic contamination. Different cultures associate words with different objects or experiences, but they are grouped into sets of meanings that are characteristic of those cultures. These sets of meanings form the basis of shared social theories, which over time become entrenched beliefs. Words then come to function as mental entities that have cognitive effects and can influence the genesis of tangible objects or sets of objects, with significant implications for on ecological and social life.

The metaphorical basis underlying the lexical motivation for the act of 'understanding' emerges from the intersection of visual experience and light, which is fundamental to primate visual perception. Expressions such as "having a vision of things", "from my point of view", "transparent reasoning", "perceiving clearly", "enlighted mind", among others, arise from this association between light, vision and understanding (Lakoff & Johnson 1980).

Mark Johnson (1987) explores the embodiment of the mind, suggesting that humans possess generalised unconscious structures known as kinaesthetic image schemas. These schemas are rooted in basic bodily experiences, such as the container schema, the part-whole schema, the source schema, the path schema, and the goal schema. They originate in bodily proprioceptive and sensorimotor experience and can be defined in terms of structural elements that have a fundamental logic and are subject to metaphorical projection. These metaphorical projections act as structuring elements for different cognitive domains (Johnson 1993). For example, in the container schema, the structural elements are the inside, the boundary, the outside, and its fundamental logic lies in the concept of inside or outside. This metaphorical projection imparts a conceptual structure to visual experience (things enter or leave the visual field), interpersonal relationships (entering or leaving a relationship), and system logic (systems contain elements), among others. Examples such as these illustrate the formation of image schemas from sensorimotor interactions, giving rise to pre-conceptual structures of lived experience, which are then projected into the imaginary. These projections are not arbitrary; on the contrary, they are carried out, in part, through metaphorical and metonymic procedures.

An intriguing observation is the largely unconscious nature of these processes of imaginary and linguistic translation, which seem to be in line with the mechanisms of the unconscious identified by Lacan 1953. In the Rome Discourse (1953), Lacan describes metaphor and metonymy as the two fundamental poles of language. He suggests that the analysis of the dynamics of the unconscious can derive from these elements because, for Lacan, the unconscious is structured like a language.

The narrative and imaginative capacities of the human brain are likely to play a crucial role in the advancement of knowledge, whether scientific or philosophical. Castoriadis (1999), for example, writes that the human imagination has the power to create 'possibilities' that did not previously exist. In anthropology, Descola (2005) became interested in the schematisation of thought and identified the binary opposition between physicality and interiority as one of them, embodied in various forms by different cultures. For Descola, the dichotomy of physicality-interiority is common to all human beings, and through combinations of their similarity and/or dissimilarity, he identified four major systems of cultural "world compositions", or ontologies: animism, where beings are similar in interiority but different in physicality; its opposite, modern Western naturalism, where beings are similar in physicality and different in interiority; a totemic cultural ontology in which beings are similar in both interiority and physicality, although divided into dissimilar groups; and finally, the opposite of this, analogism, an ontology in which each being is unique, different from all others, but is related to them by analogy.

These four ontologies, or modes of identification, proposed by Descola, are not mutually exclusive. They tend to appear in all cultures, with one of them being dominant and usually characterising thought and practice. For example, until recently most of humanity did not make much of a distinction between what is natural and what is social, as is typical of naturalism, nor did they consider the treatment of humans and the treatment of non-humans to be in entirely different spheres. Different ontologies correspond to different modes of socio-cosmic organisation, or patterns of distribution of beings in collectives, and figurative art manifests these aspects more obviously in contemporary culture, as in the case of analogism (Descola 2021).

There are societies that use analogy to establish order in the world, and analogy is familiar to us because European peoples were analogists before, especially in the Renaissance, the Middle Ages and Antiquity. At present, analogism characterises the Far East, the Amerindian populations of the Andes and Mexico, and much of Africa. For Descola (2005), analogy is the idea that all entities in the world are fragmented into a multiplicity of essences, forms, and substances, and that their contrasts are combined into a dense network of similarities that links the properties of each entity. The interiority and physicality of these objects are quite distinct, and analogy allows for a cosmology that contains, within its boundaries, all the relationships and determinations necessary for its proper functioning. Analogism can be found in Chinese divination techniques, in African groups who believe that social disorders and natural disasters are related, and in European medicine before the 17th century, which emphasised the search for analogies between symptoms, objects, and parts of the human body. The essence of analogism lies in the search for signs of similarity (Descola 2005).

Analogism appears in different areas of contemporary thought. It makes it possible to respond to the need for knowledge in the face of a profoundly changing world. This is the case, for example, with modern bioethics casuistry, which focuses on the analysis of individual cases to identify relevant ethical principles that can be applied to other similar situations. Bioethics uses analogy to transfer ethical principles well established in one context to a related, novel context. For example, the principles of autonomy, beneficence, non-maleficence, and justice, widely used in medical issues, can be adapted, and applied to ethical decisions related to scientific research, public health, or innovative treatments. Bioethical challenges related to cultural diversity and different value systems rely on analogy to identify commonalities and shared ethical principles between different cultures.

In environmental science and ecology, analogies are often used to understand the relationships between humans and non-humans—such as the application of the legal and moral status of subject to natural entities such as rivers, mountains or animals, and its philosophical, legal, aesthetic and moral implications. Technological advances such as artificial intelligence, genetic therapies, and biotechnologies use analogy to identify similar ethical issues faced in other fields and apply these considerations to guide decision-making in emerging technologies. Analogism is clearly present in digital science, artificial intelligence, and robotics through bio-inspired systems emulating the functioning of the human brain and other biological systems to create intelligent algorithms and robots. Some examples include artificial neural networks, which mimic the structure and functioning of neurons in the brain, and genetic algorithms, which are based on the process of natural selection and biological evolution to solve complex problems. There are also various bio-inspired approaches to robotics, where robots are designed to resemble animals or natural organisms in their appearance and/or behaviour. For example, some robots mimic the locomotion of insects, birds, or marine animals. Pattern recognition algorithms, for example, can be designed to find similarities and correspondences in images, similar to how the human brain processes the visual world. The concept of biofeedback seeks to create a more natural and intuitive connection between man and machine. These analogical relationships are easy to reify, and one of the cyclical recurring questions is whether digital systems and robots are conscious. Animistic inferences can easily be drawn from objects that behave as if they have agency—computers, for example, or artificial intelligence that uses language, such as ChatGPT.

Theory of mind, in primate ethology, is based on the conscious human experience that other humans have intentional inner lives and, consequently, that other primates also have subjectivity. This is the result of "anthropogenism" and not of anthropocentrism, as de Waal (2022) argues, meaning that the attribution of intentions to other primates, who are evolutionarily close to us, is based on an analogy with our own subjective experience. In cell biology, the outer membrane is often thought of analogously as a physical boundary separating the inside from the outside environment. In ecology, the web analogy is used to describe trophic relationships between different species, animals, plants, microorganisms, etc. In environmental modelling, analogous reasoning applies to the Earth's atmosphere as a greenhouse, to high biodiversity as a dense protective network for the environment, and the term atmospheric river is used to designate water-laden clouds, etc. The analogy in medicine is used, for example, in the idea of the immune system as an army fighting infections, in the conception of the heart as a pump, of the body as a machine. The analogy of the environment to human health is proposed by some scientists to make the environmental crisis closer and more personal.

Yet these analogies do not weave together a coherent socio-cosmic system, or ontology. They are fragmentary, unconnected, and do not give rise to an explicit analogical ontology in Western culture. They remain tacit and as local forms of thought. Perhaps one of the tasks of a novel contemporary epistemology would be to explore these analogical relationships and to propose a consistent foundation for them, a grid of epistemic approaches that can underpin a global and shared view, uniting ecologies and cultures in the fields of science, technology and the humanities.

## *3.4 Transversality*

Transversality suggests the idea of crossing, intersecting, or relating different fields, objects, disciplines, and methodologies in order to seek relationships and interactions between different aspects of a problem or situation. Transversal thinking integrates knowledge from different fields of study to gain a broader comprehensive and more multidimensional understanding of a problem or situation. Its aim is a broader understanding of phenomena and one of its imperatives is the unity of knowledge. Transversality allows for the preservation of disciplinary content, the development of the imaginative and integrative aspects of its themes in an exploratory conception of science that is opposed to traditional reductionist positivism, and the development of complex and dynamic knowledge in the search for meaning.

Transversal projects in the curricula of biology, medicine or psychology can bring out more clearly the implicit values of these scientific disciplines. These values are naturalistic at their core and generally include the identification of causal elements, analytical materialism, the logic of the excluded third, categorisation, and universality. These values can, for example, be re-examined in the light of ecological and humanistic values, for instance. The discussion of freedom, diversity, beauty, justice, dignity, and even truth, understood as the sense of coherence between disparate information, can and should be questioned in the context of their relationships with the new planetary and climatic situation, with other peoples and cultures, with non-humans. These inquiries show the relationship between disciplinary thinking and structural forms of conceiving reality. They make it possible to give the disciplines a sense of timeliness and urgency, an ecological density and a new anthropological meaning. They also allow different ways of thinking to be introduced in a less superficial way. Literature and visual art vignettes in climate fiction (cli-fi) that address the issues raised by scientific disciplines will enable students to perceive the reality described in a deeper and more tangible way.

Transversal thinking is particularly valuable when dealing with complex and challenging issues that cannot be adequately understood or solved by a single approach or discipline. A deeper approach to these questions would desirably affect the methods of the scientific disciplines themselves, redefining their objects and repositioning the ways of acting upon them, as the work of Barbara McClintock has shown. This is one of the most poignant contradictions of the two cultures, as they were called in the 1980s and still active: despite decades of critical thinking, and despite the problems with naturalism, scientists working in or out of laboratories generally retain the modern beliefs that underpin their sciences. Perhaps the environmental and anthropological crisis of our times has not yet been strong enough to change the physicalist paradigm, which seems to make it easier to grasp the world. An organic paradigm has not yet established itself on the scientific panorama, but all the signs are that it is in the making.

It is important to bear in mind that this paradigm shift will not only take place through logical and rational discourse, but also, and perhaps above all, through practice and through the emotion and empathy that art can generate, because both are more holistic, real and imbued with subjective understanding in their solutions. In line with cognitive studies, emotional messages generate more empathy and engagement than a simple theoretical explanation. It is this composite reality, experienced in the places where life unfolds, that we will now discuss. We may be witnessing the plural emergence of extra-modern collectives that, without denying the historical reality of modernity, recognise its internal limitations and attempt an analogical hybridisation of eco-cultural adaptations.

## 4 *Alternativas*. Building Possible Futures

According to the previous section, we may succeed in addressing climate change and the environmental crisis if we adopt different mindsets, taking into account in particular the four concepts discussed earlier: non-binary thinking, relational ontology, analogy as a way of thinking, and thematic, disciplinary, or other forms of transversality.

The documentary titled "Alternatives. Building Possible Futures", directed by Agustí Corominas and Juan del Rio, which premiered in Portugal on 16 July 2023 in Póvoa de Varzim, shows how these issues can be put into practice (Corominas and del Rio 2023).[2]

Another characteristic of this documentary is that it shows precisely the path, in most cases under way, of forms of socio-ecological worlds compositions that respond to the climate and environmental crisis. As one of the interviewees, anthropologist and

[2] See the article in the Público newspaper: https://www.publico.pt/2023/07/16/azul/noticia/alternativas-filme-dar-corda-esperanca-2057030.

environmentalist Emilio Santiago from the ‘Rompe o Círculo’ Transition Institute, puts it: “[a] new society cannot be born only from denunciation; it also needs seeds: new social relations and new institutions.”

The documentary opens with Antonio Turiel, a researcher at the Institute of Marine Sciences in Barcelona, saying that the inevitable shortage of fossil fuels will soon leave the world with two diametrically opposed alternatives: an eco-fascism in which the richest control the rest of the energy production, or a democratic and participatory model that allows the common use of renewable energy sources. “Alternatives. Building Possible Futures” explores this second path. In parallel with some of the ideological idiosyncrasies inherent to the documentary, we are interested in pointing out, among many others, three cases that illustrate the practical and feasible possibility of responding to the climate challenge, and how these alternatives arise from new ways of thinking that we spoke of earlier.

## *4.1 Goiener*

The first case is Goiener. It is a not-for-profit cooperative that trades in renewable energy in certain parts of Spain. The documentary refers to the cases of Donostia and Ordizia in San Sebastian, in the Basque Country. In general, this cooperative seeks to produce energy by adapting the technology to the specific resources of each territory, as opposed to the widely used fossil model. In other words, respecting the principle of renewability, energy can come from the sun, wind, rivers, or biomass: the choice of one of these sources depends on the resources of a given territory. This approach does not advocate a single energy transition path, but rather the most appropriate one for a given location.

To paraphrase Antonio Turiel again, this option has four advantages: it does not require high technology, it has a limited environmental impact, it distributes wealth throughout the territory, and it guarantees local employment. One of the Goiener associates, interviewed in the documentary, expresses his satisfaction at having an energy supplier other than Iberdrola (the electricity supplier company in Spain). It is a participatory and democratic model in action. But it is also a new way of thinking: binary thinking is overcome, because ecological resources are not simply to be exploited from a subject-object perspective, externalising the effects of usage. Instead, they are conceived as units of a local collective that needs to be taken care of, in the medium to long term, from a cooperative perspective. We also find that this collective of humans and non-humans implicitly performs relational ontology: for example, managing water resources from a temporal perspective, so that they do not run out in hot, dry summers, and considering future generations, requires attention, thinking and behaviours that consider both the temporal characteristics of water cycles and the local needs associated with them. The whole system is interconnected through action.

### 4.2 *Reci-Creativa*

The second case comes from Granada, Andalusia. The associative and voluntary project *Reci-Creativa*, which is part of the wider 'Movement for Transition' founded in England at the beginning of this century[3], aims, according to its promoters, to "fight against planned obsolescence", particularly of everyday household appliances. Inviting volunteers with skills and knowledge in the field, *Reci-Creativa* offers to "repair anything that breaks in our homes". The most common items are blenders and toasters, and often the cost of the item to be repaired is not more than one euro. In the documentary, a volunteer tries to open a projector for repair and comments: "Modern things are designed so that they cannot be opened", that is, they are designed to stop working, to be discarded, even if the cost of repair is minimal. This is where transversality comes into play, namely the multiple skills that make it possible to reuse objects intended to become obsolete. Moreover, these services are often available in local markets for the exchange of more general services, without the intervention of global currencies, thus promoting relational ways of thinking about the human and non-human community.

### 4.3 *Entrepatios*

The third case is related to the issue of housing: *Entrepatios* in Madrid seeks to address the looming housing crisis and the need to reduce energy consumption in the real estate sector. As Iñaki Alonso, architect and member of the *Entrepatios* cooperative explains, "(...) it is a housing model that tries to build a different relationship with the environment, between people and also with the economy".[4] Seventeen families bought plots of land and built seventeen independent, low-energy ecological houses. The building has communal areas for the whole community of residents: for example, there are only three washing machines in the communal laundry for all the residents. They organize the washing through a shift schedule for using these machines. According to one resident, there are no conflicts and everything runs smoothly. In other words, for the individualist culture we are used to, we are in fact talking about a reduction of fourteen machines: this is the sustainability path that this association is following. This example shows how a collective promotes relational ontology, intertwining it with a subtle non-binary perception of the functioning of culture and nature. The use of a common space implies connecting with the constraints of others, escaping individualism and promoting cooperation and relationships. Using fewer washing machines to reduce the impact on natural elements such as energy and water implies the notion that these are related to, and impact on, a broader whole, the effects of which then reverberate. It is the awareness that cultural practices have natural effects and vice versa that helps this collective relate to each other and to

---

[3] Cf: https://transitionnetwork.org/.

[4] Quotation from the documentary.

the larger interconnected world, thus reducing ecological impact. Undoubtedly, the mutation of practices is the concrete path to sustainability, but it is accompanied by the emergence of new ways of thinking, mostly implicit because not always explicitly articulated in consciousness. They carry within them new perceptions and different compositions of worlds.

To sum up, these and other examples cited in the documentary make us realise that it is possible to tackle the world's current environmental problems without brutal changes or fanfare. This is the great virtue of "Alternatives. Building Possible Futures": it is possible, here and now, to build these new worlds.

As Rob Hopkins, co-founder of the 'Transition Movement', says of the current model of limitless economic growth: "(...) my children grew up to a certain point, and I'm very happy with that. If they kept growing and growing, there was something deeply wrong. We created this economic model that needs to grow always. When we look at these problems, there is a solution."[5] Hopkins draws an analogy between children's growth and economic growth, and promotes an organic—not mechanical—conception of this economic growth: non-binary, in relation to the environment, and managed by internal and external dynamics. Now, this idea that we don't have to grow forever, that the sky is not the limit (a phrase so popular today), requires the conceptual skills presented in the previous section, as well as practical examples such as those presented in the documentary "Alternatives. Building Possible Futures".

## 5 Conclusion

This chapter explores the crucial intersection of climate change and evolving modes of thought, exemplified by non-binary thinking, relational ontology, analogism, and transversality. The urgency of tackling climate change requires a paradigm shift away from the conventional binary thinking that has often contributed to the exploitation of the environment and the separation of humanity from the natural world. The emergence of alternative frameworks challenges these dualisms and emphasises interconnectedness and interdependence. Relational ontology rejects reductionism and emphasises the intrinsic connections between all life forms and ecosystems. Analogism as a mode of thought highlights the importance of recognising patterns and relationships across different systems for the holistic understanding that is essential in addressing climate challenges. Transversality, which promotes sustainable practices across disciplinary boundaries and is rooted in interdisciplinary collaboration, emerges as a key strategy for navigating the Anthropocene. The *Alternativas* project serves as an exemplary model of how transversal, non-binary approaches can drive innovative solutions to climate change through collaborative efforts. As climate change accelerates, embracing these new ways of thinking will be essential for a sustainable and interconnected future.

---

[5] Quotation from the documentary.

## References


Bauman Z (2000) Liquid modernity. Polity Press, UK

Beck U (2016) The metamorphosis of the world. How climate change is transforming our concept of the world. John Wiley and Sons Ltd.

Bertalanffy L von (1968) General system theory: foundations, development, applications. N. Y.: George Braziller

Campelo A, Paz Barroso E, Barbosa F, Calheiros J, Lencastre MPA (Coord.), Farinha Marques P, Cunha P, Pinto RJ, Sá-Nogueira Saraiva R de, Estrada R (Coord.), Magalhães S, Toldy T (2022) Compor mundos. Humanidades, bem-estar e saúde Vol 1. Porto: Edições Universidade Fernando Pessoa

Castoriadis C (1999) Figures du pensable. Les carrefours du labyrinthe. Seuil, Paris

Corominas A, del Rio J (2023) Documentário alternativas. Construindo mundos possíveis. Produzido por Red de Transición—Transition Network Spain

Crutzen P, Stoermer E (2000) The "Anthropocene." Global Change Newsletter 41:17–18

de Waal F (2001) Tree of origin: what primate behavior can tell us about human social evolution. Harvard University Press, USA

De Waal F (2022) Different. Gender through the eyes of a primatologist. W. W. Norton & Company, USA

Descola P (2005) Par-Delà Nature et Culture. Gallimard, Paris

Descola P (2018) Conférence à la Fondation Louis Vuitton, Paris

Descola P (2021) *Les formes du visible. Une anthropologie de la figuration*, Paris: Seuil

Eibl-Eibesfeldt I (1989) Human ethology. Routledge, UK

Johnson M (1987) The body in the mind: the bodily basis of imagination, reason and meaning, Chicago: Chic. Press, Un

Johnson M (1993) Moral imagination. Implications of cognitive science for ethics, U.S.A.: University Chicago Press

Lacan J (1953) Discours de Rome, Paris

Lakoff G, Johnson M (1980) Metaphors we live by, USA: Chicago Un. Press

Lakoff G (1987) Women, fire and dangerous things. What Categories Reveal about the Mind. USA: Cambridge University Press

Latour B (1991) Nous n'avons jamais été modernes. Essai d'anthropologie symétrique. La Découverte, Paris

Latour B (2017) Facing Gaia. Eight lectures on the new climatic regime. Polity Press, UK

Lencastre MPA, Estrada R (2022) Compor mundos. Humanidades, bem-estar e saúde Vol 2. Porto: Edições Universidade Fernando Pessoa

Lovelock J (2007) The revenge of Gaia: earth's climate crisis & the fate of humanity. Basic Books

McClintock B Women who changed science. The Nobel Prize. https://www.nobelprize.org/womenwhochangedscience/stories/barbara-mcclintock Consulted online 2013.12.19.

Oddone N (2007) Revisitando a "epistemologia social": esboço de uma ecologia sociotécnica do trabalho intelectual. Ciência Da Informação, Brasília 36(1):108–123

Rocha GR, Rocha LFS, Keller EF (2022) Ciência e gênero, feminismo e história das ciências: entrevista com Evelyn Fox Keller. História, Ciências, Saúde. Manguinhos: Rio de Janeiro 29(3):725–735

Saraiva R (1996) Determinantes etológicos, psicológicos e culturais da personalidade. Projecto Praxis pcsh/c/out/0182/96

Wohleben P (2015) The hidden life of trees: what they feel. Greystone Books, How They Communicate. Discoveries from a Secret World

**Marina Prieto Afonso Lencastre** Retired Full Professor at the University of Porto, Full Professor at the University of Fernando Pessoa, Psychotherapist and Supervisor Specialist at the Portuguese Society of Clinical Psychology, Coordinator of the network "Composing Worlds: Humanities, Well-being and Health".

**Rui Estrada** Full Professor at the Fernando Pessoa University, Integrated Researcher at CITCEM, Coordinator of the network "Composing Worlds: Humanities, Well-being, and Health" and researcher of Permanent Observatory Violence and Crime, University Fernando Pessoa, 4249–004 Porto, Portugal.

**Juan del Río** Director of the documentary 'Alter Natives, building possible futures'.

# Eco-Anxiety and Pro-Environmental Behaviour in Adolescents

**Sergio Vila-Tojo, Cristina Gómez-Román, Paulo Vítor Lisboa, Ana Paula Monteiro, and Pedro Cunha**

**Abstract** Eco-anxiety is increasingly recognized as a significant psychological phenomenon among adolescents, reflecting their deep concern about climate change and environmental degradation. This chapter delves into the concept, the generating factors, and implications of eco-anxiety in young people, examining the dual role of eco-anxiety as both a psychological barrier and a potential catalyst for environmental action. On one hand, eco-anxiety has been observed to paralyze, leading to inaction due to the overwhelming nature of environmental crises. On the other hand, eco-anxiety can act as a catalyst for pro-environmental behaviours, motivating young individuals to engage in actions that mitigate these threats. We discuss how the perception of the severity of the environmental threat and belief in a personal capacity

S. Vila-Tojo · C. Gómez-Román (✉) · P. V. Lisboa
CRETUS, Department of Social Psychology, Basic Psychology and Methodology, University of Santiago de Compostela, Santiago de Compostela, Spain
e-mail: cristina.gomez@usc.es

S. Vila-Tojo
e-mail: sergiovila.tojo@usc.es

P. V. Lisboa
e-mail: paulo.lisboa@rai.usc.es

A. P. Monteiro
CIIE—Centre for Research and Intervention in Education, Faculty of Psychology and Education Sciences of the University of Porto (FPCEUP), Porto, Portugal

Department of Education and Psychology, University of Trás-Os-Montes and Alto Douro, Vila Real, Portugal

A. P. Monteiro
e-mail: apmonteiro@utad.pt

P. Cunha
Faculty of Human and Social Sciences, Fernando Pessoa University, Porto, Portugal
e-mail: pcunha@ufp.edu.pt

Research Centre for Health Technologies and Services (RISE-Health), Porto, Portugal



W. Leal Filho et al. (eds.), *Composing Worlds: Humanities, Health and Wellbeing in the XXI Century Towards a More Sustainable World*, World Sustainability Series,
https://doi.org/10.1007/978-3-031-87108-5_12

to effect change influence mobilization towards pro-environmental behaviour. Additionally, the chapter identifies the need for targeted interventions to support adolescents in transforming their anxiety into action. Applying strategies that enhance group identity and social norms, we explore how eco-anxiety can be transformed into a constructive force for environmental engagement among young individuals.

**Keywords** Eco-anxiety · Pro-environmental behaviour · Adolescents · Coping strategies

## 1 Eco-Anxiety in Adolescents: Concept, Contextualization, Effects

The growing recognition of climate change as a global threat underscores socioecological challenges—such as water scarcity, raising temperatures or extreme weather events–that pose significant risks to the psychological and social well-being of individuals, particularly adolescents. This age group, in the middle of an evolutionary transition, is particularly vulnerable to stressors caused by the increasing severity of environmental crises (Hayes et al. 2018).

The negative impacts of climate change are widespread, and detrimental to human lives and livelihoods. With the effects predicted to worsen, the responses of institutions and political systems to adapt to this crisis have generally not been proportionate to the threats posed, nor successful in reaching internationally agreed goals such as keeping warming within a two-degree range. In 2022, the Intergovernmental Panel on Climate Change (IPCC) Report addressed the impact of climate change on mental health for the first time, shifting the focus from its previous emphasis solely on physical health impacts. This recognition of disorders such as post-traumatic stress, trauma, depression, and anxiety when facing natural disasters or climate-forced displacement has validated what many have understood and felt for some time: that the effect of the climate crises on mental health is significant and widespread and affects young people most (Hickman 2024). Highlighting the severity of these impacts, research has shown that direct exposure to disasters associated with climate change can have negative impacts on the capacity for emotional regulation and lead to cognitive deficits, as well as behavioural and adjustment problems (Burke et al. 2018).

Many people, particularly adolescents, are reasonably concerned about this crisis and the lack of action, and are reporting associated weakening in their mental health. Several terms have emerged in an attempt to describe this personal situation, such as, solastalgia, environmental despair, ecological grief, and ecological guilt (Pihkala 2020). From this proliferation of concepts, the term "eco-anxiety" (also called climate anxiety) has gained prominence as an umbrella term that incorporates concerns about this sociopolitical and ecological problem (Hickman 2024; IPCC, 2022). It is defined by the American Psychological Association as “a chronic fear of environmental destruction” (Clayton et al. 2017).

Eco-anxiety is perceived as a unique cognitive and emotional experience and, in terms of intensity, can vary between mild, medium, significant, severe and critical climatic distress (Hickman 2024), constituting a multifaceted psychological process. In this sense, eco-anxiety is an emerging mental health crisis, which is connected to environmental, social and cultural factors that are constantly changing in today’s world (Hickman 2024). For some authors, eco-anxiety is associated with anxious-depressive symptoms in response to actual or anticipated impacts of climate change on one’s life (Hickman 2020). Some studies have shown that there may be a relationship between measures of environmental concern and characteristic symptoms of anxiety, depression, and stress issues (e.g., Helm et al. 2018; Pihkala 2020). From the perspective of Ojala et al. (2021), there are three possible explanations for the significant relationship between worry, anxiety and climate anguish and well-being: (a) negative emotions about environmental problems can cause lower levels of well-being; (b) low general well-being itself can generate negative environmental emotions; and (c) there may be no causal relationship, but there may be personal vulnerability, that is, a general tendency to experience negative affect due to personality factors.

In this sense, it should be noted that there is a certain degree of agreement regarding the fact that, while eco-anxiety negatively affects well-being (Ogunbode et al. 2022), it is not a pathological condition. Therefore, there is a need to avoid medicalization or specific treatments for this form of anxiety arising from the threat of climate change (Brophy et al. 2023). So, eco-anxiety it is a specific form of anxiety (distinct from disorders such as generalised anxiety) related to the distress caused by our perceptions and knowledge of negative environmental changes (Coppola 2021; Hogg et al. 2021).

Research, activists, non-governmental organizations, and media coverage have expanded our understanding of the negative effects on well-being produced by this threat (Ogunbode et al. 2022; Pihkala 2020). Studies have shown how the increased awareness of climate crises contributes to a spectrum of negative consequences such as distress, confusion and anxiety. It highlights the impact of these environmental challenges on the mental well-being and behaviour of children and young people (Clayton 2021; Helm et al. 2018; Hickman 2019; 2020). They are increasingly manifesting emotional and relational problems, stemming from fears about the uncertainty of their future in a changing climate. This collective body of evidence underscores the need to address the negative effects that climate change has on well-being, especially among younger populations, who face the challenge of inheriting a threatened planet. Along with sadness and anxiety, adolescents report experiencing feelings of betrayal, abandonment, and rejection by people in power who they expected to protect them (Hickman et al. 2021; UNICEF 2022). In this regard, one of the questions that has been at the centre of the current discussion on eco-anxiety is whether it paralyzes or inhibits young people’s pro-environmental behaviour or whether, on the contrary, it represents a motivating factor for action against environmental threats (Sangervo et al. 2022). Regardless of the situation generated by eco-anxiety, the research suggests that psychological, emotional, and social support can be very helpful in supporting the well-being of children, young people, and adults seeking to sustain conscious awareness of our context of living on a suffering planet (Aston 2023).

In order to better understand the effects of eco-anxiety, this chapter will discuss how it can act as both a barrier and a catalyst for pro-environmental behaviour in adolescence. We will explore the complex nature of eco-anxiety, highlighting its ability to constrain but also to serve as a trigger that prompts action. In addition, we will propose strategies for transforming eco-anxiety into a constructive agent, seeking to turn worry and anxiety into motivating forces for positive change.

## 2 Eco-Anxiety as a Psychological Constrain of Pro-Environmental Behaviour

Every day, adolescents are susceptible to news reports across traditional media and social networks highlighting the environmental risk and the adverse effects of climate change, threatening their well-being (Hase et al. 2021).

These environmental awareness messages to which adolescents are exposed follow a sequential logic: (a) increase the perception of the environmental problem in order to (b) produce a cognitive discomfort (eco-anxiety) that leads them to (c) act pro-environmentally. Considering that people try to reduce or eliminate any negative experience in their lives, the main hypothesis behind these messages is that eco-anxiety would lead to the active search for solutions to environmental problems, or at least to perform pro-environmental behaviours that would help to alleviate this cognitive discomfort.

However, perceptions of the environmental problem do not necessarily lead to greater eco-anxiety, nor is adopting pro-environmental behaviours the alternative necessarily chosen to alleviate eco-anxiety (Helm et al. 2018; Shao and Yu 2023). While this sequence serves as a basic structure, simplifying the issue to just this sequence or considering it in isolation is insufficient and may even be counterproductive. It is also necessary to assess other aspects such psychological distance, threat perception, attribution of causality and other behavioural alternatives that the adolescents might consider.

It should be noted that adolescents may recognize the environmental problem but, at the same time, perceive it from a psychological distance. Along with the information about climate change via different media, adolescents confront several vital issues which are critical in their stage of life. These years are particularly important for the development of their personality, for individual and group identity, for the establishment of certain interpersonal relationships among peers and other academic and professional concerns (Lisboa et al. 2024; Tanti et al. 2011).

Adolescents will not necessarily ignore the problem but will prioritise the issues that concern them. Our brain is programmed to attend to the present, not to what lies outside our event horizon (Gifford 2011), which may lead adolescents to foreground immediate events over future ones. In this sense, it is possible that, while recognising the problem, they may consider that climate change affects areas far away from where they live (geographical distance) or that its negative consequences will not occur for a long time (temporal distance). They may also have a strong optimistic bias (Weinstein 1980) and therefore perceive a low probability that climate change will affect them or their loved ones (social distance) (Costa-Font et al. 2009). Ultimately, the more distant the problem of climate change is perceived as being, the less likely adolescents are to take action to address it.

In this sense, the protection-motivation theory proposes that the relevance and priority given to a problem does not depend so much on the recognition of the problem, but on whether or not it is perceived as threatening (Rogers and Prentice-Dunn 1997). This requires a perception that the situation is serious, that its effects are fatal and that it may affect them directly or people close to them. Such an assessment will trigger a range of emotions such as fear, dread or worry (Slovic 2000). The combination of these perceptions and emotions in assessing and anticipating the environmental problem will form the basis of what we call eco-anxiety, arising in response to climate change stressors. Eco-anxiety may be even more pronounced if an exercise of internal attribution of responsibility is carried out, which would be associated with experiencing emotions of guilt for the environmental situation (Hogg et al. 2021).

Adolescents who are in such a situation (high threat perception and internal attribution of responsibility) but who do not behave pro-environmentally may experience a strong cognitive dissonance that needs to be resolved (Festinger 1957). In this respect, excessive eco-anxiety may be associated with an immediate need to seek relief, and lead to thoughts and/or behaviours that are not associated with increased pro-environmental behaviour (Heeren et al. 2022). Alternatively, adolescents may choose other options to resolve dissonance. One of these options is to transfer the blame for the climate crisis to other people or institutions, acting as scapegoats. Carrying out this exercise of external attribution of responsibility can be effective given the stage of life in which adolescents are. Although adolescents begin to have independence in some aspects of their lives, they are still under the supervision of their parents, who assume a large part of the responsibility for their needs. In this sense, adolescents may perceive that environmental damage has been mostly induced by previous generations and that it should be those generations that seek solutions. This displacement of responsibility may contribute to an alleviation of associated guilt and a reduction in pro-environmental behaviour. In doing so, adolescents would gain a way to justify and exonerate their environmentally irresponsible actions (Bandura 2016).

Also, adolescents could obtain a neutralization of the eco-anxiety, naturalizing the phenomenon. Few people today deny climate change, but a significant number of people argue that natural cycles are the main cause, and that anthropocentric influence is limited (Poortinga et al. 2018). This fatalistic conception could lead adolescents to think that their contribution to the problem is small, even insignificant, and to assume the "fate" of the planet. Acceptance of the problem as irresolvable would lead to inaction and potentially further minimization of the consequences (Bandura 2016).

In short, an excess of eco-anxiety accompanied by a feeling of lack of control or not knowing how to act can trigger a series of thoughts and actions contrary to the pro-environmental path. In this sense, it is necessary to explore the conditions under which eco-anxiety functions as a facilitator rather than a barrier to pro-environmental behaviour. To this end, we must focus on which variables affect the coping strategies employed by adolescents to reduce this eco-anxiety in a proactive, action-oriented, and constructive manner.

## 3 Eco-Anxiety as a Motivator for Pro-Environmental Behaviour

As mentioned previously, eco-anxiety can constrain pro-environmental behaviour through feelings of paralysis and denial; however, it also has the potential to serve as a powerful motivator for action (Pihkala 2020). The distress induced by eco-anxiety, characterized by a deep concern for the environmental crises facing our planet, can drive individuals to find reassurance and purpose in contributing to environmental causes, thereby transforming their anxiety into actionable change (Lutz 2023). Several studies have shown that individuals experiencing high levels of eco-anxiety are more predisposed to participate in a range of pro-environmental behaviours, from recycling and reducing energy consumption to actively supporting environmental causes (Clayton et al. 2017; Ojala 2012; Swim et al. 2019). Indeed, the relationship between eco-anxiety and sustainable behaviour and climate activism is especially strong among adolescents (von Gal et al. 2024). To further explore the idea that eco-anxiety can act as a catalyst to raise awareness and turn concern into action, it is essential to explore the psychosocial mechanisms that underpin this transformative process.

It has been mentioned previously that environmental awareness messages follow a sequential logic to activate cognitive dissonance by exposing people to the discrepancy between their environmental concerns and their behaviour. While it is true that this strategy does not always work and is not for everyone, its intention is to act as a booster for pro-environmental behaviour based on the idea that by seeking to reduce dissonance, people will tend to find ways to increase the coherence between their personal values and actions. Precisely, one way that adolescents could implement to increase consistency would involve adjusting their actions (making them more pro-environmental) to align with their values (Festinger 1957; Izuma and Murayama 2013; Pihkala 2022). This process not only alleviates eco-anxiety, but also strengthens their engagement with sustainable living practices.

One factor that may facilitate the reduction of dissonance through pro-environmental behaviour, rather than through belief change (e.g., denial, external attribution), is the perception of efficacy. Efficacy refers to the belief in one's capability to successfully carry out specific actions (Bandura 1977). In the context of eco-anxiety, various forms of efficacy, such as the efficiency of actions, self-efficacy perception, and collective efficacy, can influence how individuals deal with environmental concerns (Hussey 2023). For example, self-efficacy perception, which involves confidence in one's ability to make a difference, can empower individuals to cope with eco-anxiety and participate in pro-environmental actions (Innocenti et al. 2023). On the contrary, without this perception of self-efficacy, the experience of eco-anxiety may induce feelings of helplessness and lead to inaction (Pikhala 2020).

Also important in translating eco-anxiety into behaviour is the belief in the collective capacity to drive change. Perceived collective efficacy can also significantly contribute to addressing eco-anxiety by fostering a sense of community and shared responsibility in addressing environmental issues (Cheng et al. 2014). Nevertheless, it is not enough to simply believe that one is capable of taking action (self-efficacy) or that, together with others, it is possible to effect change (collective efficacy); it is equally important to perceive that the actions performed are effective and efficient in addressing environmental challenges directly (response efficacy) (Bostrom et al. 2019). Thus, efficacy may act as a moderator in the relation between eco-anxiety and pro-environmental behaviour, leading people to channel their eco-anxiety into actions, such as supporting environmental causes (Becht et al. 2023; Innocenti et al. 2023; Whitmarsh et al. 2022).

Beyond efficacy, another element that fuels action is the perception of injustice (Vilas et al. 2016). Specifically among adolescents, who face the worst part of the long-term consequences of environmental degradation, injustice appears as a drive toward seeking solutions and engaging in climate activism (Haugestad et al. 2021). This sense of injustice is amplified by the belief that the impacts of climate change will affect them without their having contributed significantly to the problem (O'Brien et al. 2018; Fine 2022). The emotional response to perceived injustices related to environmental degradation, especially among younger generations who inherit the consequences of past actions, often manifests as anger or frustration (Lorenzini and Rosset 2024). These strong emotions can result in external attribution of the problem, but when this displacement of responsibility is directed towards achievable goals (or at least perceived as such), the anger or frustration can be channelled through climate activism and social demand for change (Gregersen et al. 2023; Kurth and Pihkala 2022; Kollmuss and Agyeman 2002), transforming eco-anxiety from a paralyzing concern into a motivation to seek ways to address this perceived injustice.

Furthermore, the recognition of environmental injustices can unite individuals and communities. This unity is essential for the effectiveness of environmental activism, as it amplifies the call for change and reflects the widespread support for sustainability and equity in addressing climate issues. Adolescents, in particular, can play a key role in this collective action. Their unique position as inheritors of the planet's future gives their voices particular weight, and their engagement can inspire others to take notice and join in the efforts to demand and create meaningful change. It points to the potential of eco-anxiety to support a collective environmental identity, motivating sustained action (Mackay et al. 2021). The realization that one is not alone in one's eco-anxiety can foster a powerful collective identity among individuals sharing these concerns. This sense of unity transcends individual differences, cultivating a resilient, motivated community dedicated to proactive environmental management (Thomas et al. 2019). Central to this collective identity is the shared purpose and understanding of the impact of combined efforts promoting a sense of empowerment and shared responsibility (Horowitz 2017).

The intersection of eco-anxiety and the mobilizing power of emotions through perceived injustice, a sense of collective identity and efficacy, and cognitive dissonance highlight the complex psychological landscape of environmental activism. By understanding how these elements interact, we can better help adolescents and others affected to channel their eco-anxiety and emotional responses into productive activism. This not only helps reduce the psychological burden of eco-anxiety, but also contributes to building a more equitable and sustainable future.

## 4 Transforming Eco-Anxiety: From Barrier to Action

The previous sections have explored how eco-anxiety can both paralyze and drive adolescents toward pro-environmental behaviour. Acknowledging the dual nature of eco-anxiety allows us to establish the groundwork for approach to strategies aimed at empowering adolescents to channel their eco-anxiety into constructive environmental action oriented to address contemporary socioecological challenges associated with climate change.

However, to exploit this potential, we need to go beyond traditional awareness campaigns. This involves developing messages and educational programmes with a dual focus. Firstly, highlighting the urgency of environmental issues is necessary. However, an overemphasis on the negative aspects can discourage people from engaging in pro-environmental behaviours (Mathers-Jones and Todd 2023). Hence, messages should not only address these urgent issues, but also show tangible ways in which adolescents can contribute to solutions, thereby reinforcing their sense of agency and collective efficacy. In addition, it is important to consider those broader narratives that embrace the sense of injustice and collective capacity for change that adolescents can identify with. Precisely, strategies that focus on reinforcing biospheric values, pro-environmental identity or seek to align the social norm to the environmental issue may contribute to pro-environmental behaviour being perceived as an adequate and effective form of coping.

In this sense, the development of biospheric values is key in strengthening environmental self-identity, particularly during adolescence—a critical period for value formation and change (Daniel and Benish-Weisman 2019; Bacchini et al. 2023). This type of values points to the importance of environmental protection as a personal life goal, fostering a self-concept as an environmental protector (van der Werff et al. 2013a).

In this regard, strategies should be designed to promote biospheric values within adolescent communities and integrate them into their identities. Connecting with nature plays a key role in this process, as it directly influences the development and reinforcement of biospheric values (Martin and Czellar 2017). Educational programmes offering experiential learning in natural settings, such as participating in conservation efforts, engaging in outdoor education, or simply spending reflective time in nature, have proven effective (Chawla 2020). These activities deepen individuals' connections to the environment and enhance their commitment to its protection (Mackay and Schmitt 2019; Whitburn et al. 2020). Additionally, developing community-based environmental projects and platforms for adolescents to express and share their environmental values and past actions is positive (Fanghella et al. 2019). These experiences not only strengthen individual biospheric values but also highlight shared environmental concerns and commitments within peer groups (Bouman and Steg 2022). This would be countering the misperception of isolation in environmental action, addressing the feelings of helplessness associated with eco-anxiety, and strengthening their identity as contributors to environmental conservation. These experiences can be particularly impactful for adolescents, who are in an intense period of changing values with a view to stabilization (Bacchini et al. 2023; Daniel and Benish-Weisman 2019) and, therefore, for whom such encounters can shape lifelong values and attitudes towards the environment.

However, it is noteworthy that opportunities for these nature-based experiences have been in decline (Richardson et al. 2022), especially among younger generations that are living in increasingly urbanized contexts (Puhakka and Hakokängäs 2024) and whose contact with nature has been rapidly reducing (Richardson et al. 2020). This trend highlights the importance of creating intentional spaces and opportunities for adolescents to connect with the natural world (Nisbet et al. 2019; Richardson et al. 2021). By actively engaging with the environment, adolescents not only would confront and manage their eco-anxiety but also would begin to see themselves as key players in the fight against environmental degradation. In this regard, it is important to note that schools, families, and community organizations are critical in facilitating these experiences, promoting group interactions that not only foster a direct connection to nature but also build community around shared environmental values (Chawla 2020).

In addition to the problem of the reduction of direct contact with nature, a challenge arises in the form of societal underestimation of biospheric values. Research indicates that people often believe that others value and care about the environment less than they actually do, leading to a potential undervaluation of their environmental actions and priorities (Bouman and Steg 2019; Bouman et al. 2020). This belief can lead to a self-enhancement bias (Krueger 1998), in which one believes one is doing more than others, and in some instances, thinking that others are doing little or nothing. In such cases, some adolescents may believe that peers are taking advantage of their efforts to protect the environment. This perception of the presence of free riders in the group may cause the adolescent to feel frustrated or perceive collective ineffectiveness, heightening eco-anxiety but also risks diminishing their pro-environmental behaviour to avoid feeling like the only "fool" making an effort (see *sucker effect*; Kerr 1983).

Moreover, beyond perceptions, free riders can be a reality and put the collective effort at risk. To prevent these profiles from appearing, the creation of solid social norms in adolescence, where personal norms and habits are still underdeveloped, can be very useful, especially for those whose personal norms are weak (de Groot et al.,

2021) Normative influence is related to the need to maintain a positive social image, the approval of others and the avoidance of social sanctions, but they also serve as a guide as to what is right and wrong. Hence, social norms can contribute to developing personal norms—the individual's sense of moral obligation to perform an action (Schwartz 1977)—in this early life stage. They are where social expectations and individual moral obligations converge to guide behaviour, which may be particularly relevant in preventing free-rider behaviour. The interaction between personal and social norms in influencing behaviour reflects how social influence and individual moral judgments combine in sustainability decisions (Bamberg et al. 2007). All together play a key role in forming one's environmental identity, highlighting the interconnection between individual beliefs and collective values (van der Werff et al. 2013b).

In this regard, adolescents, with their sensitivity to the actions and attitudes of their peer group (Collado et al. 2017), can be significantly influenced by a social normative context that is favourable to environmental issues and provides guidance for pro-environmental behaviour. Emerging environmental problems challenge long-established behavioural patterns that are transmitted from generation to generation. In this sense, adolescents are paying attention to what opinion they have and how relevant others act. The influence of descriptive norms, which detail the behaviours of the majority, often carries more weight than injunctive norms, which prescribe what ought to be done, particularly for those who may not strongly identify with a specific group (Bertoldo and Castro 2016). This insight points to the necessity of fostering a supportive community environment where environmental actions and values are not only observed but are actively endorsed and shared among peers, further enriching the foundations for a robust environmental identity. The adoption of pro-environmental behaviours, therefore, is not just a matter of individual choice but is also deeply rooted in the social environment of their lives.

Considering the above, participatory approaches that engage adolescents in dialogue, decision-making, and action can be particularly effective. By involving them in environmental projects that address local issues, we can provide real-life contexts where their actions have visible impacts. Such experiences enhance self-efficacy and strengthen connections to the environmental movement through shared values and collective identity. Furthermore, incorporating the concept of environmental self-identity into these strategies may amplify adolescents' motivation to act. Reflecting on past pro-environmental actions and understanding the role of social norms in behaviour reinforce adolescents' dedication to environmental causes, acknowledging that environmental identity evolves through peer interactions, experiences, and societal expectations.

Ultimately, transforming eco-anxiety from a barrier to a catalyst for action among adolescents requires a multifaceted strategy that addresses psychological, social, and experiential dimensions. By fostering a sense of efficacy, emphasizing the collective potential for impact, and encouraging a pro-environmental identity, we can empower adolescents to cope with their eco-anxiety constructively. In doing so, we not only alleviate the psychological discomfort of eco-anxiety but also mobilize a new generation committed to addressing contemporary socio-ecological challenges and to creating a more sustainable world that improves the well-being of people and care for the planet.

**Acknowledgements** This work was co-funded by the Portuguese Foundation for Science and Technology, IP (FCT), under the multi-year funding awarded to CIIE [grants no. UIDB/00167/2020 and UIDP/00167/2020].

## References

Aston N (2023) The experience of eco-anxiety of individuals who have been through an ecopsychology support group for facing the climate crisis: a grounded theory inquiry. University of East London. https://doi.org/10.15123/uel.8v4qz

Bacchini D, Affuso G, Aquilar S, Dragone M, Esposito C (2023) Values across adolescence: a four-year longitudinal study. The predictive role of community violence and parental acceptance-rejection. Eur J Pers 37(6):686–704. https://doi.org/10.1177/08902070231167199

Bamberg S, Hunecke M, Blöbaum A (2007) Social context, personal norms and the use of public transportation: two field studies. J Environ Psychol 27:190–203. https://doi.org/10.1016/j.jenvp.2007.04.001

Bandura A (1977) Self-efficacy: toward a unifying theory of behavioral change. Psychol Rev 84(2):191–215. https://doi.org/10.1037/0033-295X.84.2.191

Bandura A (2016) Moral disengagement: how people do harm and live with themselves. Worth Publishers. https://psycnet.apa.org/record/2015-43532-000

Becht A, Spitzer J, Grapsas S, van de Wetering J, Poorthuis AMG, Smeekes A, Thomaes S (2023) Feeling anxious in a warming world: climate anxiety and adolescents' pro-environmental behavior. https://doi.org/10.31234/osf.io/a9k5h

Bertoldo R, Castro P (2016) The outer influence inside us: exploring the relation between social and personal norms. Resour Conserv Recycl 112:45–53. https://doi.org/10.1016/j.resconrec.2016.03.020

Bostrom A, Hayes AL, Crosman KM (2019) Efficacy, action, and support for reducing climate change risks. Risk Anal 39(4):805–828. https://doi.org/10.1111/risa.13210

Bouman T, Steg L (2022) Engaging city residents in climate action: addressing the personal and group value-base behind residents' climate actions. Urbanisation 7:S26–S41. https://doi.org/10.1177/2455747120965197

Bouman T, Steg L, Zawadzki SJ (2020) The value of what others value: when perceived biospheric group values influence individuals' pro-environmental engagement. J Environ Psychol 71:101470. https://doi.org/10.1016/j.jenvp.2020.101470

Bouman T, Steg L (2019) Motivating society-wide pro-environmental change. One Earth 1:27–30. https://doi.org/10.1016/j.oneear.2019.08.002

Brophy H, Olson J, Paul P (2023) Eco-anxiety in youth: an integrative literature review. J Ment Health Nurs 32(3):633–661. https://doi.org/10.1111/inm.13099

Burke SE, Sanson AV, Van Hoorn J (2018) The psychological effects of climate change on children. Curr Psychiatry Rep 20(5):1–8. https://doi.org/10.1007/s11920-018-0896-9

Chawla L (2020) Childhood nature connection and constructive hope: a review of research on connecting with nature and coping with environmental loss. People Nat 2:619–642. https://doi.org/10.1002/pan3.10128

Cheng CCJ, Cl Y, Sheu C (2014) The link between eco-innovation and business performance: a Taiwanese industry context. J Clean Prod 64:81–90. https://doi.org/10.1016/j.jclepro.2013.09.050

Clayton S (2021) Climate change and mental health. Curr Environ Health Rep 8:1–6. https://doi.org/10.1007/s40572-020-00303-3

Clayton S, Manning C, Krygsman K, Speiser M (2017) Mental health and our changing climate: impacts, implications, and guidance. In: American psychological association and ecoAmerica

Collado S, Staats H, Sancho P (2017) Normative influences on adolescents' self-reported pro-environmental behaviors: the role of parents and friends. Environ Behav 51(3):288–314. https://doi.org/10.1177/0013916517744591

Coppola IG (2021) Eco-anxiety in "the climate generation": is action an antidote? Environmental studies electronic thesis collection. https://scholarworks.uvm.edu/envstheses/71

Costa-Font J, Mossialos E, Rudisill C (2009) Optimism and the perceptions of new risks. J Risk Res 12(1):27–41. https://doi.org/10.1080/13669870802445800

Daniel E, Benish-Weisman M (2019) Value development during adolescence: dimensions of change and stability. J Pers 87(3):620–632. https://doi.org/10.1111/jopy.12420

Fanghella V, d'Adda G, Tavoni M (2019) On the use of nudges to affect spillovers in environmental behaviors. Front Psychol 10:00061. https://doi.org/10.3389/fpsyg.2019.00061

Festinger L (1957) A theory of cognitive dissonance. Stanford University Press. https://psycnet.apa.org/record/1993-97948-000

Fine JC (2022) Language and social justice in US climate movements: barriers and ways forward. Front Commun 7:920568. https://doi.org/10.3389/fcomm.2022.920568

von Gal A, Fabiani G Piccardi L (2024) Climate change anxiety, fear, and intention to act. Front Psychol 15:1341921. https://doi.org/10.3389/fpsyg.2024.1341921

Gifford R (2011) The dragons of inaction: psychological barriers that limit climate change mitigation and adaptation. Am Psychol 66(4):290–302. https://doi.org/10.1037/a0023566

Gregersen T, Andersen G, Tvinnereim E (2023) The strength and content of climate anger. Glob Environ Change 82:102738. https://doi.org/10.1016/j.gloenvcha.2023.102738

de Groot, JIM, Bondy K, Schuitema G (2021) Listen to others or yourself? The role of personal norms on the effectiveness of social norm interventions to change pro-environmental behavior. J Environ Psychol 78:101688. https://doi.org/10.1016/j.jenvp.2021.101688

Hase V, Mahl D, Schäfer MS, Keller TR (2021) Climate change in news media across the globe: an automated analysis of issue attention and themes in climate change coverage in 10 countries (2006–2018). Glob Environ Change 70:102353. https://doi.org/10.1016/j.gloenvcha.2021.102353

Haugestad CAP, Skauge AD, Kunst JR, Power SA (2021) Why do youth participate in climate activism? A mixed-methods investigation of the #FridaysForFuture climate protests. J Environ Psychol 76:101647. https://doi.org/10.1016/j.jenvp.2021.101647

Hayes K, Blashki GA, Wiseman J, Burke S, Reifels L (2018) Climate change and mental health: risks, impacts and priority actions. J Ment Health Syst 12(1):1–12. https://doi.org/10.1186/s13033-018-0210-6

Heeren A, Mouguiama-Daouda C, Contreras A (2022) On climate anxiety and the threat it may pose to daily life functioning and adaptation: a study among European and African French-speaking participants. Clim Change 173(15):1–17. https://doi.org/10.1007/s10584-022-03402-2

Helm SV, Pollitt A, Barnett MA, Curran MA, Craig ZR (2018) Differentiating environmental concern in the context of psychological adaption to climate change. Glob Environ Change 48:158–167. https://doi.org/10.1016/j.gloenvcha.2017.11.012

Hickman C (2020) We need to (find a way to) talk about… eco-anxiety. J Soc Work Pract 34(4):411–424. https://doi.org/10.1080/02650533.2020.1844166

Hickman C (2024) Eco-anxiety in children and young people—a rational response, irreconcilable despair, or both? Psychoanal Study Child 77(1):356–368. https://doi.org/10.1080/00797308.2023.2287381

Hickman C, Marks E, Pihkala P, Clayton S, Lewandowski RE, Mayall EE, Wray B, Mellor C, van Susteren L (2021) Climate anxiety in children and young people and their beliefs about government responses to climate change: a global survey. Lancet Planet Health 5(12):e863–e873. https://doi.org/10.1016/S2542-5196(21)00278-3

Hickman C (2019) ediChildren and climate change: exploring children's feelings about climate change using free association narrative interview methodology. In: Hoggett P (ed) Climate psychology: on indifference to disaster, pp 41–59. https://doi.org/10.1007/978-3-030-11741-2_3

Hogg TL, Stanley SK, O'Brien LV, Wilson MS, Watsford CR (2021) The hogg eco-anxiety scale: development and validation of a multidimensional scale. Glob Environ Change 71:102391. https://doi.org/10.1016/j.gloenvcha.2021.102391

Horowitz J (2017) Who is this "we" you speak of? Grounding activist identity in social psychology. Socius 3. https://doi.org/10.1177/2378023117717819

Hussey D (2023) Exploring the impact of actionable guidance and feedback on pro-environmental behaviors, climate anxiety, and environmental efficacy. Proc Hum Factors Ergon Soc Annu Meet 67(1):1735–1740. https://doi.org/10.1177/21695067231195912

Innocenti M, Santarelli G, Lombardi GS, Ciabini L, Zjalic D, Di Russo M, Cadeddu C (2023) How can climate change anxiety induce both pro-environmental behaviours and eco-paralysis? The mediating role of general self-efficacy. Int J Environ Res Public Health 20(4):3085. https://doi.org/10.3390/ijerph20043085

Intergovernmental Panel on Climate Change (2022) Climate change 2022: impacts, adaptation and vulnerability. Contribution of working group ii to the sixth assessment report of the intergovernmental panel on climate change. Cambridge University Press. https://doi.org/10.1017/9781009325844

Izuma K, Murayama K (2013) Choice-induced preference change in the free-choice paradigm: a critical methodological review. Front Psychol 4:1–12. https://doi.org/10.3389/fpsyg.2013.00041

Kerr NL (1983) Motivation losses in small groups: a social dilemma analysis. J Pers Soc Psychol 45(4):819–828. https://doi.org/10.1037/0022-3514.45.4.819

Kollmuss A, Agyeman J (2002) Mind the gap: why do people act environmentally and what are the barriers to pro-environmental behavior? Environ Educ Res 8(3):239–260. https://doi.org/10.1080/13504620220145401

Krueger J (1998) Enhancement bias in descriptions of self and others. Pers Soc Psychol Bull 24:505–516. https://doi.org/10.1177/0146167298245006

Kurth C, Pihkala P (2022) Eco-anxiety: what it is and why it matters. Front Psychol 13:981814. https://doi.org/10.3389/fpsyg.2022.981814

Lisboa PV, Gómez-Román C, Guntín L, Monteiro, AP (2024) Pro-environmental behavior, personality and emotional intelligence in adolescents: a systematic review. Front Psychol 15:1323098. https://doi.org/10.3389/fpsyg.2024.1323098

Lorenzini J, Rosset J (2024) Emotions and climate strike participation among young and old demonstrators. Soc Mov Stud 23(1):39–55. https://doi.org/10.1080/14742837.2023.2178406

Lutz PK, Zelenski JM, Newman DB (2023) Eco-anxiety in daily life: relationships with well-being and pro-environmental behavior. Curr Res Ecolog Soc Psychol 4:100110. https://doi.org/10.1016/j.cresp.2023.100110

Mackay CML, Schmitt MT, Lutz AE, Mendel J (2021) Recent developments in the social identity approach to the psychology of climate change. Curr Opin Psychol 42:95–101. https://doi.org/10.1016/j.copsyc.2021.04.009

Mackay CM, Schmitt MT (2019) Do people who feel connected to nature do more to protect it? A meta-analysis. J Environ Psychol 65:101323 https://doi.org/10.1016/j.jenvp.2019.101323

Martin C, Czellar S (2017) Where do biospheric values come from? A connectedness to nature perspective. J Environ Psychol 52:56–68. https://doi.org/10.1016/j.jenvp.2017.04.009

Mathers-Jones J, Todd J (2023) Ecological anxiety and pro-environmental behaviour: the role of attention. J Anxiety Disord 98:102745. https://doi.org/10.1016/j.janxdis.2023.102745

Nisbet E, Zelenski JM, Grandpierre Z (2019) Mindfulness in nature enhances connectedness and mood. Ecopsychology 11(2):1–11. https://doi.org/10.1089/eco.2018.0061

O'Brien K, Selboe E, Hayward BM (2018) Exploring youth activism on climate change: dutiful, disruptive, and dangerous dissent. Ecol Soc 23(3):42. https://doi.org/10.5751/ES-10287-230342

Ogunbode CA, Doran R, Hanss D, Ojala M, Salmela-Aro K, van den Broek KL, Bhullar N, Aquino SD, Marot T, Schermer JA, Wlodarczyk A, Lu S, Jiang F, Maran DA, Yadav R, Ardi R, Chegeni R, Ghanbarian E, Zand S, Karasu M (2022) Climate anxiety, wellbeing and pro-environmental action: correlates of negative emotional responses to climate change in 32 countries. J Environ Psychol 84:101887. https://doi.org/10.1016/j.jenvp.2022.101887

Ojala M (2012) Regulating worry, promoting hope: How do children, adolescents, and young adults cope with climate change? Int J Environ Sci Educ 7:537–561

Ojala M, Cunsolo A, Ogunbode CA, Middleton J (2021) Anxiety, worry, and grief in a time of environmental and climate crisis: a narrative review. Annu Rev Environ Resour 46(1):35–58. https://doi.org/10.1146/annurev-environ-012220-022716

Pihkala P (2020) Anxiety and the ecological crisis: an analysis of eco-anxiety and climate anxiety. Sustainability, 12(19):7836. https://doi.org/10.3390/su12197836

Pihkala P (2022) Eco-anxiety and pastoral care: theoretical considerations and practical suggestions. Religions 13(3):192. https://doi.org/10.3390/rel13030192

Poortinga W, Fisher S, Bohm G, Steg L, Whitmarsh L, Ogunbode C (2018) European attitudes to climate change and energy. Topline results from Round 8 of the European Social Survey

Puhakka R, Hakoköngäs E (2024) Adolescents' experiences in nature: sources of everyday well-being. J Leis Res 55(2):250–269. https://doi.org/10.1080/00222216.2023.2204346

Richardson M, Dobson J, Abson DJ, Lumber R, Hunt A, Young R, Moorhouse B (2020) Applying the pathways to nature connectedness at a societal scale: a leverage points perspective. Ecosyst People 16(1):387–401. https://doi.org/10.1080/26395916.2020.1844296

Richardson M, Hamlin I, Butler CW, Thomas R, Hunt A (2021) Actively noticing nature (not just time in nature) helps promote nature connectedness. Ecopsychology 14(1):8–16. https://doi.org/10.1089/eco.2021.0023

Richardson M, Hamlin I, Elliott LR, Mathew P (2022) Country-level factors in a failing relationship with nature: nature connectedness as a key metric for a sustainable future. Ambio 51:2201–2213. https://doi.org/10.1007/s13280-022-01744-w

Rogers RW, Prentice-Dunn S (1997) Protection motivation theory. In: Gochman DS (ed) Handbook of health behaviour research I: personal and social determinants, Plenum Press, pp 113–132. https://psycnet.apa.org/record/1997-36396-006

Sangervo J, Jylhä KM, Pihkala P (2022) Climate anxiety: conceptual considerations, and connections with climate hope and action. Glob Environ Change 76:102569. https://doi.org/10.1016/j.gloenvcha.2022.102569

Schwartz, SH (1977). Normative influences on altruism. In: Berkowitz L (ed) Advances in experimental social psychology, Academic Press, vol 10, pp 221–279. https://doi.org/10.1016/S0065-2601(08)60358-5

Shao L, Yu G (2023) Media coverage of climate change, eco-anxiety and pro-environmental behavior: experimental evidence and the resilience paradox. J Environ Psychol 91:102130. https://doi.org/10.1016/j.jenvp.2023.102130

Slovic P (2000) The perception of risk. Earthscan Publications. https://psycnet.apa.org/record/2001-01329-000

Swim JK, Geiger N, Lengieza ML (2019) Climate change marches as motivators for bystander collective action. Front Commun 4. https://doi.org/10.3389/fcomm.2019.00004

Tanti C, Stukas AA, Halloran MJ, Foddy M (2011) Social identity change: shifts in social identity during adolescence. J Adolesc 34(3):555–567. https://doi.org/10.1016/j.adolescence.2010.05.012

Thomas EF, Zubielevitch E, Sibley CG, Osborne D (2019) Testing the social identity model of collective action longitudinally and across structurally disadvantaged and advantaged groups. Pers Soc Psychol Bull 46(6):823–838. https://doi.org/10.1177/0146167219879111

UNICEF (2022) UNICEF Annual Report 2021. Protecting child rights in a time of crises. https://www.unicef.org/reports/unicef-annualreport-2021

van der Werff E, Steg L, Keizer K (2013a) The value of environmental self-identity: The relationship between biospheric values, environmental self-identity and environmental preferences,

intentions and behaviour. J Environ Psychol 34:55–63. https://doi.org/10.1016/j.jenvp.2012.12.006
van der Werff E, Steg L, Keizer K (2013b) It is a moral issue: the relationship between environmental self-identity, obligation-based intrinsic motivation and proenvironmental behaviour. Glob Environ Change 23(5):1258–1265. https://doi.org/10.1016/j.gloenvcha.2013.07.018
Vilas X, Alzate M, Sabucedo JM (2016) Mobilized citizenship: Motives, emotions, and context. Divers Perspect Psicol 12(2):167–181
Weinstein ND (1980) Unrealistic optimism about future life events. J Pers Soc Psychol 39(5):806–820. https://doi.org/10.1037/0022-3514.39.5.806
Whitburn J, Linklater W, Abrahamse W (2020) Meta-analysis of human connection to nature and proenvironmental behavior. Conserv Biol 34(1):180–193. https://doi.org/10.1111/cobi.13381
Whitmarsh L, Player L, Jiongco A, James M, Williams M, Marks E, Kennedy-Williams P (2022) Climate anxiety: what predicts it and how is it related to climate action? J Environ Psychol 83:101866. https://doi.org/10.1016/j.jenvp.2022.101866

# Algorithms as a Journalistic Tool for Scientific Information Accessibility

**Ana Gabriela Frazão-Nogueira, José Miguel Túñez-López, Jesús Segarra-Saavedra, and Patrícia Weber**

**Abstract** Four years ago, the Trusted News Initiative (BBC 2020) joined forces to combat disinformation surrounding the COVID-19 vaccine. At the same time, First Draft revealed that 84% of content on Facebook and Instagram was misinformation about the vaccine. In 2018, a study identified 16 media outlets, 13 news agencies, and 21 leading news organizations that were transitioning to algorithmic information control mechanisms. In the meantime, Project 2061, from the American Association for the Advancement of Science (AAAS), has been operational for 35 years. For the general public, science is conveyed through the lens of journalism, where the encoding process is more pronounced because it occurs between non-peers. It is therefore evident that the quality of journalistic content, particularly in the audiovisual domain, and the competence in its transmission are of paramount importance. In order to ensure the implementation of its basic quality, namely the know-how to communicate science, it is essential to implement the aforementioned quality. In the field of media, trust is synonymous with credibility. Since the advent of software capable of producing texts in natural language, technical and technological knowledge has been imposing solutions that should facilitate the development of a

This publication is part of the project "Creación de una plataforma generadora de chatbots mediante IA para la comunicación del valor público del PSM "(PDC2023-145,885-I00), financed by MCIN/AEI/10.13039/501100011033 and by the European Union's "NextGeneration EU"/ PRTR.

A. G. Frazão-Nogueira (✉)
Universidade Fernando Pessoa/ICNova—Instituto de Comunicação da Nova, Porto, Portugal
e-mail: ana@ufp.edu.pt

J. M. Túñez-López
Universidade de Santiago de Compostela, Santiago, Spain
e-mail: miguel.tunez@usc.es

J. Segarra-Saavedra
Universidade de Alicante, San Vicente del Raspeig, Spain
e-mail: jesus.segarra@gcloud.ua.es

P. Weber
Universidade Fernando Pessoa/I3ID, Porto, Portugal
e-mail: pweber@ufp.edu.pt



W. Leal Filho et al. (eds.), *Composing Worlds: Humanities, Health and Wellbeing in the XXI Century Towards a More Sustainable World*, World Sustainability Series,
https://doi.org/10.1007/978-3-031-87108-5_13

new anthropology of rescue, taking into account AI's role as a screening tool and a builder of bridges and contents towards human ecology. This study, based on the narrative bibliographic review (Rother 2007), has as its primary objective an understanding of how the competencies of this algorithmic journalism, which Túñez et al. (2019) have termed "artificial journalism," contribute to the construction of scientific literacy and the dissemination of information. It also aims to facilitate further research on this system, which effectively translates function and textuality into journalistic information.

**Keywords** Scientific communication · Quality journalism · Algorithmic journalism · Informative accessibility and literacy

## 1 Introduction

Health literacy is a crucial factor in the prevention and reduction of health risks, as well as in the optimization of access and involvement of individuals and populations in the manipulation of the complex and intertwined network of inputs that establish human health. Conversely, the knowledge that different human groups develop about their environments and resources is a consequence of the more or less intrusive adaptation processes that they develop when coming into contact with a dynamic of increasing geographic, cultural, and technological awareness. This process, which gradually becomes more complex, needs to be better understood in order to include and involve human beings as part of a network of interdependence among themselves and with the planet.

In light of this more complex and holistic vision, communication strategies are also necessary to guarantee this approach to diversity that brings together, transmits, and activates a kind of knowledge that is responsible for promoting health accountability towards well-being, equity, and sustainable development. This is in line with the 2022 World Health Day's motto, "Our planet, our health," which aligns with the awareness raised in 2008, "Protecting health from climate change," and in 2010, "Urbanization and health."

It can be argued that the attack on science is not a transient phenomenon. By unifying science as an entity rather than a system, it becomes possible to encourage and strengthen this front line. In social terms, scientific literacy facilitates the distinction between science and pseudoscience, thereby rendering a fundamental understanding of scientific explanations for facts viable. It also fosters rational thinking and encourages the exercise of citizenship. In this context, in which the media is no longer confined to editorials or vehicles, but rather to in-depth research into social dialogue between groups of distinct knowledge, journalism does not combine with authoritarian scientific classicism. We would then transition from a thick, impermeable wall of science at the top and bottom to an osmotic one that transforms and influences, thereby ensuring. Furthermore, the media must transition from merely disseminating science to actively communicating it (Costa Bueno 2010). This communication must take into account the osmotic relationship between science and another sphere of knowledge, another (meta) language, and an epistemology that requires a new mediator of information. This mediator should not merely be a diffuser that shapes the general public, but rather a creative artisan of the transformative dialogue between science and everyday knowledge (Medina 1994, p. 180). This mediator must bridge the gap between science and everyday knowledge, considering the dynamics of Postman's media ecology.

The term "primary communication" of Epstein's circuit (2002, p. 61) is then assumed to be the sole element of this transmissibility. In this scenario, in which the circuit is the media, the improvement process interconnects the producer (p1) (scientist) and the producer (p2) (journalist/s), in a relationship between the normative (p1) and descriptive (p2) contexts (Frazão-Nogueira et al. 2021, pp. 9–10). This will result in a double knowledge from which an interpretive/explanatory content will be derived.

As Opdahl et al. (2023) observe, the need for quality journalism has become more pressing than ever, given the necessity for reliable and accurate media outlets that can disseminate information to the public and help to counterbalance the pervasive and rapid spread of disinformation. Thus, when an event gives rise to both veracity and falsity, the equilibrium between form and content represents a crucial determinant in gauging and averting the perils of message obsolescence, particularly in view of its utility. In addition, the JournalismAI report, based on a survey of 71 news organizations in 32 different countries, identifies efficiency as the primary motivation for introducing AI in newsrooms. It identifies six areas where AI could potentially impact "organizations ethics and editorial policy and practice" (Becket 2019, pp. 8–9), namely, combating algorithmic bias, misinformation and filter bubbles, enhancing editorial decisions and transparency, as well as balancing human and artificial intelligence.

Conversely, contemporary scientific knowledge is the result of a complex set of socio-institutional processes that extend beyond the boundaries of the Academy and exert a profound influence on it. In fact, the paradigm of the relationship of co-production between science and society (Jasanoff 1996) supports this argument, demonstrating that in the knowledge society, the commitment to science and research is (or should be) collective (Martins et al. 2015).

In the context of the communication of science by the media, the concept of functionality plays a pivotal role in the occurrence of an event that is translated into a "link between lexias" (Luna 2009). This translation system is established by the journalist through a convergence that is verified at the level of adequacy of meaning to the nature of the information, thereby making it into widespread information that "becomes useful and adaptable knowledge" (Frazão-Nogueira 2020, p. 10). It can be argued that journalism, in particular audiovisual journalism, is an indispensable component of an architecture that facilitates communication. Scientific journalism, in particular, is a valuable tool for increasing literacy. However, it is important to recognize that trust in information is contingent upon a mutual effort between journalism and science.

## 2 Methodology

Freixo (2018, p. 190) posits that.

> "[...] the frame of reference, whether theoretical or conceptual, guides the formulation of research questions or hypotheses and determines the perspective of the study itself [...] the resulting bibliographic citations also help to define both the context of the problem and its meaning and scientific importance".

From this point, a narrative bibliographic review is employed, which, by not using explicit and systematic criteria for the search and critical analysis of literature, is "appropriate to describe and discuss the development or 'state of the art' of a given subject, from a theoretical or contextual point of view" (Rother 2007). This enables a conceptual, critical, and interpretative analysis of the topic. The work is structured around four scenarios that have already been identified. These form the basis for a critical discussion on the importance of algorithmic journalism in scientific dissemination. It is also important to note that, in the absence of a narrative bibliographic review, it is necessary to refer to the sources used in this text. This is in order to respect the authors and to maintain the credibility of the text.

Four years ago, the Trusted News Initiative (BBC 2020) joined forces with other entities to combat the spread of disinformation related to the COVID-19 vaccine. In November of that year, First Draft, a non-governmental gatewatcher alliance, verified that 84% of content on Facebook and Instagram about the virus was misinformation (Smith et al. 2020, 31). In the meantime, reports of deaths caused by misinformation (Lemos 2021) reveal the significant consequences of an infodemic (Freire et al. 2021) about COVID vaccines. Prior to this, in 2018, a group of researchers (Túñez-López et al. 2018) identified 16 media outlets, 13 news agencies, and 21 world-leading news organizations that were undergoing a transition from oligopolistic to algorithmic information control mechanisms. In the meantime, Project 2061: The Education for a Changing Future initiative, spearheaded by the AAAS (Myron Atkin et al. 1997), has been in operation for over three decades. Its objective is to foster scientific literacy among all citizens, with an emphasis on early childhood education.

From this awareness of science, a new epistemology emerges, seemingly well-rooted in human reality and in a humanized environment. This leads to the examination of the impact of media on human perception, understanding, emotion, and value. It also considers the ways in which our interaction with media affects our chances of survival. (Postman 1970, p. 161) Furthermore, this awareness gives rise to the concept of inter-, trans-, and post-disciplinarity, which pertains to the interplay between scientific and everyday knowledge. This represents the final crucial element in the anthropology of rescue.

However, the challenging period of the global pandemic highlighted the necessity for journalism to cultivate scientific literacy. It is important to recall that, despite the potential for scientists to communicate directly with the public through the media, the efforts to filter journalistic content did not prevent the dissemination of a vast quantity of misinformation and disinformation. Furthermore, it was challenging for specialists to explain the risk–benefit weightings or expose the normal scientific degree of uncertainty without risking their credibility. Overall, despite their laudable efforts, the general inability of journalists to mediate the scientific discourse to the public eroded its importance and exhausted audiences in their efforts to understand it, reducing it to white noise.

An essential step towards fostering collective intelligence and intellectual independence is to acknowledge that an individual who is literate in information is one who recognizes the need for information, organizes it for practical application, integrates new information into an existing body of knowledge, uses information for problem solving, and learns to learn (Lenox apud Tarapanoff et al. 2002). Consequently, the implications of how scientific content is communicated and how this new journalism influences literacy and the construction of scientific information accessibility are two of the most contemporary major issues on information.

The primary objective of this chapter is to elucidate the manner in which the competencies of automated journalism facilitate the development of scientific news literacy. This, in turn, paves the way for the exploration of new avenues of inquiry into this system that effectively translates the journalistic function and textuality, offering a window into reality.

## 3 Science for the Laymen

The Lasswell model, which has been in use since 1948, remains a current and relevant framework for understanding the relationship between science and society. Indeed, it is widely recognized that the social function of the media extends beyond the mere provision of information. As Lasswell (2007, p. 228) notes:

> "The communication process in society performs three functions: (a) surveillance of the environment, disclosing threats and opportunities affecting the value position of the community and of the component parts within it; (b) correlation of the components of society in making a response to the environment; (c) transmission of the social inheritance".

Conversely, "discrete and digital communication through the double articulation of natural languages" exerts predominantly descriptive and cognitive functions, linguistic dimensions related to the referential or representative functions which aim to inform objectively as its core is among primary communication (PC). Secondary communication (SC) combines the aesthetic and factual functions. It can constitute elements of attraction for an audience that is not captive but voluntary (Epstein 2002, pp. 152–153).

This communication process examines the communicative structure from the perspective of the media, rather than from the perspective of the educational needs of the receiver. This "reading of the world" or "consciousness-strainer" role described by Miranda et al. (2015) differentiates the media's educational function from that of the school in terms of knowledge development. It is, however, crucial to recall that, as Tenreiro-Vieira and Vieira elucidate (cited in Façanha and Alves 2017, pp. 43–44),

> "Science education from a scientific literacy perspective can help individuals understand the world's problems [...]. It can also encourage people to use information and ways of thinking, including thinking critically, for enlightened and rational decision-making, for problem solving and for active and responsible participation in a democratic society."

From the perspective of content, digital natives exhibit distinct patterns of information consumption. In addition to experiencing "news fatigue," they also tend to avoid mainstream media due to concerns about its political bias and trustworthiness. With regard to scientific content, there is also the issue of the lack of language. It is commonly assumed that the information presented by the media is fundamental to the development of social knowledge. As Peters (2013, p. 14,107) explains: Nevertheless, the societal function of the media is not limited to the provision of information. The prevailing assumption is that if a piece of information passes the journalistic filter, it must be relevant. Consequently, it is anticipated that scientific journalism will disseminate knowledge with significant personal implications.

It is evident that the media cannot be compelled to prioritize science. However, it is crucial to recognize that this institution bears a responsibility to cultivate awareness. In this context, the significance of its "educational function" is also demonstrated by the serious crises of acceptance of scientific knowledge or, on the other hand, the absence of its relevance in individual decisions that affect everyone, such as in issues as climate change, migration, AIDS, or, as never before, public health, given the advent of the COVID-19 pandemic. In light of the aforementioned considerations, the necessity for journalists versed in scientific disciplines, particularly those pertaining to health, is becoming increasingly apparent. This is particularly evident in Costa's paradigm shift from the dissemination of scientific content to the communication of such information.

The concepts of attention, interest, perception, and sensitivity to information (Howard and Sheth 1969, p. 477) must be expanded upon. Additionally, the correspondence levels, memorization for selective message retention, and media responsibility in a context defined by Epstein (2002, pp. 61–62) must be understood. Finally, the process of improvement that brings this knowledge to the public is the process of communication, which has as a condition monosemy vis-à-vis the target audience. However, within a polysemic, metalinguistic, and referential language, that is, within a "secondary communication" (SC), meaning, in the media scenario, one dialogue, two actors.

While the contents are legitimized by the aforementioned language, which must align with the logic of the media to garner attention (Peters 2013, p. 14,103), and by professionals tasked with managing the communication between science and public opinion, scientists are, in fact, communicators in both of these arenas. These arenas are structured by different institutions and governed by different rules, yet the scientists remain objective, rigorous, and committed to the pursuit of truth. In contrast to this prevailing system, both will experience a crisis of acceptance by the non-scientific public. This will manifest in two distinct ways: firstly, in relation to the public's understanding of scientific orientation, and secondly, in relation to the practice of scientific journalism itself.

## 4 Knowledge Competences

*The Proximal Development Zone* (ZDP) is a concept developed by Vygotsky (cited in Schroeder et al. 2010, p. 22) which presents the idea of the transformations that occur through the intentional action of the teacher, promoting progress that would not happen spontaneously. The relationship between teacher and student is an asymmetric process, but of reciprocity: a dynamic of intellectual evolution that should allow a gradual transfer of responsibility from the first to the second. Schroeder also elucidates that, according to Vygotsky, this process is comprised of two levels: one corresponding to "real development," where mental functions are matured, leading to autonomy, and the other to "potential development," where the student still requires assistance in the consolidation of this knowledge.

On the other hand, in accordance with Schroeder, the journalistic discourse can be regarded as a "reference of social reality" (apud Quintanilha 2019, p. 21), a medium through which public opinion is shaped. The journalist, through their words, writing, and visuals, influences and produces public opinion. They act as a conduit for information, establishing connections between individuals and ideas. In a communication society where truth and falsehood are dichotomous, the journalist plays a crucial role in disseminating information and shaping public opinion.

However, the issue of meaning pertains to the generation of interpretation from the linguistic device, and what we may refer to as the zero degree of these "conventions of meaning" (Eco 2001, p. 74), which are essential to the realization of the reference processes. In this sense, a communicative action is not communication. In fact, the question of meaning is linked to the level of literacy, as well as the circumstances and accumulated knowledge and memory to which the "mechanisms of codification, retention and recovery" (Costa n/d) are applied. These mechanisms are dependent, in particular, on the type of content, environmental, social, physiological, motivational and emotional factors, as well as the type of content storage, that is, the synaptic and selective process of filtering data, which allows to preserve the information encoded in short or long-term memories. The application of the dialectic between memory and the perception of information enables individuals to not only comprehend a specific system of representation but also to express themselves through it. As a tool of interpretation and meaning, this structured exercise of coding and decoding capabilities is crucial (Calado 1994, p. 35). In light of the fact that only comprehensive reading allows for communication, it is evident that there are two distinct meanings (or uses) associated with the term "reading." "The two meanings of 'reading' are 'deciphering' (readability) and 'comprehension' (comprehensibility)" (Calado 1994, p. 33).

It is evident that learning and memory are interdependent processes. The various stages involved in the acquisition and retention of information, including attention, capture, fixation, maintenance, retrieval, and transmission, are closely linked to one another and influenced by a number of factors, including the social context and emotional influences. As Pinto (2001, p. 13) notes, perception plays a crucial role in determining how items are retained or stored. Moreover, the accuracy and reliability of the results obtained in the coding and recovery phases are contingent upon the degree of alignment between the indicators used. It can be argued that the manner in which scientific information is disseminated and perceived by the general public, as well as by journalists, has a significant impact on the formation of opinions. This is particularly evident in the context of scientific controversies, where the perception of problems within the scientific community can influence the way in which information is interpreted and subsequently shared.

Consequently, communication is impossible without the transmission of acquired knowledge, and decoding is similarly impossible without the use of referential knowledge. This is why literacy is a tool that is imperative for all individuals. Although we have adopted a more straightforward definition, the concept of "literacy" is not linear. For instance, it can be read in two dimensions, indicating that when considered in isolation, it is related to the comprehension of knowledge. However, when viewed in its social context, it is related to the application of this knowledge, as it is "a capacity for processing information in individual, social, and professional daily life" (Soares apud Bruno 2010, p. 11). Moreover, it is possible to consider the comprehensive and multi-dimensional design of Dionysus, which encompasses Multiliteracy, defined as "the different literacies and specificities of the languages that cross them" (Bruno 2010, pp. 12–13). Additionally, it considers how people utilize their functional literacy in a given context, and how these capacities relate to the needs, values, and practices of society.

'Information literacy' is therefore to be considered as being on a par with 'knowledge literacy'. In the context of contemporary journalism, this translates into the manner in which each journalist or user constructs their online experience, through the process of decoding and seizing codes and conventions inherent to the use of their digital skills. In essence, it operationalizes the DIKW pyramid model in the context of digital convergence. In other words, it shifts the focus from mere information to knowledge, insights, and action. Ultimately, it is the knowledge, insights, and actions that are of consequence. (I-SCOOP n/d).

It can be argued that being informed is not the same as knowing. The process of teaching and learning is a two-way street, with both the teacher and the learner playing an active role. Literacy is not an inherent quality; rather, it is a process of acquisition. In the context of scientific content, the media play a pivotal role in transforming scientific knowledge by linking it to phenomena, events, questions, and concerns that exist beyond the immediate scientific system but are reflected in it. This transformation goes far beyond the simple dissemination of information. Journalism's role is to maximize the comprehensibility of content relevant to an audience of non-scientists. It does this by articulating formats with content and audiences, and by recognizing that persuasion is based on the capacity of the informative context versus the referential. From this perspective, it is necessary to consider how the media can communicate science in a way that ensures access to adequate information when it is required.

## 5 Algorithms as a Journalistic Tool

The dissemination of misinformation about science can lead to misunderstandings. The social relationship with contradictory and primordial mistrust about the disruption of the status quo, as well as prejudice about the unknown, make unmasking and clarifying this misinformation difficult. This, in turn, makes it necessary to develop a better and more comprehensive understanding of the most effective strategies to correct disinformation. This is particularly important when conflicts over beliefs, values, and interests are central to individual decisions with a global impact.

Thus, the function of scientific journalism is to transform this content into models and formats that are readily comprehensible, influencing the individual's perception of science and its capacity to access, understand, and utilize scientific information in a critical manner. Nevertheless, while both journalism and science are based on these transient truths, the process is characterized by immediacy. In the implicit and tacit reading contract on scientific content, namely those in health crises such as the current pandemic, the public expects journalism to "present the risks before we have direct contact with them" (Geraldes 2001). A reading profile and a typology of content that leads journalists to often see themselves as "advocates for behavior change or motivators to action, which extends beyond the typical role of information provider" (Tanner et al., 2015, p. 361). The aforementioned reading profile and typology of content lead journalists to frequently perceive themselves as advocates for behavior change or motivators to action, extending beyond the typical role of information provider.

Conversely, it is also necessary to guarantee literacies in the mastery of the skills necessary for the construction of the elements that induce the critical understanding of discourses by both primary and secondary observers. This is to create the learning conditions that build the ability to know how to use, in a logic of functional and effective multiliteracy, the data available to achieve a greater and more productive convergence in the approach to the object (Franciscato 2002, p. 17).

In parallel, the current phase is characterised by the fact that technological development is focused on the creation of content and the writing of news based on algorithms to be generated by computers (Túñez-López et al. 2020, p. 19). Nevertheless, in this sequence of automation and optimization of data through a tangible value (contents) and intangible chain (primary and secondary knowledge), the subsistence of the content ecosystem corresponds to the degree of 'digital viscosity', which is "the ability to attract users, to make them stay, and to make them return on a recurring basis" (Hindman apud Quintanilha 2019, p. 58). This can be defined as fostering credibility through trust.

From the perspective of content management, in an era of hyper-information flows, journalistic algorithms serve a dual function. On the one hand, they facilitate the screening, processing, and dissemination of relevant information more effectively (Quintanilha 2019, p. 83). On the other hand, they benefit the public by increasing the ability to correlate meanings appropriately to a potential reader.

It is evident that despite the limitations of transparency as a solution to the problem of algorithmic power (Diakopoulos 2015, p. 403), the potential benefits of journalistic automation are significant. Automation can enhance the transparency of content, while also empowering journalists to pursue further research on issues that require deeper investigation. However, given that they are also susceptible to bias and error, publishers must ensure a certain degree of transparency and algorithmic accountability, particularly in the case of critical and controversial content. It is also important to note that, "news algorithms cannot function without structured data and the way data is chosen, evaluated, and 'cleaned'—standardized, normalized and validated, some would say manipulated—is, or at least should be, a process subject to journalistic decision" (Lindén 2017, p. 71).

In this context, where algorithmic journalism represents both a threat and an opportunity, there has never been a greater fear that technological advances will interfere with the role of the journalist, rather than with the media itself. For this reason, it is imperative that detailed descriptions of how the software carries out the writing process, which is modeled by the prioritization, classification, association, and filtering, become a constant. This is necessary in order to make the validation work more precise. This new requirement is reflected in a renewed framework of tasks for journalism performed by the professional (Quintanilha 2019, p. 88).

However, identifying quality information is a significant challenge in itself. This has led to the emergence of the concept of "junk news," which is defined as empty information that is not based on facts but rather on misleading narratives (Burger et al. 2019). In fact, recommendation algorithms have the potential to shift the battlefield of the economy of attention, to the point that giants like Facebook hire journalists to train their algorithms in the transmission of news (Nuñez 2019).

The process of selecting, writing, editing, positioning, scheduling, repeating, and otherwise massaging information to become news (Shoemaker et al., 2008, p. 73) is a complex one. The role of gatekeeper that journalistic algorithms perform is neither manipulative nor censoring. However, assuming this role is manipulative and censoring is neither unreasonable nor surprising. It is counterintuitive. Algorithms are neither valueless nor do they exist in isolation. They are part of algorithmic assemblages that include myriad actors, actants, activities, and audiences. Consequently, they are imbued with logics that are only sometimes reflective of journalism's (Zamith 2019, p. 1).

In essence, the deployment and utilisation of these non-human, unavoidable and disruptive information actors represent a power that is designed to respond to the overwhelming immediacy of the situation. Moreover, Lindén (2017, p. 65) posits that.

> "there is an algorithm for journalism, but even though some parts of news work—'low journalism'—will be automated, there is reason to believe that more demanding forms—'high journalism'— will benefit from man-machine combinations that ideally would have the journalist in charge."

In other words, a new high-quality journalism is emerging, which, favored by the man–machine-data triad, not only guarantees the continuity of news making but also expands its effectiveness through the automation of exhaustive editorial routines. This dynamic is already integrated in different media outlets, such as The Guardian. The New York Times, Reuters, The Associated Press, or, in the case of Portugal, Agência Lusa, exemplify the integration of technology into journalism. According to Pedro Camacho, Director of Innovation at Agência Lusa (cited in Nunes 2019), this process is "always verified by journalists, as long as faith in technology is not an unwavering thing."

In other words, a new form of high-quality journalism is emerging, which is favored by the man–machine-data triad and not only ensures the continuity of news production but also enhances its effectiveness through the automation of comprehensive editorial processes. This approach has already been integrated into various media outlets, including The Guardian. The New York Times, Reuters, The Associated Press, or, in the case of Portugal, Agência Lusa, exemplify the integration of technology into journalism. According to Pedro Camacho, Director of Innovation at Agência Lusa (cited in Nunes 2019), this process is "always verified by journalists, as long as faith in technology is not an unwavering thing."

In contrast, Thurman et al. (2017) posit that algorithmic journalism could prove advantageous for the industry, as it may enhance information consistency by expanding depth, breadth, specificity, and immediacy of available information. Dalen (2012, pp. 648–649) views it as "an opportunity to make journalism more human," suggesting that "machine-written news may prompt journalists to prioritize tasks that only humans can perform".

The atomization of data has accelerated the fragmentation of journalistic texts and documents into discrete information, which can then be recombined by algorithms and automation into adaptable and scalable news products (Thurman et al. 2019, p. 986). In the complex and intricate media evolutionary system, the changes sustained in digitalization and information automation have a direct impact on the representative, social, structural, intertextual, and ideological contents, which in turn interfere with the system's dynamics and impact on the public's eagerness to stay informed. One can readily observe that the sheer volume of data in the "Panama Papers" or the "Football Leaks" is indicative of a distinction between data and information. Indeed, the quantity of data in question has the potential to compromise the quality of journalism. However, intelligent recombination and reuse of content within the norms inherent to the news-making process can facilitate the implementation of a system that frees journalists from non-creative tasks.

This new man–machine technology has been the subject of considerable debate among journalists. Aljazairi (2016, p. 11) suggests that "human journalists should still prescribe the specific sources from which robots should glean their data to prevent them from including irrelevant sources." This is because the ability of algorithmic software to produce content more quickly, even in different languages, formats, and by combining multiple content sources, while also detecting and combating its own interference, presents a challenge to the traditional role of the journalist. In this context, it falls upon the professionals engaged in the creation of new forms of journalism to work closely with algorithm developers to create algorithms for specific purposes. Additionally, they must learn to investigate the decision-making processes of algorithms critically. Furthermore, they must develop the skills necessary to work with information in this symbiotic future.

## 6 Conclusions

In essence, the journalistic algorithm is perceived as both a blessing and a curse. In the context of information immediatism and the prevalence of autocratic governance, journalists may be susceptible to the pitfalls of poor reporting. However, the advent of robot or algorithmic journalism offers a potential solution to this problem (Aljazairi 2016, pp. 23–26). Conversely, literacy is an instrument that predisposes humans to interact with emerging social situations and to acquire new knowledge and skills that make them known (Vigotsky apud Bruno 2010, p. 18).

In summary, journalism represents an effective instrument for scientific dissemination. However, in the context of the contemporary paradigm crisis, neither science nor journalism is secure or unshakable in its effectiveness. In light of these considerations, it is essential to identify the most effective strategies for communicating science in a manner that ensures accessibility, literacy, and credibility.

> "Major challenges for practice and research in science communication include understanding the converging influences on science communication; engaging formally with the public about science; understanding the special complexities of communicating science in the face of public controversy; and communicating science in a complex, dynamic, and competitive communication media environment" (National Academies of Sciences, Engineering, and Medicine 2017, pp. 88–89)."

It is evident that the role of the communication professional is to facilitate the identification of media that are inclusive and capable of disseminating information to the public in its entirety (Silva 2017, p. 59). The advent of new technologies has the potential to streamline daily practices. However, it is crucial to recognize that the issue of accessibility and information literacy is inextricably linked to broader concepts of behavior and citizenship. In other words, if the general public is better informed about scientific progress, they can not only make more informed decisions on a personal level, but also influence decisions that affect their towns and nations (Zerzan 2019).

The integrative perspective is of fundamental importance. In order to adopt a new stance, journalism must acknowledge its role as a social mediator. Conversely, perception is a multifaceted process that occurs simultaneously on multiple levels. It is carried out on a mental and emotional level, as well as on a logical and intuitive level. However, what are the interactions between the various elements involved in the communication of science, including specificities, structures, elements, and processes, and how do they affect the understanding, response, and utilization of science by different audiences? Furthermore, it is necessary to consider the potential role of digital technologies in the effective dissemination of scientific information. In conclusion, how can scientific communication be adapted to meet the needs, concerns, and actions of the various audiences, taking into account their attitudes, characteristics, and perceptions.

The concept of "digital viscosity" is revisited here in light of the potential for digital media to provide more information, more quickly, and more closely than ever before. Kotler (Martini et al. 2019, p. 361) posits that "the key to generating customer loyalty is to provide high perceived value." Customer satisfaction, as defined by Martini et al. (2019, p. 359), is an emotional response to experiences related to the service that corresponds to their expectations and, therefore, of perceived quality. In addition, the SERVQUAL key (Pelissari et al. 2012) can be employed to assess the relationship between content and the public. This includes the dimensions of tangibility, reliability, responsiveness, assurance/guarantee, and empathy. It is also necessary to consider the scientific content in the media and its recognized value in the context of an age of information disruption. The hypothesis is that journalism and communication must act strategically in the circulation of knowledge in society, from the place of production of this knowledge, that is, scientific institutions.

In the intersection and tension of the contents of both universes emerges this third perspective—scientific journalism—which interweaves the first two (i.e., specialized journalism and science), through the dialectic between critical and investigative reflection on specific contents that correspond to the requirements of informative quality, maximizing the knowledge that science provides us.

As Pavlik (2000, p. 229) notes, journalism has historically been influenced by technological developments. This is because the communication act is only effectively developed when the emitter and receiver are involved. The motivation of the receiver can be influenced by a number of factors, including meeting expectations, aligning with perception indicators, such as memory, selective perception based on emotional associations, and interpretive, significant, and digital factors.

According to Postman (2000, p. 10), 'a medium is a technology within which a culture grows'. This interconnection can either imply the future complete success of the medium, or the actual crisis of degeneration of the dissemination process, which is enshrined both in science and in the so-called scientific journalism. The latter is the translation and dissemination of scientific content to the general public. Consequently, the efficacy of this accessibility initiative hinges on the communicator's objectives. To this end, it is imperative to consider additional measures of the quality and effectiveness of scientific communication, beyond merely grasping the role of context in rendering scientific information comprehensible. It is equally crucial to assess the audience and opinion leaders' profiles with regard to the generation of decision-making impact. In essence, the concept of successful communication is contingent upon the perspective of both the communicator and the target audience. For instance, when the audience seeks information from science, the communicator may also seek to persuade. This discrepancy in objectives can be elucidated through the lens of social dataset analysis, which offers a more comprehensive understanding of the audience by identifying and addressing the underlying motivations and perceptions that influence their engagement with scientific information.

This approach aligns with the insights of Gillespie (2016), who posits that the abstractions commonly encountered in scientific discourse, such as those described by Striphas (apud Gillespie 2016, p. 23), can be effectively addressed through the integration of social data analysis. The "black boxes" of Steensen (2019, p. 187) could be useful in predicting trends and behavioral flows, as well as their effect on the understanding and perception of science, particularly in relation to societal impact. This could be achieved by retrieving the question of credibility, quality, readability, and "likeability," especially from the Expectation-Confirmation Theory (Bhattacherjee apud Graefe and Haim 2018). In terms of the science's "news value," which also in journalism converges on the "inexpectability" of fact (Loose and Lima 2014, p. 14), this will be proportional to its importance in terms of belonging or having social implications and consequences. However, if one considers the social implications of research, as discussed by Medina and Greco (1994, p. 12), three key questions emerge:

> "First, it's necessary to choose emergency topics that point to solutions; second, it's necessary to mobilize multiple areas of scientific knowledge, since social impasses cover several or all specialties; third, emergencies are localized and the protagonists of the situation have something to say about their translation and possible changes —after all, what is at stake is not science and technology in the abstract, but the happiness of the local user of this or that specialized knowledge."

The Internet is a dynamic platform of content, and scientific journalism represents one of its interfaces. In this context, it is necessary to consider the potential shift in the communication of scientific information with 'news value' from a primary to a secondary channel. This shift is directed towards a lay public and may present challenges in terms of the perception of this 'competent knowledge' by the media. Conversely, the dissolution of this asymmetry of knowledge requires the involvement of a disseminator that has "a good degree of scientific literacy" (Loose and Lima 2014, p. 8). In order to translate into a natural language, the specific (meta)language, it is necessary to avoid any tendency to skid between the lack of reliability and/or the lack of competence. In this objective relationship, both scientific and journalistic syntaxes are employed to seek truth and to exercise technical and technological control.

"Scientific journalism is thus one of the tools for the growing literacy in society" (Frazão-Nogueira and Túñez-López 2021, p. 13) and the algorithms facilitate the integration of PC and SC since "one of the important differences is that automated journalism does not work directly on reality defined by facts but rather on a reality codified mainly in data" (Túñez-López et al. 2020, p. 20). Nevertheless, there is a mutualist effort between journalism and science that ensures the credibility of information. This is evidenced by the assertion that "when a medium offers quantity, quality, and diversity in both the set of information and the sources that have provided it, it is more credible and its influence is greater" (Edo 2002, p. 111).

Therefore, scientific journalism signifies a profound comprehension of the subject matter through coherent and accessible information structures, integrated into a media ecology that is mindful of the accessibility and cognitive relationship between media, information, society, and the target audience, and conducive to fostering literacy. In this regard, we cite Guzmán (2019, p. 1188), who asserts that "these efforts must go beyond people's assessments of the quality of the product created during the automation process and directly examine how people understand technology within this particular communicative role." In a study conducted by Graefe et al. (2018), in which 986 individuals evaluated two articles on credibility, readability, and journalistic expertise, the impact of these factors can be observed.

> "Varying the declared source had small but consistent effects: subjects rated articles declared as human written always more favorably, regardless of the actual source. Varying the actual source had larger effects: subjects rated computer-written articles as more credible and higher in journalistic expertise but less readable" (Graefe et al. 2018, p. 1)."

For the sake of argument, it is worth recalling another technological advance that has had a profound impact on the world: the Gutenberg press. It is evident that there is currently no desire to resume pre-press time, particularly in view of the use of algorithms in journalism, as assumed by Nuno Santos Silva, Secretary of State for Cinema, Audiovisual and Media, at the conference "Artificial Intelligence and the future of journalism—AI will take possession of the 4th power?" "Let us not seek to impede or delay its adoption on the grounds of disruption. Rather, let us endeavor to guide this process in accordance with the fundamental values of the press, including truth, rigor, and pluralism" (LUSA, 2014).

It is an irrefutable fact that the human environment and the maintenance of human health are interdependent. This interdependence imposes tension, demands, and, consequently, challenges to contemporary societal and socio-ecological dynamics. In light of the aforementioned considerations, the principal strategy for health literacy can be said to correspond to communication, encompassing its various forms, including audiovisual communication. In essence, journalism is the only reliable source of information that can facilitate a deeper understanding of the nuances of global events. It plays a pivotal role in translating complex information into accessible knowledge, fostering enhanced comprehension and enlightened citizen behavior. In other words, journalism, now enhanced by AI, is the sole instrument for the accessibility and literacy of scientific information.

In this context, in which the health of humans and of the planet is socially determined, the impact on improving both will only be achieved through literacy. This is because an increase in the holistic awareness of the individual will systematize habits and decisions that will bring about—and return—a socio-ecological and human ecology perception. An expansion of and within the value of the ecosystem as a co- and inter-participative entity, thereby establishing it as a somatic heritage, in which the human being is neither distinct, separate, nor superior, but rather one of the equally important components of the planet's biocultural heritage.

## References

Aljazairi S (2016) Robot Journalism: threat or an opportunity. Ph.D. diss., Örebro University. https://bit.ly/35O6jSY
BBC (2020) Trusted News Initiative (TNI) to combat spread of harmful vaccine disinformation and announces major research project. BBC, December 20. https://bbc.in/3uqaOxg
Beckett C (2019) New powers, new responsibilities: a global survey of journalism and artificial intelligence. London School of Economics. https://blogs.lse.ac.uk/polis/2019/11/18/new-powers-new-responsibilities/
Bruno AS (2010) Ler com Arte Estudo de Caso com Adultos em Processo de Alfabetização. PhD diss., Universidade de Lisboa. https://bit.ly/3rpGm4r
Burger P, Kanhai S, Pleijter A, Verberne S (2019) The reach of commercially motivated junk news on Facebook. PLoS ONE 14(8):e0220446. https://doi.org/10.1371/journal.pone.0220446
Calado I (1994) A Utilização Educativa das Imagens. Porto Editora, Porto
Costa P (n/d) A memória e a sua influência no processo de aprendizagem. NetSaber-Artículos. https://bit.ly/3HsIM82
Costa Bueno W (2010) Comunicação científica e divulgação científica: aproximações e rupturas conceituais. Informação & Informação 15(1):1–12
Dalen AV (2012) The algorithms behind the headlines: how machine-written news redefines the core skills of human journalists. J Pract 6(5–6):648–658. https://doi.org/10.1080/17512786.2012.667268
Diakopoulos N (2015) Algorithmic accountability. Journalistic investigation of computational power structures. Digit Jlsm 3(3):398–415. https://doi.org/10.1080/21670811.2014.976411
Eco U (2001) Semiótica e Filosofia da Linguagem. Instituto Piaget, Lisboa
Edo C (2002) Del papel a la pantalla: la prensa en Internet. Comunicación Social, Badajoz
Epstein I (2002) Divulgação Científica: 96 verbetes. Pontes, Campinas
Façanha AAB, Alves FC (2017) Popularização das Ciências e Jornalismo Científico: possibilidades de Alfabetização Científica. Amazônia. Revista de Educação em Ciências e Matemática 13(26):41–55. https://doi.org/10.18542/amazrecm.v13i26.4283
Franciscato CE (2002) Limites teóricos e metodológicos nos estudos sobre a noticiabilidade. Anais do 11° Encontro Anual da Compós. Galoá, Campinas. https://bit.ly/35Pa8Hy
Frazão-Nogueira AG, Túñez-López JM (2021) O Jornalismo Visual e a Infografia como ferramentas para a literacia científica e mediática. In: Gradim A, Moura C, Ramalho R (coords) Comunicar ciência num mundo em mudança. pp 9–26
Freire N, Cunha IC, Ximenes Neto FR, Machado MH, Minayo MC (2021) A infodemia transcende a pandemia. Ciência & Saúde Coletiva 26(09). https://bit.ly/349bnRv
Freixo M (2018) Metodologia Científica: Fundamentos, Métodos e Técnicas. 5th edn. Edições Piaget, Lisboa
Geraldes E (2001) Jornalismo e legitimação científico-tecnológica: o caso da erradicação da poliomelite no Brasil. XXIV Intercom, Campo Grande. https://bit.ly/3uqWJ2v
Gillespie T (2016) Algorithm. In: Peters B (ed) Digital keywords: a vocabulary of information society and culture, pp 18–30. https://bit.ly/3snCn7B
Graefe A, Haim M, Bastian H, Brosius HB (2018) Readers' perception of computer-generated news: credibility, expertise, and readability. Journalism 19(5):595–610. https://doi.org/10.1177/1464884916641269
Guzmán AL (2019) Prioritizing the audience's view of automation in journalism. Digit J 7(8):1185–1190. https://doi.org/10.1080/21670811.2019.1681902

Howard J, Sheth J (1969) The theory of buyer behavior. J Am Stat Assoc 65(331):1406–1407. https://doi.org/10.2307/2284311
I-SCOOP (n/d) The DIKW model for knowledge management and data value extraction. https://bit.ly/2IVhey6
Lasswell H (2007) The structure and function of communication in society. In: İletişim Kuram ve Araştırma Dergisi, pp 215–228. https://bit.ly/3fpneAs
Lemos V (2021) Coronavirus: la epidemióloga que perdió a su padre porque "él preferiría creer en los mensajes de WhatsApp". BBC News Brasil. https://bbc.in/3ur25Li
Lindén CG (2017) Algorithms for journalism: the future of news work. J Media Innov 4(1):60–76. https://doi.org/10.5617/jmi.v4i1.2420
Loose E, Lima MDV (2014) A Comunicação Científica sob a Ótica de Isaac Epstein. Ação Midiática. https://doi.org/10.5380/am.v0i7.37202
Luna D (2009) Links como articuladores do sentido na narrativa do hipertexto jornalístico do portal JC Online. Anais Hipertexto, pp 1–8
LUSA (2014) Inteligência artificial deve "servir" o jornalismo, diz Comissão Europeia. Observador. https://bit.ly/3sDMNAv
Martini K, Sangging IW, Merta M (2019) The influence of customer satisfaction in terms of service quality and perceived value. Atlantis Press. https://doi.org/10.2991/teams-19.2019.57
Martins M, Baptista MM, Araújo E, Latif L (2015) Ciência e conhecimento: políticas e discursos. Revista Lusófona de Estudos Culturais 3(2): 7–17. https://doi.org/10.21814/rlec.111
Medina C, Greco M (orgs) (1994) Saber plural: o discurso fragmentalista da ciência e a crise de paradigmas. ECA, São Paulo
Medina C (1994) Epistemologia Pragmática e Saber Plural. In: Medina C., Greco M. (orgs) Saber Plural: o discurso fragmentalista da ciência e a crise de paradigmas, pp 177–187
Miranda C, Schaefer R, Reis Medeiros V (2015) O Jornalista como mediador de consciência e operador da sociedade. In: Ontopsicologia Ciência Interdisciplinar, pp 197–215. https://bit.ly/3rx4UZw
Myron Atkin J, Bianchini JA, Holthuis NI (1997) The different worlds of project 2061. In: Raizen SA, Britton ED (eds) Bold ventures. pp 131–245. https://doi.org/10.1007/978-94-009-0341-8_3
National Academies of Sciences, Engineering and Medicine (2017) Communicating science effectively: a research agenda. The National Academies Press. https://doi.org/10.17226/23674
Nunes F (2019) Robôs também já escrevem notícias em Portugal. Estão a estagiar na Lusa. Eco https://eco.sapo.pt/especiais/robos-tambem-ja-escrevem-noticias-em-portugal-estao-a-estagiar-na-lusa/
Nuñez M (2019) Facebook joins other tech giants in employing journalists to curate news. Forbes https://bit.ly/3GxHR4V
Opdahl AL, Tessem B, Dang-Nguyen DT, Motta E, Setty V, Throndsen E, Tverberg A, Trattner C (2023) Trustworthy journalism through AI. Data & Knowl Eng 146:102182. https://doi.org/10.1016/j.datak.2023.10218
Pavlik J (2000) The impact of technology on Journalism. J Stud 1(2):229–237. https://doi.org/10.1080/14616700050028226
Pelissari AS, Solis DR, Ianagui EC, Gonzalez IVDP, Setubal FMR (2012). Aplicação e Avaliação Do Modelo Servqual Para Analisar a Qualidade Do Serviço. https://doi.org/10.6020/1679-9844/2301
Peters HP (2013) Gap between science and media revisited: scientists as public communicators. PNAS 110(3):14102–14109. https://doi.org/10.1073/pnas.1212745110
Pinto AC (2001) Memória, cognição e educação: Implicações mútuas. In: B Detry, F Simas (eds) Educação, cognição e desenvolvimento: Textos de psicologia educacional para a formação de professores. pp 17–54. https://bit.ly/3B0ZUiC
Postman N (1970) The reformed english curriculum. In: Eurich AC (ed) High school 1980: the shape of the future in american secondary education, pp 160–168. https://nysgs.org/page-1075266

Postman N (2000) The human of media ecology. Proc Media Ecol Assoc 1:10–16. https://bit.ly/35UpDxV
Quintanilha TF (2019) Desafios no Jornalismo e sustentabilidade no sector da Imprensa escrita. PhD the. Instituto Universitário de Lisboa. https://bit.ly/35Uq3nZ
Rother ET (2007) Revisão sistemática X revisão narrativa. Acta paul. Enferm 20(2):V–VI
Schroeder E, Ferrari N, Pedrosa Maestrelli S (2020) A Construção dos Conceitos Científicos em Aulas de Ciências: a teoria histórico-cultural do desenvolvimento como referencial para análise de um processo de ensino sobre sexualidade humana. ALEXANDRIA Revista de Educação em Ciência e Tecnologia 3(1):21–49. https://bit.ly/3uCfcZY
Shoemaker P, Vos T, Reese S (2008) Journalists as gatekeepers. In: K Wahl-Jorgensen, T Hanitzsch (eds) The handbook of journalism studies, pp 73–87. Routledge. https://bit.ly/35Jt6z7
Silva MA (2017) Jornalismo e Acessibilidade Comunicacional: Estratégias para a inclusão de pessoas com deficiência visual através dos dispositivos móveis. PhD diss. Universidade da Beira Interior, Covilhã. https://bit.ly/3GIAOXn
Smith R, Cubbon S, Wardle C (2020) Under the surface: Covid-19 vaccine narratives, misinformation & data deficits on social media. First Draft. https://bit.ly/3WkgkwU
Steensen S (2019) Journalism's epistemic crisis and its solution: disinformation, datafication and source criticism. Journalism 20(1):185–189. https://doi.org/10.1177/1464884918809271
Tanner A, Friedman D, Zheng Y (2015) Influences on the construction of health news: the reporting practices of local television news health journalists. J Broadcast Electron Media 59(2):359–376. https://doi.org/10.1080/08838151.2015.1029123
Tarapanoff K, Suaiden E, Oliveira C (2002) Funções sociais e oportunidades para profissionais da informação. DataGramaZero. Revista de Ciência da Informação 3(5). https://bit.ly/3J7y1Z1
Thurman N, Lewis S, Kurnet J (2019) Algorithms, automation, and news. Digit J 7(8):980–992. https://doi.org/10.1080/21670811.2019.1685395
Thurman N, Dörr K, Kunert J (2017) When reporters get hands-on with robô-writing. Professionals consider automated journalism's capabilities and consequences. Digit JIsm 5(10):1240–1259. https://doi.org/10.1080/21670811.2017.1289819
Túñez-López JM, Toural-Bran C, Cacheiro-Requeijo S (2018) Uso de bots y algoritmos para automatizar la redacción de noticias: percepción y actitudes de los periodistas en España. El Profesional De La Información 27(4):750–758. https://doi.org/10.3145/epi.2018.jul.04
Túñez-Lopez JM, Toural-Bran C, Valdiviezo-Abad C (2019) Automatización, bots y algoritmos en la redacción de noticias. Impacto y calidad del periodismo artificial. Rev Lat Comun Soc 74:141–1433. https://doi.org/10.4185/RLCS-2019-1391
Túñez-López JM, Toural-Bran C, Frazão-Nogueira AG (2020) Impacto de la robotización en la elaboración de noticias. La percepción de los periodistas ante el Robot Journalism. In: Ruiz-Alba N, Oliveira JS (eds) Nuevas fórmulas del ejercicio periodístico, pp 13–31
Zamith R (2019) Algorithms and Journalism. J Stud. https://doi.org/10.1093/acrefore/9780190228613.013.779
Zerzan A (2019) Why should scientists communicate clearly with the public? Voices Magazine https://bit.ly/3L8vIXz

**Ana Gabriela Frazão-Nogueira** has a PhD in Audiovisual and Journalism from the University of Santiago de Compostela and is a professor at the Fernando Pessoa University (UFP) where she teaches mainly in the areas of Radio and Journalism. Responsible for the Radio and Hypermedia Laboratory at that institution, she also attended a postgraduate degree in Marketing and Strategic Communication, at UFP, and another in Cinema, from the Aristotle University of Thessaloniki (Greece). Member of ICNova, her areas of research are Infographics and Visual Journalism, Information Literacy and Accessibility and Automation of journalistic content. Besides, she has a particular interest in issues of Information Continuity and a particular passion for Interviews and the use of Voice as an informative tool.

**José Miguel Túñez-López** is Professor of Journalism and professor of Communication Management in the Department of Communication Sciences at the University of Santiago de Compostela, as well as an advisor and professor to various European and Latin American organizations and universities on communications management. Doctor in Journalism from the Autonomous University of Barcelona, he is part of the Novos Medios Research Group of the USC, since December 2007, and IP-1 of the VALCOMM project. He is the author of more than a hundred reports, articles and books on communication. He has been Dean of the Faculty of Communication Sciences at USC (2004–2009), Director of the International Doctoral School in Arts, Humanities, Social and Legal Sciences at USC. He is currently Deputy Director of the USC International Doctoral School, EDIUS (Coordinator of the Social and Legal Sciences area). Queen National Journalism Award and partner, among others, of the Spanish Association for Communication Research (AE-IC), Spanish Journalism Society (SEP), Latin Society of Social Communication (SLCS) and the Latin Platform of Communication Magazines (PLATCOM).

**Jesús Segarra-Saavedra** is a senior lecturer at the University of Alicante (UA). Extraordinary Doctorate Award (2019), PhD in Audiovisual Communication and Advertising (2016) and Degree in Advertising and Public Relations (2005) by UA. Since 2018, lecturer and professor in the Department of Communication and Social Psychology at the UA. He has been lecturer and professor at UNIR (2016–2020), UMH (2018), UCJC (2019–2020) and UEMC (2019–2020) and foreign visiting professor at UTADEO (Colombia, 2019). Postdoctoral researcher in the groups "Comunicación y públicos específicos" (Compubes) at UA and "Comunicación y sociedad digital" (Coysodi) at UNIR, among others. He is co-editor of Revista Mediterránea de Comunicación, deputy editor of journal Health and Addictions, co-editor of IROCAMM. International Review of Communication and Marketing Mix. He is also a member of the scientific committee of Revista Latina de Comunicación Social, Revista de la Asociación Española de Investigación en Comunicación (RAE-IC), Miguel Hernández Communication Journal.

**Patrícia Weber** is a professor at Fernando Pessoa University (UFP), Porto–Portugal. With a master's degree from UNISINOS, Patrícia Weber completed her Ph.D. in Information Sciences, in Journalism and Media Studies, at UFP. Her work emphasis on Communication Sciences, specifically Research Methodology in Communication, Journalism and Radio. In Brazil, she supervised more than 70 scientific research carried out by students in the 1st cycle and participated in more than 70 1st cycle juries, where also coordinated the radio nucleus of the Experimental Communication Agency of UNISINOS and the EaD centers of that institution. Patrícia was also a journalist and worked on radio stations, newspapers, websites and communications agencies.

# Respect for Nature Virtues, Rules, and Climate Emergency

**Maria José Varandas**

**Abstract** Considering the current challenges brought forth by the Anthropocene crisis and claiming that technological solutions, although necessary, are not sufficient, this paper emphasizes the need for an ethical approach that goes beyond the inter-human walls, equating life in its broadest meaning and appealing to the moral excellence of human agency. In this sense, grounded in Taylor's ethical approach, we contend that the practical effectiveness of regulative principles of environmental action also demands the agent's moral strength. In our view, the development of a virtuous character, outlined by a wide range of virtues along with rules and principles, and forged in pedagogical programs committed to environmental citizenship, is the right way to respect nature and overcome the climate emergency.

**Keywords** Environmental ethics · Virtue ethics · Life-centered-ethics · Inherent worth · Agents and moral subjects · Moral excellence · Climate emergency

## 1 Introduction

The focus of ethical reflection is on moral choices, on what makes an action morally right or wrong, in the framework of community interactions settled by principles and rules that subsume duty and the obligation to act in accordance with it. This means that in practical terms, the freedom of the moral subject is intrinsically linked to the freely commitment to respect the moral norm, whose aim is the community good.

Since Aristotle, ethics has been considered to have a theoretical component that refers to the ultimate ends and values of actions and a practical component that provides guidance for action. For the Greek philosopher the Supreme Good or ultimate end of active life is Happiness (Aristotle, 1934: 1.7.[5]) whose path is routed on the virtuous conduct that models a good character (*ethos*). In the 18th century, Hume and Kant determined the further development of the conceptual foundations

M. J. Varandas (✉)
Center for Philosophy of Faculty of Arts and Humanities of the University of Lisbon, Lisbon, Portugal



W. Leal Filho et al. (eds.), *Composing Worlds: Humanities, Health and Wellbeing in the XXI Century Towards a More Sustainable World*, World Sustainability Series,
https://doi.org/10.1007/978-3-031-87108-5_14

of ethics and moral agency. Whereas morality is defined and upheld by reason in Kant (1785), moral agency is grounded in sympathy and pleasure for Hume (1739). Contemporary ethics oscillates between these tendencies, sometimes teleological, sometimes deontological, sometimes utilitarian calculating the maximum pleasure for the greatest number (Stuart Mill 1863).

Today we are witnessing a profound crisis, which is at the same time ecological and anthropological and not unrelated to the legacy of modernity with its optimistic belief in the unlimited power of human reason, which has evolved throughout the dystopian vertigo of total control over nature's forces. Ironically, the utopian myth of Bacon's *Regnum Hominis,* architected in his book *New Atlantis* (1626), woke up in a present reality seriously threatened by the apocalyptic risk of the future of the humanity. Today, the features presented by climate change have become an inter-generational, global, and collective, problem. The impending extreme climatic events, biodiversity loss, and food and water shortages call for a vast deployment of political and technological means, with the National Program for Climate Change (PNAC 2020–2030) advancing the urgent goal of promoting the transition to a low-carbon economy. Nevertheless, for an effective change, the epistemological framework of modernity in its ethical and anthropological dimensions must be profoundly rethought. Beyond technological solutions, the climate emergency needs a structural change in the attitudes and behaviors of all agents involved. So, one can ask: how ethics is called upon to respond to the current challenges, insofar as the Anthropocene crisis is, in essence, an ethical problem, that is, a problem that concerns human action? Among others two questions arises.

Knowing that climate change threatens humans, nonhuman animals and nature in potentially devastating ways (Gardiner 2012) is it still possible to understand the concept of moral community in the strict inter-human sense? What is expected of moral agency today?

Despite repeated warnings in recent decades from the scientific community, the IPCC (Intergovernmental Pannel on Climate Change), the IPBES (the Intergovernmental Science-Policy Platform on Biodiversity and Ecosystem Services) and other instituitions as well as intensive research demonstrating the harmful effects of climate change (e.g. Willis and Shonil 2009; Habibullah et al. 2022), the 21st century surpasses the levels of environmental disruption of previous centuries worsening the degree of ecological changes that threaten the future of human beings, with devastating consequences for ecosystems and unprecedented losses in biodiversity (IPBES 2019). Due to the global dimension of climate emergency, the challenges it demands for large-scale concerted action, the contradictions inherent to its multiple and different aspects, and its specific complexity, presents itself as a true "moral storm"[1] (Gardiner 2006; 2011), which strongly appeals to the ethical categories of

[1] "The peculiar features of the climate change problem pose substantial obstacles to our ability to make the hard choices necessary to address it. Climate change is a perfect moral storm. One consequence of this is that, even if the difficult ethical questions could be answered, we might still find it difficult to act. For the storm makes us extremely vulnerable to moral corruption." Gardiner (2006: 398).

responsibility, prudence, justice, imposing a profound reflection on the (moral) value of non-human beings and the type of obligations towards them.

Such a critical time calls for an environmental philosophy that goes beyond intra-human walls and perceives human actions in the broad context of the natural world, pursuing the need to operate a shift in the ethical paradigm and to found moral agency according to respect for nature. This call was already clearly formulated by Paul Taylor in 1986 with the book *Respect for Nature. The Theory of Environmental Ethics,* when the ethical debate on environmental issues, which had begun in the seventies, was dominating the field of applied ethics and was proving to be particularly fruitful, with a significant number of publications aimed at deconstructing the anthropocentric paradigm, redefining the status of the human being, and reflecting on the value of nature.[2]The emergence of environmental ethics stems from this specific reflective context and is defined by the inquiry and the in-depth understanding of the ethical relationship between humans, the natural environment, and non-human beings, identifying the set of values that should govern these relationships, and seeking to determine the norms that should regulate their interactions.

Among distinctive environmental ethics approaches, we will focus on Paul Taylor's biocentrism and his proposal of a radical reconfiguration of the human mental universe, which encompasses an ontology, an anthropology, and an ethics, asserting itself as made up of three interconnected components: a belief system, an ultimate moral attitude, and a set of rules of duty and standards of character. In a time of dramatic impoverishment of planetary life, we think it is enlightening to re-visit this approach and retain its principal message—life on Earth matters. To understand and preserve its value, rules and virtues are needed as standards of correct environmental action and should be contextualized and promoted trough social-political-educational programs (Carvalho et al. 2023) toward a constructive future where respect and nature come across.

## 2 The Biocentric Outlook of Nature

As the word Biocentrism specifies, this modality of ethics presents itself as a life-centered ethics, inspired by Albert Schweitzer, whose work *The Philosophy of Civilization* (1923) postulates an ethics of reverence for life with respect to the ontological principle inscribed in every living being: the will-to-live.[3] Paul Taylor substantiates Schweitzer's inspiration with the conceptual axes of Kantian and Aristotelian ethics, combining the ethical normativity of rules and principles of action with the

[2] For an overview of environmental ethics' founding texts, modalities, and main themes, see Zimmerman/Callicott (eds.) 1998. Environmental Philosophy: From Animal Rights to Radical Ecology. Prentice Hall, New Jersey.

[3] "The world is indeed the grisly drama of will-to-live at variance with itself" Albert Schweitzer ([1923] 1987: 313).

virtues that should characterize the agent, along with the assumption of biocentric egalitarianism[4] justified by the inherent worth of each singular organism.

Inherent worth is a concept taken from Kantian ethics, resulting from the distinction that the Germanic philosopher establishes between means and ends, from which he attributes moral status and then inherent worth to the being that is an end-in-itself, i.e., that has intrinsic value. For Taylor, the reasons that underlie the extension of the concept of inherent worth to every singular living being have its justification and consistency in biology. In fact, as any singular organic existence fights for survival and defends its physical integrity as an end, it has, according to Taylor, a *telos* inscribed in its own being (Taylor 1986:14–33), which is the pursuit of its life against death. In this sense, both its somatic integrity and survival have their own value that entirely belongs to its entity, and therefore, according to the author, this gives the organic being an inherent worth that requires respect from moral agents. Paradoxically, Taylor's declared non-anthropocentrism is grounded on Kantian anthropocentric reasoning, by appropriating a fundamental dichotomy from Kantian ethics (1785:Sect. 2), with which the German philosopher delimits the universe of moral beings and makes the division between humans (ends-in-themselves) and non-humans (means). Taylor, using the same moral categories, extends the kingdom of ends-in-themselves to all singular living beings, defending that each organism is a dynamic unit of self-preservation and self-realization, and its existence constitutes the realization of the *good-of-its-own* (Taylor 1986:75), imposed in its nature as an instinct or, in a Spinozian sense, *conatus,*[5] to fully develop its biological faculties. This makes every biological entity valuable as a teleological center of life (Taylor 1986:7), pursuing an ultimate end (*telos*)—the preservation of its organic identity and its well-being, "striving to preserve itself and to realize his own good in its unique way" (Taylor 1998:78).

In a life-centered ethics, it is the very fact of being alive that defines the concept 'end-in-itself', and consequently any organism is worthy of moral concern.[6] Take a plant, says Taylor, like any other organism, a plant is a spontaneous system of life, of self-preservation, with a control program. Plants have no goals; they are not subjects-of-a-life, they do not suffer. Even so, a plant grows, reproduces, resists death, and struggles to maintain its botanical identity. For that reason, the life of the plant per se is an end-in-itself and its identity has an inherent worth (Taylor 1986:75–76). Assuming an attitude of respect towards nature, states the philosopher, consists of seeing wild plants and animals in natural terrestrial ecosystems as bearers of moral standing. Which implies that they cannot be used as means to someone's ends, as they are moral subjects.

---

[4] "Humans are members of the Earth's Community of Life in the same sense and on the same terms in which other living things are members of that Community." Taylor (1986: 99).

[5] "Each thing, as far as it can by its own power, strives to persevere in its being." Spinoza ([1677] 1989: Book III, proposition 6).

[6] "The good of the biotic community can only be realized in the good lives of its individual members." Paul Taylor (1986: 70).

## 3 The Moral Community of Life

Stressing that the status of moral agent presupposes the possession of deliberative capacities, the ability to judge right and wrong, and the ability to commit oneself morally to one's own decisions (Taylor 1986:14), Taylor recognizes that, ultimately, the normativity of moral principles will be tributary to the character of the agent, or, to put it another way, if the principles are regulators of action, their full practical realization will only be possible through the agent's virtuous way of being, as a being whose conduct follows standards of virtue or, negatively, of vice. "Such standards (virtue and vice) apply only to beings who are moral agents" (Taylor 1986:16).

On the other hand, the moral subject does not manifest such capabilities, but, as an end-in-itself he possesses a moral status that obliges moral agents to the duty of preserving and not affecting him. In the human moral community, the baby, the insane person, or someone in coma, do not exhibit either the capacity for deliberation and decision or standards of virtue; however, the fact that they belong to the moral community through their inherent worth as ends in themselves implies respect for them owed by moral agents, which derives from the duty not to affect their lives and from the responsibility to protect and care for them. Hence, the human moral community shows that, although all human beings are moral subjects, worthy of respect and consideration, not all human individuals are moral agents. By extending the scope of the moral community beyond the human sphere and including all living beings in it, with the justification that each organic being is the bearer of its own good, i.e., its life, and consequently has an inherent worth, from this perspective each living individual has the status of a moral subject imposing obligations to the agent. Which means that in Taylor's environmental ethics, the moral community considered is composed of individual living beings, belonging to the terrestrial biota, in which only one class of beings (humans) exhibits the attributes of moral agency.

> From the perspective of a life-centered theory, we have prima facie moral obligations that are owed to wild plants and animals themselves as members of the Earth's biotic community. We are morally bound (…) to protect or promote their good for *their* sake (…) Their well-being as well as human well-being is something to be realized as an *end-in-itself.*
>
> Paul Taylor (1998:72)

Beyond principles and imperatives, the unlimited responsibility imputed to the agent requires, as Taylor recognizes, the robustness of character that the attitude of respect consummates. Respect, one's ultimate moral attitude (Taylor 1986:46) is entirely rooted in the agent's willingness to follow the normative principles and standards of conduct implied in the assumption of a biocentric worldview. This disposition reflects their character. A good character, Taylor tells us (1986:87), with virtues such as honesty, benevolence, and compassion, includes, in addition to the deliberative aspect, a practical aspect that compels the agent to act in accordance with their moral judgement. As Taylor says, "The practical aspect of virtue is the counterweight to weakness of will" (1986:87).

This brings us to the crux of our reasoning:

To what extent is virtue a prerequisite for action within the broad context of the natural world?

## 4 Virtues and Environmental Ethics

A common reader will certainly come across news of occurrences that, e.g., sometimes report the care for non-human animals, sometimes exalt the aesthetic appreciation of nature, sometimes exemplify sustainable ways of managing energy resources, but also, negatively, report on acts of pollution of rivers and soils, and of maleficence against animals.

These examples come in relation to the definition of virtue ethics, whose development was inspired by the claim of Margaret Anscomb (1958) to return to ancient Aristotelian virtues.[7] In general, virtue ethics seeks to answer a simple question, namely, "What kind of person is behind this action?" (Hill 1983). Thus, in this ethical approach, we can see a shift from the reflective axis of action (is the action correct?), to the type of character required from the moral agent in each practical context (has the person demonstrated the good character necessary for the situation in question?). If the context in question is the natural world, what we ask is what type of virtues are environmentally justified, or what type of character satisfies the demands placed by ecological reality. This presupposes a close connection between virtue ethics and environmental ethics, and in this sense, when it comes to the modality of environmental ethics in question, the type of criteria that defines the universe of moral consideration influences the requirement of appropriate virtues. The different approaches that seek to outline a general theory of virtue in an ecological context indicate a series of traits that define an environmentally virtuous character—humility, prudence, wisdom, solidarity, and compassion, among others. Our description of this subject will closely follow Ronald D. Sandler's argumentation on his thesis:

> An adequate ethical theory must provide an ethic of character (…) Environmental Ethics is simple ethics as it pertains to human-environment interactions and relationships. So, an adequate environmental ethics likewise requires not only an ethic of action—one that provides guidance on what attitudes and dispositions we ought and ought not to do to the environment—but also an ethic of character—one that provides guidance on what attitudes and dispositions we ought and ought not to have regarding the environment.
>
> Sandler (2005:2)

[7] For further information we refer to the work of Rosalind Hursthouse, *On Virtue Ethics* (1990), in which she defends the thesis of the centrality of the virtuous character in the ethical understanding of action. The fully virtuous human being has, according to the author, a set of correct rules to respond positively to the circumstances of the action, acting in a correct way because his character leads him to act in that way and not another. In the virtuous human being, the motivation to act correctly does not depend on the calculation of the benefits of correct action or on a reason that thinks what should be done, but rests on the self-motivation to do what is right. Hursthouse's delineation of virtuous character is rooted in the Aristotelian theory of virtue, enriched with a biological, social, and rational contextualization that extends the field of application of Virtue Ethics to the environment.

Ultimately, as Sandler affirms, any environmentally correct or incorrect action refers to the character of the agent, that is, to his inner willingness to value one or other reasons that influence their action and emotion in specific circumstances. Inasmuch as virtue is an acquired character trait that responds excellently to the world, i.e. in the right way and for the right reason[8] the virtuous character lies in one's disposition to respond correctly to the world in accordance with its goods and values. Based on this assumption, Sandler affirms the need for environmental virtue ethics in the context of action. Quoting Sandler (2005:3), two questions arise here:

> What attitudes and dispositions constitute environmental virtue? What is the role of an ethics of character in the context of environmental ethics?

Sandler defines environmental virtue as the appropriate disposition or appropriate character trait in human interaction with the environment.[9] This means that an environmentally virtuous person is willing to respond, emotionally and practically, to the environment and non-human individuals in a way that strives for excellence. The notion of excellence embodies a series of virtues that positively contribute to the well-being of the community. Applying these traits to a community enlarged to non-human entities, there is no qualitative distinction in the expression of virtues, whatever the reference community, as they will be equivalent to those that characterize inter-human relationships. Compassion, gratitude, and gratification, for example, that someone expresses towards other humans do not have a morally relevant difference if their field of application concerns non-human animals or natural entities.[10] The notion of moral excellence points to what is commonly called a "good person", that is, a person whose natural and proper disposition seeks, throughout his life, with sensitivity and wisdom, to promote the good functioning of social groups to which they belong and encourage others to maintain balanced and good-natured relationships with that group. When the community of belonging is the ecological community (as in Leopold's land ethic), the disposition of beneficence that contributes to good ecological functioning will be constitutive of environmental virtues. Sandler (2013) reminds us that this environmentally virtuous character is clearly illustrated by persons recognized for their exemplarity, such as John Muir, Henry Thoreau, Aldo Leopold, or Rachel Carson, among others, whose lives are a representative model of environmental commitment, care, and preservation. Muir revered the prodigality and

---

[8] " (…) Virtue is the observance of the mean, in point of excellence and rightness it is an extreme", Aristotle Nichomachean Ethics: 2.6 [17].

[9] "[Virtue] It is a disposition to respond to considerations in the world in excellent ways", Sandler (2013).

[10] "When a character trait is justified as a virtue at least in part by environmental goods and values—be they instrumental goods (e.g., natural resources, ecosystem services, or recreational opportunities), intrinsic values (e.g., aesthetic or spiritual values), or inherent worth (i.e., the value of environmental entities in and of themselves)—it is an environmentally justified virtue. Care for living things, appreciativeness of natural beauty, and moderation in use of natural resources are examples of environmentally justified virtues. Nonmaleficence toward nonhuman living things, insensitivity to natural beauty, and profligacy in the use of natural resources are some corresponding vices" Sandler (2013).

aesthetic presence of nature.[11] Thoreau called for simplicity and frugality as paths to a harmonious relationship with the natural world.[12] Aldo Leopold did not understand a relationship with the land without love and respect.[13] Rachel Carson wrote in *The Sense of Wonder* (1965), a true hymn to the cultivation of environmental virtues: sensitivity, humility, reverence, and wisdom.

In each one of these personalities, what underlies their action is a strong affective and binding disposition to natural reality, which informs and models a good environmental character, revealed throughout their entire life.

According to Sandler, in order to determine environmental virtues, it is important to take into account their context of inscription, because if the criterion for moral consideration is the ability to feel pleasure or pain, the required character trait is one that has the virtue of promoting pleasure and reducing pain, whereas if the context to be considered morally is the biotic or ecological community, the virtues in question are those that promote the health and integrity of ecosystems (Aldo Leopold) or the integrity and flourishing of the biotic community (Paul Taylor):

> For a utilitarian (…) environmental ethic such as Peter Singer's, on which the criterion for moral standing is the capacity to experience pleasure and pain, the virtue of compassion and the vice of cruelty are central, and a character trait is a virtue to the extent that it is conducive to promoting pleasure and the absence of pain (e.g., considerateness and benevolence) (...) For a Kantian biocentric ethic … such as Paul Taylor's, on which all living things are regarded as having inherent worth, the virtue of respect for nature and the vice of nonmaleficence toward living things are central, and a character trait is a virtue to the extent that it is conducive to allowing living things to pursue their own goods unconstrained by human activity (…) On a communitarian (…) environmental ethic such as Aldo Leopold's, on which the biological community is of primary importance, the virtue of ecological sensitivity and the vice of hubris will be central, and character traits will be virtues to the extent that they promote the health, integrity, or flourishing of the biotic community (e.g., love and gratitude)
>
> Sandler (2013)

Thus considering, the author declares that the set of rules and principles postulated by a theory of environmental action, although it is a necessary condition, will not be sufficient, especially when we are in the presence of different moral universes, sometimes conflicting with each other. It requires an appropriate disposition, a good character, in which sensitivity, wisdom, prudence, and knowledge coalesce, assessing the most favorable course of action. In this sense, the author infers that environmental virtues are effectively present in environmental action, concluding that an ethics of environmental virtues involves both positive requirements for environmentally appropriate behavior in accordance with the rules and principles of environmental action, and standards to guide and to promote the moral improvement of the agent. As Sandler (2013) states,

> Environmental virtue ethics is a vibrant area of environmental ethics (…) Work is also being done on the roles of virtue in responding to particular environmental problems (e.g., global

[11] J. Muir, "A New View of the High Sierra," in Carlson and Lintott (2008: 45–56).

[12] Thoreau, "Walking," in Carlson and Lintott (2008: 54–63).

[13] "When we see land as a community to which we belong, we may begin to use it with love and respect." Aldo Leopold (1949: Foreword).

warming and environmental injustice) and to environmental challenges more generally. In addition, there is continued interest in whether virtue ethics provides a viable and preferable alternative environmental ethic to those supported by other normative theories. What appears to be settled, however, is that environmental virtue ethics is an indispensable component of environmental ethics.

It seems consistent to us that environmental ethics, in whatever form, integrates, even if implicitly, environmental virtues, considering that the wider and more extensive the moral community is, the more it needs the robustness of the agent's character. In a community composed of a myriad of beings with specific interests and often antagonistic to each other, the dimension and scope of what is morally at stake require the strength of the agent's character to put into practice an action that decisively overcomes the imbalances. This means that a moral agency regulated by moral principles of universal applicability needs as a necessary condition the agent's judicious common sense and appropriate environmental virtues.

Referring back to biocentrism, which aims for maximum amplitude of action defending and respecting life in all its forms throughout the agent's robust moral force, we consider that, analogously, the magnitude of the climate emergency, too fast and too global, also demands of the agents a persistent constancy in the attitude of respect for nature, or, in other words, for planetary life. According to Paul Taylor this attitude must be *expressed* or *embodied* in the agent´s character:

> We are now in position to see how a person's character in matters of environmental ethics can express or embody the attitude of respect for nature (…) Virtues give one the steady ability to do the right thing with the right aim and for the right reasons in situations where moral failure (…) are not uncommon. In such circumstances virtues are needed for doing what is right and refraining from what is wrong.
>
> Taylor (1986:87–88)

If character, in matters of environmental ethics, *embodies* respect for nature and virtues are *needed* for environmentally correct action, the attitude of respect for nature is the *ultimate attitude* of the theory of environmental action,[14] i.e. the foundation of all acts and attitudes that imply commitment to an ethical system, understood and accepted as valid, both in terms of its normative principles and of the willingness to adhere to them (*ibid.*: 92). Accordingly, these principles subsume both standards of good character (the character of an environmentally committed person), and rules of correct action. The rules stated—the rule of non-maleficence, the rule of non-interference, the rule of fidelity, the rule of restitutive justice, which are, nowadays, core principles commonly used in bioethics (Beauchamp et al. 2012), specify the morally relevant properties of the action and come accompanied by a set of virtues because,

> Living the right kind of life in relation to the domain of nature and life on Earth is not simply a matter of having our choice and conduct conform to a set of rules and priority principles. The development of certain character traits is equally essential.
>
> Taylor (1986:198)

[14] "The attitude of respect for nature is not only a moral one, but an ultimate one", Taylor (1986: 92).

These traits encompass a series of virtues such as conscientiousness, integrity, courage, temperance, perseverance, constancy, benevolence, compassion, sympathy, and care, among others, configuring an ideal of moral excellence that compels the agent to the duty of cultivating these traits in the sense of constant moral improvement,[15] even if its full realization is presented as an enlightenment horizon for the enhancement of character. Such character standards are, as the philosopher states, normative guides for practical life, valid criteria of moral merit and in this sense constitute a fundamental part of normative ethics, as important as the rules of duty.

Paul Taylor's biocentrism raised multiple objections, and even though the focus of this paper is on the need to bring together virtues and rules toward a climate emergency action grounded in the respect for life in general, we believe it is instructive to recall Rosalind Hursthouse and her objections to Taylor's concept of 'respect for nature'.

To Hursthouse (2007:163–168), 'respect for nature' should be understood as a virtue that cannot be acquired by a simple decision after a few appropriate and enlightening readings. Insofar as it is not plausible to acquire the virtue of respect from one moment to the next, or that someone adopts a correct attitude, a disposition to act well, just by rational decision, Hursthouse asserts that the willingness to act well is the result of appropriate education and continued training of moral character since childhood.[16] Although we think this is indeed the desirable path to an overall environmental citizenship, we must argue that the intense dissemination of ecological literacy in public policies and the constant warnings from the scientific community about the global threats of climate change (Ripple et al. 2020) have been, in fact, producing cyclical changes in attitudes and behaviors all over the world, promoting an environmentally positive behavior. The decision not to eat meat, for example, whether based on sentience criteria, environmental criteria, or both, is now adopted by a progressively increasing number of agents. In most cases, it is about people who, after a lifelong experience of habits centered on a carnivorous diet, decide to become vegans from one day to the next because information about animal suffering or ecological crises affected them and gave rise to a radical change of attitude. Furthermore, Hursthouse's considerations about a moral universe centered on virtue acquired through pedagogical modeling from childhood, make it unfeasible to include someone who, for whatever reason, didn't receive an appropriate moral education in his childhood, nor does it seem to admit that a non-virtuous person can, through rational clarification and decision, start to act correctly or that a virtuous person can, at a given moment, for imponderable reasons, act malevolently. We are of the opinion, with Taylor, that the spreading of ecological literacy along with an ethics of respect for nature, positively informs human agency, promoting the change of behaviors and attitudes. Another relevant counterargument points to the atomist

[15] The idea of moral improvement through the persistent practice of virtue was expressed by Aristotle as follows: "The virtues therefore are engendered in us neither by nature nor yet in violation of nature; nature gives us the capacity to receive them, and this capacity is brought to maturity by habit.", Nicomachean Ethics: 2. 1 [3].

[16] "These dispositions and capacities have to be inculcated, from childhood, in the moral training of character", Hursthouse (2007: 163).

and individualist worldview of the biocentric ethics. We subscribe to this objection, since Taylor's approach draws a closed line between the animate and the inanimate, and this restrains the principle of respect for nature, conflicting with a holistic view of nature. This view states the interdependence between the biotic and the abiotic and is justified by ecological sciences and proved by countless facts, e.g., the relation between climate change and loss of biodiversity (Willis et al. 2009; Habibullah et al. 2022) which highlight an undeniable fact: Nature is a network of intricate interdependencies and interconnections of all its components.

At this point we must note two important concepts emerging in environmental literacy grounded in recent scientific research that break down the human-nature dichotomy and that bring us closer to the idea that life is a continuum involving all natural entities, including humans. That is, the concepts of 'One Health' and of 'Criature's Dignity'. The latter results from the research of the Swiss organization ECHN (The Federal Ethics Committee on Non-Human Biotechnology), an independent expert committee working with Basel University in the field of non-human biotechnology and gene technology appointed by the Swiss Federal Council and mandated to advise the federal authorities from an ethical perspective. Reading the scientific report about the moral value of plants (Federal Ethics Committee of Non-Human Biotechnology 2008), we find in it the main concepts of Taylor's ethics:

> The Federal Constitution has three forms of protection for plants: the protection of biodiversity, species protection, and the duty to take the dignity of living beings into consideration when handling plants (…) Previous discussion within constitutional law relates the term Würde der Kreatur («dignity of living beings») to the value of the individual organism for its own sake (…) All members of the ECNH are unanimous that plant communities also always have an instrumental and a relational value, but that this does not exclude their having inherent worth. As moral objects that should be protected for their own sake.

The goal of ECHN's scientific research in the non-human domain is to offer an ethical and scientific underpinning for legislation and permit applications of essential importance in the fields of gene technology and non-human biotechnology. Climate Change is also a relevant subject of its scientific reports, which serve as the architectural framework within which laws, rules, and best practices are developed, substantiating the legal sphere and public policies. This is just one of many instances that clearly demonstrate the scientific community's efforts to influence and support decision-making that respects nature, trying to stop the escalation of environmental degradation and its impacts on ecosystems and public health, affirming the interconnection between humans and nature, and putting forth a new paradigm in which humans are part of nature and not beings apart from it. An understanding that builds up the concept of One Health, as suggests the One Health Global Network (2012–2015): "One Health recognizes that the health of humans, animals and ecosystems are interconnected. It involves applying a coordinated, collaborative, multidisciplinary and cross-sectoral approach to address potential or existing risks that originate at the animal-human-ecosystems interface". We must recall here the Land Ethic's environmental philosopher, Aldo Leopold, when he states that the health of the land is an imperative for action, insofar as the health of an ecosystem is also the health of all its component parts, and infers that as planetary citizens, it is our duty to preserve and respect it for our own good and the good of all natural beings, because we are all interconnected in this planetary life.

The reasons presented in this article, seeking to link Taylor's environmental ethics with the challenges of the Anthropocene, underlie his claim for 'moral strength', as it postulates an ideal of excellence composed by a set of virtues and defines the moral agent as a morally exceptional and exemplary character. Implicitly, this ideal seeks to model a social ethos constitutive of public policies and pedagogical programs, inasmuch as it is the virtuous character that guarantees the persistence of the will in correct action.[17] Even if scientific warnings and environmental literacy are essential tools for influencing people's behavior, they are not enough on their own to bring about a systemic shift in the way that people interact with natural realities. In our view, this requires that public policies recognize the need for pedagogical training beginning in childhood for a widespread and structural change of attitudes in the relationship between humans and nature,[18] because, even if scientific research joins environmental ethics in the appeal to respect nature, in the end, it is in the moral agent that this appeal is expressed. Hence, the widespread modeling of a good character that faces and overcomes 'moral corruption' (Gardiner 2006) presents itself as a crucial complementary task to be adopted by public policies.

## 5 Conclusion

As emphasized in the previous sections, our attention to Paul Taylor's environmental ethics has been less for what it purports to be—a theory of environmental action—but more for what it subsumes—a theorization of environmental duties and virtues underlying respect for nature.

We consider that, among the different modalities of environmental ethics, none other than biocentrism prescribes such a complete set of environmental duties, simultaneously appealing to the virtuous character of the moral agent and outlining its ideal portrait in detail. The relation between virtue and duty is relevant in moral theory and, therefore, considerable in the current historical moment, in which bioecological changes have reached alarming and dangerous levels for life in general. In our opinion, the normative apparatus needs to be embraced by a character that seeks to improve the quality of life in Earth, acknowledging that we are living amid a perfect moral storm and running the risk of 'moral corruption' in judgments and choices that undermine constructive solutions to tackle environmental concern. Climate change involves serious ethical issues,[19] especially in its global, intergenerational,

[17] "Virtue is, therefore, the moral strength of man's will in the fulfillment of his duty" Kant, ([1797]1996): II, Introduction to the Doctrine of Virtue, XIV.

[18] To this regard we remark the work of Stephen Kellert and Peter H. Kahn, explained in Varandas (2016): 201–219.

[19] "Even our best theories face basic and often severe difficulties addressing basic issues such as scientific uncertainty, intergenerational equity, contingent persons, nonhuman animals and nature. But climate change involves all of these matters and more." Gardiner (2006: 407).

and ecological dimensions. Despite challenges owing to underdeveloped theories and pragmatic issues, there is a strong consensus concerning the need for, and the overall shape of, serious action and the relevance of key ethical values, such as prudence, fairness, and responsibility (Gardiner 2012). To this, we must add that the ethical component in the thinking of this time raises questions that go beyond the universe of the climate change agenda in terms of understanding its causes—the responsibility of technoscience in past emissions—and in relation to solutions—adaptation and mitigation strategies. We wonder if the development of such solutions stems from an anthropocentrism that excludes important modalities of environmental ethics, such as biocentrism and ecocentrism, or not. If policies integrate principles and themes of environmental ethics—e.g., precaution and the theme of future generations (Hans Jonas 1984)—wouldn't it be crucial to equate all the contributions that pursue the value of respect for planetary life? Since the normativity of ethics ultimately concerns the character of the agent, as it is in the individual that this normativity is expressed, the conceptual framework of society in its different dimensions—ethical, legal, and political–, should consider ways of promoting an effective shift in the moral agency towards an active and integrated participatory attitude in the broad context of climate change.

As said before, in addition to consistent information and practical rules that lead to cyclical changes in human agency, it will be necessary an effective structural change, not simply cyclical change, that manifests itself in the character of the agent to adopt an environmentally correct conduct capable of resiliently facing the threats of a climate emergency. This critical time need an extended ethical framework that brings into account a deeper comprehension of the practical aspects of human behavior in all its dimensions—not only culturally and socially but also individually. Principles, rules, and standards of character should be integrated into a unified framework that encourages and develops the best practices to restrain environmental vices and maleficent actions that threaten planetary life and, consequently, the future of humanity. This unified framework should guide public policies, legislation, and direct educational systems, reverting to pedagogical programs in which the cultivation of virtues enables the assumption of righteous principles and rules of positive environmental action.

To the extent that the good of the terrestrial community concerns all members and all the relationships that bind it together, Paul Taylor's environmental ethics deserves particular attention. Not only for its rules, such as those of beneficence, non-maleficence, justice, and similar, that are already adopted in different ethical contexts and not just in bioethics, but mainly for its horizon of reflection as it reveals an ideal of human excellence to be sought and pursued, especially at this crucial moment in which the accentuated crisis of the Anthropocene imposes an inclusive and comprehensive ethical reflection on moral choices that ought to model not only adequate public policies but also the good character of the agent.

## References

Aristotle (1934) Nicomachean ethics. In: Rackham H (ed) Loeb classical library. Harvard University Press. https://www.perseus.tufts.edu/hopper/text?doc=Perseus%3Atext%3A1999.01.0054%3Abook%3D1

Anscombe M (1958) Modern moral philosophy. Philosophy 33(124). https://files.meetup.com/328570/Anscombe-20Modern%20Moral%20Philosophy.pdf

Beauchamp TL, Childress JF (2012) Principles of biomedical ethics. Oxford University Press

Carlson A, Lintott S (2008) Nature Aesthetics and Environmentalism. From Beauty to Duty. Columbia University Press

Carvalho L, Dourado L, Varandas MJ (2023) Conceptions of Portuguese science teachers on the concept of ecoethics. Edu Sci 13(11):1089. https://doi.org/10.3390/educsci13111089

Gardiner SM (2006) A perfect moral storm: climate change, intergenerational ethics and the problem of moral corruption. Environ Values 15(3):397–413. https://doi.org/10.3197/096327106778226293

Gardiner SM (2011) A perfect moral storm: the ethical tragedy of climate change. Oxford University Press https://philarchive.org/archive/GPEROA

Gardiner SM (2012) Ethics and global climate change. https://www.nature.com/scitable/knowledge/library/ethics-and-global-climate-change-84226631/

Habibullah S, Din BH, Tan SH et al (2022) Impact of climate change on biodiversity loss: global evidence. Environ Sci Pollut Res 29:1073–1086. https://doi.org/10.1007/s11356-021-15702-8

Hill Jr T (1983) Ideals of human excellences and preserving natural environments. Environ Ethics 5(3):211–24. https://doi.org/10.5840/enviroethics19835327

Hume D ([1739], 1985) A treatise of human nature. Penguin Books Ltd, London. https://oll.libertyfund.org/title/bigge-a-treatise-of-human-nature

Hursthouse R (1990) On virtue ethics. Oxford University Press

Hursthouse R (2007) Environmental virtue ethic. In: Walker RL, Ivanhoe PJ (eds) Working virtue: virtue ethics and contemporary moral problems. Oxford University Press, pp 155–172. http://hettingern.people.cofc.edu/Environmental_Philosophy_Sp_09/Hursthouse_Environmental_Virtue_Ethics.pdf

Jonas H (1984) The imperative of responsibility: In search of an ethics for the technological age. The University of Chicago Press

Kant I ([1785] 2002) Groundwork of the metaphysics of morals. In: Wood AW et al (ed) translated JSTOR, Yale University Press. https://www.jstor.org/stable/j.ctt1njjwt

Kant I ([1797]1996) The metaphysics of morals. In: Gregor MJ (ed) Cambridge University Press. https://philpapers.org/rec/KANTMO-4

Leopold A (1949) A sand county almanac and sketches here and there. Oxford University Press

Mill S (1863) Utilitarianism. Parker, Son and Bourne, London. https://socialsciences.mcmaster.ca/econ/ugcm/3ll3/mill/utilitarianism.pdf

Ripple W, Wolf C, Newsome T, Barnard P, Moomaw W (2020) World scientists' warning of a climate emergency. BioScience 70(1): 8–12. https://www.researchgate.net/publication/337679925_World_Scientists'_Warning_of_a_Climate_Emergency

Sandler R (2005) Introduction: environmental virtue ethics. In: Sandler R, Cafaro P (2005) Environmental virtue ethics. Rowman & Littlefield Publishers, Lanham, MD, pp 1–12

Sandler R (2013) Environmental virtue ethics. In: The international encyclopedia of ethics. Wiley Online Library https://doi.org/10.1002/9781444367072.wbiee090

Schweitzer A ([1923]1987) The philosophy of civilization. Campion CT Transl. Prometheus Books, Buffalo, New York

Spinoza B ([1677] 1989) Ethics. Elwes RHM Transl. Prometheus Books, New York

Taylor PW (1986) Respect for nature. In: A theory of environmental ethics. Princeton University Press. https://doi.org/10.1016/0140-1750(88)90072-3

Taylor PW (1998) The ethics of the respect for nature. In: Zimmerman ME et al (eds) Environmental philosophy: from animal rights to radical ecology. Prentice Hall, New Jersey, pp 71–85

Varandas MJ (2016) A natureza: solo de conjunção da ética e da estética. NEA, Säarbrucken
Willis K, Shonil B (2009) Biodiversity and climate change. Science 326(5954): 806–807. https://doi.org/10.1126/science.1178838

## *Public Reports:*

Federal Ethics Committee of Non-Human Biotechnology (2008) The dignity of living beings with regard to plants. https://www.ekah.admin.ch/inhalte/_migrated/content_uploads/e-Broschure-Wurde-Pflanze-2008.pdf Federal Ethics Committee on Non-Human Biotechnology: https://www.ekah.admin.ch/en/homepage-1
IPCC (2023) Climate change 2023: synthesis report. In: Core Writing Team, Lee H, Romero J (eds) Contribution of working groups I, II and III to the sixth assessment report of the intergovernmental panel on climate change. IPCC, Geneva, Switzerland, pp 35–115. https://doi.org/10.59327/IPCC/AR6-9789291691647
IPBES (2019) Global assessment report on biodiversity and ecosystem services. Intergovernmental Science-Policy Platform on Biodiversity and Ecosystem Services (IPBES), Germany. https://doi.org/10.5281/zenodo.3553579
One Health (2012–2015). https://www.onehealthglobal.net/; https://www.onehealthglobal.net/what-is-one-health/

**Mª José Varandas** Degree in Philosophy, pre-Bologna Master's and PhD in Philosophy of Nature and the Environment from the Faculty of Arts and Humanities of the University of Lisbon. Member of the Philosophy Center at the Faculty of Arts and Humanities (FLUL). Founder, president, and vice president of the Society for Environmental Ethics. Invited professor in PhD seminars and environmental ethics courses in higher education institutions. Author of publications in the field of environmental ethics (articles and books). Organizer and co-organizer of conferences on environmental issues. Speaker at various academic and public institutions. Member of the Scientific Council of the Portuguese Congress of Philosophy (2014–2018). In 2009, she won the Gulbenkian Environment Prize for her book *Ambiente: Uma Questão de Ética.*

# Socioecological Management and Social Participation

# Mountainous Environment in the Central Region of Mucajaí, Roraima, Brazil: Citizen Science and Geotouristic Approaches

**Ana Sibelonia Saldanha Veras, Waqas Ahmed, Nelson Azevedo Barros, and Maria Alzira Pimenta Dinis**

**Abstract** Geotourism consists of a form of interpretation and awareness practiced in ecosystems geographically located in environments of expressive geological and geomorphological aesthetics, associated with citizen science, which is also inclusive, with the strong involvement of ordinary citizens who collect data from observation of natural environments, streamlining scientific research and generating scientific knowledge, preparing citizens to manage the diverse demands of remote areas. The balance lines of these domains can be severely impaired by global warming, for which human beings have been blamed in this Cenozoic era, destabilizing access to recreation and guided tours. Through a methodological approach based on an international literature review focusing on the northern Amazon, the main objective of this chapter is to demonstrate the relevance of sustainability practiced by citizens from the local reality of mountain geosites, establishing the relationship with the municipality of Mucajaí, Roraima, Brazil. The results can be considered useful references for the scientific community, contributing to the sum of local strategies, with global repercussions, for the development of citizenship, so as not to contribute to the increase in the climate emergency experienced today. To this end, it will be necessary to promote

A. S. S. Veras (✉)
Faculty of Science and Technology, University Fernando Pessoa, Porto, Portugal
e-mail: anasibelonia@gmail.com

W. Ahmed
Department of Operations and Supply Chain, NUST Business School, National University of Sciences and Technology, Islamabad, Pakistan

N. A. Barros · M. A. P. Dinis
Fernando Pessoa Research, Innovation and Development Institute (FP-I3ID), University Fernando Pessoa (UFP), Porto, Portugal
e-mail: nelson@ufp.edu.pt

M. A. P. Dinis
e-mail: madinis@ufp.edu.pt

M. A. P. Dinis
Marine and Environmental Sciences Centre (MARE), University of Coimbra, Coimbra, Portugal

 
W. Leal Filho et al. (eds.), *Composing Worlds: Humanities, Health and Wellbeing in the XXI Century Towards a More Sustainable World*, World Sustainability Series,
https://doi.org/10.1007/978-3-031-87108-5_15

behaviors that avoid episodes of great repercussion and irreparable damage to the communities involved in conducting, encouraging and rescuing the sense of preservation, thus promoting ecological knowledge, and contributing to a more sustainable world.

**Keywords** Geotourism · Global warming · Mountain ecosystem · Sustainability · Mucajaí · Citizen knowledge

## 1 Introduction

Recent years have witnessed the evident exploitation of natural resources in the Central Region of Mucajaí, Roraima, Brazil, which was seen in actions induced by the recurrent removal of rock materials and plant species on steep slopes of hills and mountains (Gomes 2022; Veras et al. 2022a). Anthropogenic actions in landscapes make them fragile, which is added to strong extreme rains, as mentioned by Santos et al. (2023), long-lasting events and dangerous intensities, e.g., landslides and floods, putting at risk lives in territories worthy of recreation and leisure in mountainous nature, ecosystems that serve as true lessons in interaction with citizens (Gastauer et al. 2022; Veras et al. 2022a, b). The geological scenario stands out in the Central Region, essential for humanity and biodiversity, satisfying the curiosities and investigations of researchers and lovers of natural areas when interpreting the geological and geographical phenomenon, combining leisure, education, and preservation, i.e., the sustainability of remarkable places (Leng et al. 2023). In this scenario, geotourism is consolidated by the number of geoparks made official by UNESCO, it is supported by the phenomenon of tourism, one of the largest industries that prioritizes geographic reliefs, also in terms of global growth (Mosgrove 2023; Štrbaet al. 2023).

Research contributing to Citizen Science (CS) witnesses relevant advances in projects led by professional scientists, offering opportunities to ordinary citizens to participate at a simplistic level in the collection of data, practices and techniques that solve research. Such practice provides a new vision between scientists and non-scientists to connect in a complete evolution, particularly because it does not require the use of a fixed protocol but rather the commitment of the citizens (Sanabria et al. 2022; Haklay et al. 2023). It allows voluntary support, mutual and conscious citizen collaboration, as well as participation in projects and in various areas of knowledge, with benefits for regions lacking large resources. Still, there are people disengaged in this context (Batchelder et al. 2023). By involving the citizen in an active process, CS acts towards advancement based on the opinion of the resident and their experiences, a scope of diversified participation that helps to emphasize sustainable territories (Lima 2023). Furthermore, carrying out a scientific project with partnerships in strategic rural areas encourages and empowers the target audience to develop skills for local solutions and for the protection of nature and habitats, e.g., currently focusing on climate as a major concern for everyone in this era (Weiner et al. 2022; Rosado Gonzalez et al. 2023).

Therefore, aligning geotourism and CS actions means to capture the enthusiasm of citizens surrounding geosites as a worthy partner, strengthening CS and encouraging confidence in choosing the researcher's role, promoting interaction and awareness about geological heritage and disseminating the fulfilment of the 2030 Agenda goals in mountain environments affected by climate change (Palmer et al. 2022).

Agricultural explorations in the Central Region of Mucajaí include agricultural fires that often go out of control and result in forest fires, damaging ecological and microclimatic areas. The strong influence of the *El Niño* phenomenon, responsible for severe droughts in the Amazon, affects ecosystems (Da Silva et al. 2008; Fortes et al. 2023). Therefore, challenges come together in order to mitigate the growing anthropogenic impacts in all terrestrial biomes worldwide (Coratzaet al. 2023, Bussard & Reynard 2023, Mori et al. 2023).

Based on these assumptions, the objective of this chapter is to demonstrate the magnitude of sustainability practiced by citizens, i.e., CS, based on the local reality in mountain geosites, establishing the relationship of the topic with the municipality of Mucajaí, in Roraima, Brazil, translating it into the urgency of a greater commitment of everyone in relation to what activities are detrimental to the environment.

## *1.1 Global Warming Emergency and Geotourism*

Throughout the twenty-first century, and according to Covey & Megonigal (2019), concern about global climate change caused by greenhouse gas emissions, the burning of fossil fuels, in addition to the impact of human activities in terrestrial ecosystems, such as deforestation and industrial advancement, have been reflected on very serious factors, spread of pests and diseases, phenomena that have been presaged by several scientists (Raihan 2023; Hadiet al. 2023).

Nowadays, the effects of global warming have been felt by everyone in many parts of the Earth's surface, since the increase in the planet's average temperature changes the sea level due to the melting of the polar ice caps, as stated by Qiao& Zhang (2023), which can cause the disappearance of islands and flooding in densely populated coastal cities (Barnes et al. 2023). In this sense, cities located on the coastal strip deserve attention due to the melting of ice in Antarctica, which causes rising sea waters and altered resurgence regimes, intensifying storms, loss of life and displacement of people (Pallathadka 2023; Rabelo et al. 2023). In studies by scientists Volk et al. (2023), there is a forecast of a greater frequency of maximum climatic events, such as tropical storms, floods, extreme heat, drought, and hurricanes, with serious consequences for populations and ecosystems, which could lead to the extinction of animal and plant species.

Brazil, a country of great continentality, has an extensive coastal strip, i.e., the Northeast Region with the states: Rio Grande do Norte, Bahia, Piauí, Alagoas, Ceará, Paraíba, Pernambuco, Sergipe, Maranhão; the North Region with the states: Pará and Amapá, the Southeast Region with the states: Espírito Santo, São Paulo and Rio de Janeiro; and the South Region, with the states: Paraná, Santa Catarina and Rio Grande do Sul (IBGE 2018). Following this perspective, the Intergovernmental Panel on Climate Change (IPCC), accordingly Hartz (2023), views several biomes on the planet and states that the risk of climate change potential is determined by vulnerability and exposure, human interaction with the natural environment, and the anthropic actions generated (Masson-Delmonte et al. 2021).

Among the consequences of climate change, Volk et al. (2023)state that the effects are in the forest region, which neutralize and get the carbon (C) from the atmosphere, an important chemical element, the fourth essential for life on planet Earth, essential for living beings and an important component in organic life, found in the compositions of natural vegetation and in the process of photosynthesis (Correa et al. 2023). In the global context, geotourism innovation is guided by the maximum postulate of interpreting landscapes of geological prominence (Oksanen 2023; Zhanabayev et al. 2023). Similarly, this region of high scientific value offers significant information that requires protection from climate change and, at the same time, contributes to sustainable local development, using geoconservation and geoeducation strategies, valuing residents in local activities and their active role in the sum of the 2030 Agenda and its 17 Sustainable Development Goals (Giudice 2022; Reinhart et al. 2023). Thus, the use of geodiversity encompasses abiotic elements and their pre-existing characteristics of rock with added value increasingly relevant to global economy, as they are raw materials, in addition to composing the Earth's landscape (Alvarenga 2023). These resources are part of the global challenge, part of the ecosystem with an impact on the quality of life and the anthropogenic exploitation, which harms biodiversity, i.e., ecology (Ioannou et al. 2023).

## 2 Mountain Ecosystem in the Context of Sustainability

The mountain ecosystem is a topic that has been much discussed in the last twenty years as it considers the maintenance of the landscape to be a crucial point for sustainable development (Dax &Tamme 2023). Geoecosystems are home to very vulnerable endemic birds, animals and plants, i.e., existing living organisms, suppliers of goods and services, such as fresh water, of great regional importance (Machate et al. 2023). Due to the quality of its attractiveness to a very particular audience, it is important to understand the relevant commitment to sustainability through tours and research, as well as learning to interpret local geology and geomorphology and to avoid human action, encouraging zeal and the objective of sustainable development (SDG) 15 and 17, linked to the level of commitment, adopting the carrying capacity so as not to overload these survival niches of many individuals (Veras et al. 2021; Veras et al.2022a, b).

As one of the six Brazilian biomes, the Amazon Forest is a very biodiverse ecosystem, however, it is threatened daily by climate change and anthropogenic activities; The components of these biomes are complexes of giant trees and other variations that have resilience, providing environmental services, such as air quality, essential for the planet (Signori-Müller et al. 2023). In addition to these biomes at the local level, community members are involved in providing information for various projects, which often do not have technology available and do not master more complex areas, knowing only mobile technologies, requiring deeper engagement on technological tools to disseminate and promote SDG 15 (Elegbede et al. 2023).

## 3 Citizen Science and Its Role in Contemporary Times

As a result of demographic expansion, human beings advance towards the vicinity of habitats of non-domesticated and wild animals, altering landscapes. These incursions open windows for interaction with ecology, i.e., the relationship between species and humans, however, this practice has created a high risk for natural environments (Tyagi et al. 2023). Recently, however, CS has made a connection with society as it gives credibility to citizens living in outskirts and rural environments; This citizen helps with collaborative information in the search for sustainability, an attitude that translates into the empowerment and innovation of this collaborator in research as it is linked to social and environmental dynamics (Butkevičienė et al. 2021).

CS, including scientists, citizens, participants, users, volunteers, collaborators and co-researchers, only comes into effect effectively with citizen engagement, which helps in identifying landscapes and cataloging geosites, thus characterizing themselves as global citizens, with personal and professional opportunities through commitment to the causes of the earth's surface in which they live (Haklayet al. 2023; O'Donnell& Brundage 2023). The irreversible damage resulting from climate change affects the nexus of ecology, therefore, the ecosystem services of biodiversity and geodiversity, social integration in local contexts, have public participation as great allies, with volunteers investigating and providing low-cost data to be analyzed by scientists, e.g., about the start or end of fires, still widely used in landscape units (Purdy et al. 2023).

In this sense, the important sciences that stand out are Geology, Geography, Geochemistry, Paleontology and Glaciology, and Sociology is also concerned with applying its norms to modernity in the interventions of social movements (Carvalho et al. 2021). They adhere to the recent CS, not as precise as the systematic scrutiny of the (IPCC) according to Hartz (2023). It focuses on plural and comprehensive assessments of climate change McGowan (2023), outlining scenarios to be known and felt by everyone. Anchored in popular wisdom, where society supports contemporary academic postulates, in other ways the CS is still disconnected in many places around the world, however, in Europe, the United States and Australia, academic movements are robust (Pelacho et al. 2021; Smaniotto & Passani 2023). To this end, the citizen's commitment is embodied in information with rich details about the locations, for example, the consumption and behavior of species, the occurrence of monuments and changes in ecosystems, as observed by Frigerio et al. (2021).

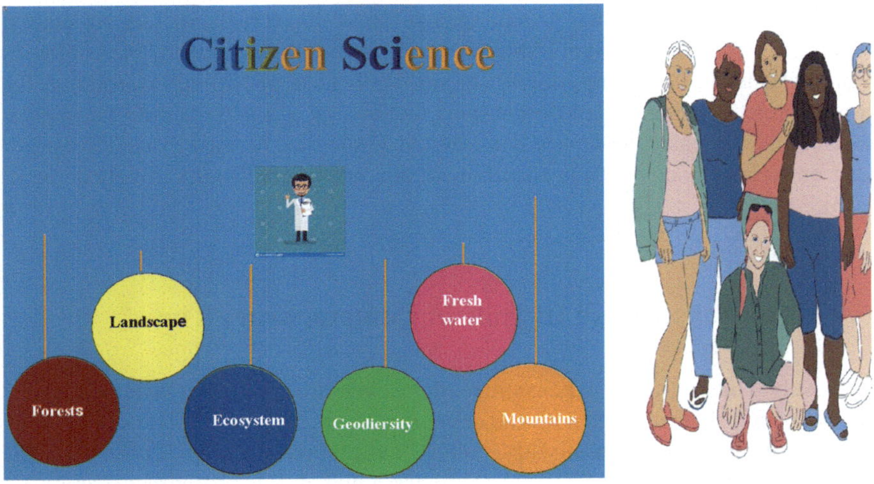


**Fig. 1** Profile of the geoturism CS network in articulation with ordinary citizens, independent from academia

Figure 1 demonstrates how CS involves different areas of science, contributing to scientific studies.

In Brazil, the interdisciplinarity and challenges of this practice are increasing, e.g., the Brazilian Citizen Science Network (*Rede Brasileira de CiênciaCidadã* - RBCC), launched in March 2021 in a workshop, deepened by the discussion on the recognition of CS by scientists in the social, ecological and technological, a moment that gave rise to embryonic projects on the topic of sustainability, standardization of science, data quality, and fundraising (Haklay et al.2023; Smaniotto & Passani 2023). According to De Castro (2023), CS mirrors the involvement of brilliant minds, a way of giving visibility to a new generation of observatories that communicate in a simple and clear way, sustaining the rigor of research with data from ecological and geological environments on a local scale, actively in decision-making, understanding the relevant role of facilitating the interaction of researchers and citizens prone to the topic being researched, making scientific knowledge more accessible (Hartz 2023). Much of the so-called human progress and its achievements deteriorate biodiversity and impact the climate, associated with the Industrial Revolution, population growth and the increased consumption of natural resources originating from the Anthropocene, which bring consequences to life and nature, as McGowan (2023) points out, referring to the description of studies on this topic and the geological force on the earth at a planetary level; This began with irreversible geochronological developments of unparalleled magnitude and which effects are still completely unpredictable (Batchelder et al. 2023). This precise diagnosis has a variety of consequences for the Social Sciences. Thus, social scientists need to frame the situation referred to as the

Anthropocene and analyze it in anticipation of future social processes, e.g., the structural, institutional, normative, and individual consequences of human-caused climate change, rapid loss of biodiversity, acidification of oceans and land degradation.

# 4 Methodology

The literature review methodology was used in this study, by means of a selection of scientific articles from Web of Science, Scopus, repositories, mctic.gov.br, unicef.org, Roraima state fire department, and other relevant existing collections on the Google Scholar platform. Environmental factors, i.e., natural and anthropogenic, were also used to verify the fires. Also included are the last years of studies on CS in its different phases, which contribute innovatively, providing and integrating new reliable data, monitoring compliance with the SDGs; since, even though people are not scientists, they are in a place where they can voluntarily inform about the eccentricities of mountain geosites and all the aspects that can occur if there is no contribution from each citizen related to climate change (Fraisl et al. 2022; UN 2018; Purdy et al. 2023). As for the context, it covers the critical situation of the burning season, a practice widely used in the Amazon Region, also involving frequent logging and cattle raising through the exploitation of pastures (Coutinho 2022).

## *4.1 Citizen Participation in Research in Government Projects*

The municipality of Mucajaí, in partnership with the State Government, supports and registers activities that necessarily have a positive impact on local citizens, with a focus on Amazonian environmental sustainability, especially in areas such as family farming, the forestry sector, and small businesses, aiming to optimize land use (Roraima 2020). Citizen participation requires registration with the municipality's programs, such as: Rural Environmental Registry (*Cadastro Ambiental Rural* - CAR) of the State Prevention Committee, Fire Control and Forest Fire Fighting in Roraima, all with policies to direct education and fire control actions and policies, which strengthen partnerships with the City Hall of the municipality of Mucajaí.

## *4.2 Information Acquired from Institutions Operating in the Area*

Institutions such as the State Environmental Foundation (*FundaçãoEstadual do Meio Ambiente* - FEMARH), entities from the Municipality of Mucajaí City Hall (*Prefeitura do Município de Mucajaí* - PMM) and the Municipal Secretariat for the

Environment were contacted weekly, in the month of August 2023, to identify the actions and partnerships developed for the mountain area, supported by Bartuśet al. (2023) statements, in which measures are necessary for implementation, conducive to geotouristic tours, such as those aimed at students, leisure use for residents of the region and actions for conservation in order to guarantee and protect the scenic spots of the abiotic universe for those who depend on biodiversity (Veras et al. 2020; Cai et al. 2023).

## 4.3 Actions that Include the Mountain Environment

Supported by article 20 of Resolution n. 217 of December, 1948 and Declaration of the Rights of Workers and Other People Residing in Rural Areas (Hartz 2023). Social groups are considered to be all those that have a common interest, feeling of belonging and group identity, as these shape the evolution of a place, define products to be cultivated and activities to be practiced in order to add resources to the local economy and sustainability for Mother Earth (UN 2018). These rural environments are the home base to small and medium-sized farmers, rural settlers, traditional communities, that is, CS with potential for scientific advancement, involvement in the conservation of the territory from which they derive their livelihood, with small animal husbandry, and much to share for health and ecosystem management (Fortes 2023). In this way, different knowledge can be used when conducting geotourism and designing activities, evaluating studies from a sustainable perspective. Notably, CS expanded on issues related to the visibility of projects involving public, private and voluntary managers, an aspect of great value in an era in which sustainability is needed (Guillén Martin & Mateo Medero 2023).

## 4.4 Location of the Study Area

The iconic destination is located in the Central Region of Mucajaí, state of Roraima, Brazil, a young city founded in 1951 as an agricultural colony and, officially, by Federal Law No. 7,009, of July 1, 1982, (Brazil 1982). It is a geotourist region, with temperatures ranging from 23° to 28° C, it has a topographic range of 70 m in relation to sea level. The months with the highest rainfall are May, June, and July, with an average altitude of 75 m above sea level (Veras, 2014, Farias Filho 2015; Veraset al. 2020). Mucajaí occupies lands whose lithological substrate is represented in the Central Guiana Domain (*Domínio Guiana Central* - DGC), therefore, they support landscapes that display mountain ranges, hills aligned on an attenuated to slightly undulating surface over the extensive area defined as the *Rio Branco–Rio Negro* Pediplane, all of these characteristics are found in the Amazon Biome (Beserra Neta 2008). As for ethnic-racial characteristics, it has five categories: white, black, *pardo*, indigenous, and yellow. Many of these citizens are organized in agricultural colonies,

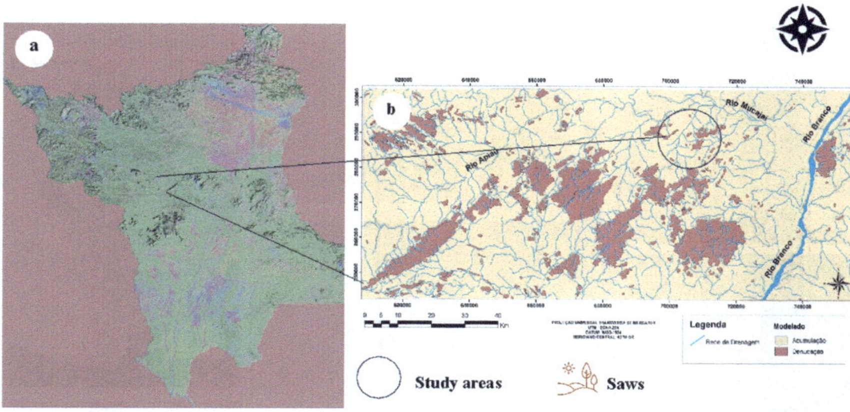


**Fig. 2** Study area. **a** State of Roraima. **b** Geomorphological map highlighting the mountainous part

with the common use of cleaning the land with fire, which can trigger large fires, reflecting on the climate change and the application of geotourism (Veraset al. 2022a, b; Drükeet al. 2023). The region has an area of 12,337.851 km2, a demographic density of 1.46 ha/km2 and a schooling rate of six to 14 years of around 92.6%; revenues from external sources total 93,7%; all these particularities are included in this immensity of territory, according to IBGE (2023). Figure 2 indicates the location of the research area.

## 5 Results and Discussion

The results of this study provide valuable records that can help understand the important role of CS at the local level of the Mountain Region. Greater dissemination and involvement on the part of Higher Education Institutions (HEIs) is suggested in raising awareness regarding the need to co-opt young people, less young people and tourism enthusiasts, i.e., geotourism in the robust task force in favour of nature preservation in recognition of the impact that Homo sapiens has on the planet (Manono et al. 2023). Such as other remote regions, it must receive alerts and awareness talks about climate change, so as not to ignore the effects of this reality experienced in areas whose choreography is not at all beautiful, contrasting the location, e.g., the increase in solid waste (Borges et al. 2023) left in the surrounding areas, resulting in mountain scenery, pastures and vegetation, altering the pattern, distribution and alteration in the physiologies of species, temperature and precipitation, as well as reducing the abundance of animals, increasing pests and insects (Coutinho 2023; Volk et al. 2023).

**Fig. 3** Climate change evidence in the research area: **a** The landscape in its fullness and well preserved, denoting harmony in nature, **b** Burning mountains photographed in flight (Author of imagens a, b and c: Jorge Macedo), **c** Recording the fire danger in close proximity to a thatched house with a high fire risk, **d** Spiral of the Anthropocene geological era, i.e., *Anthropos* means humans and *Kainos* means new. It was widely discussed by scientists and non-scientists, who call for sustainability and environmental commitment actions (Adapted from McCowan, 2023)

In the Amazon, the use of fire is a very common practice to open new areas for agricultural or livestock exploration, as well as to keep these areas free of trees and bushes. However, this practice creates an opportunity for the fire to spread towards the forest, a factor triggered by dry vegetation and strong winds (Farias Filho 2015). All these events and the constant burning in rural lots tend to be serious during the dry season, adding to the *El Niño* climate phenomenon, a climate episode that warms the Pacific Ocean, causing devastating effects in different regions of the world, resulting in severe damage to the environment, in addition to this, the felling of trees and, consequently, the atmosphere is unsuitable for geotouristic tours (Veras et al. 2021; Luo et al. (2023),. All these episodes cause enormous damage to fauna, flora and the citizens who inhabit these distant lands (Fig. 3).

## 6 Implications of This Research at Different Levels

The fate of small and more remote environments in the immense Amazon is fundamental to resolving on-site crises as they comprise living and non-living elements, inherently connected, with no possibility of separating themselves from geodiversity. However, geodiversity is the stage, the substrate of living elements that need to be preserved to maintain their functionality (Manono et al. 2023). Among the abiotic elements referenced, the mountains and hills, i.e., geosites called *inselbergs*, interpreted as geoheritage, make up the greatest expressiveness and, according to Veras et al. (2022b), deserve a more conservative look and attitudes, in the sense of measures of protection and monitoring, with the community as an essential factor in this surveillance. This study, with a view to contributing to CS, has a high potential to contribute to tackling problems of human conduct in practice and use in the mountainous areas in search of the SDGs, for which everyone is responsible, listed below.

### 6.1 Mountains and Hills

Mountains and hills have functions that manifest themselves as freshwater deposits and energy supply, thus, their socio-environmental role is considered essential, with their vegetation and evapotranspiration, which can be changed according to human activities (Veras et al. 2022a, b). The low mountain biome in Mucajaí is highly relevant, being one of the highest points, with the possibility of building viewpoints to view the rural landscape and economy focused on agricultural production, small animal farming and pasture formation; Furthermore, the scenic beauty present is an object of advertising and a means of entertainment. However, mountains and hills are subject to impacts, resulting in ecological damage that can harm the ecosystem.

### 6.2 Interpretive Panels

Interpretive panels should include information about the plateau, central regions, soils, stratigraphic units, including the Holocene era, and environmental care. The panels constitute a means of visual communication for the visitor at the tour site, even when led by a local guide, which enriches the local attraction with the information to be passed on to the visitors, with students being important passers on of this information to other society groups.

### 6.3 Building a Link Between Geodiversity and Biodiversity

It is worth highlighting that geodiversity is the support for biodiversity, with geodiversity being the holder of diverse features and landscapes (geological and geomorphological) that are supported by the scenic beauty of the northernmost Amazon Biome. These reliefs shelter biodiversity, provide vegetation that develops in the rock walls, providing water supply to an entire natural chain, therefore, it is the interface and connection of Natural Science and Geosciences.

## 6.4 Listening to Traditional Leaders and People is Promoting Empowerment

The working understanding of traditional leaders and people, who are resilient citizens, constitutes a challenge for management policies, considering that these people have lived in harmony with nature for hundreds of years, in the sense of recognizing the skill with which they treat forests, therefore, these people are those who best contribute to climate balance. The Amazon Forest, in greater proportion in the Northern region of Brazil, is a place that houses magnificent biodiversity and geodiversity, as well as the habitat of a significant number of ethnicities willing to pass on this cultural knowledge to the world (Santos 2021; Veras et al. 2022a, b).

## 6.5 Management of Protection Areas

In view of the large territorial dimension, the signage of heritage protection areas in the Central Region has not yet been completed with the appropriate indication of signs of sites permitted for use by tourism, i.e., geotourism. For properties and users, setting fires as a cleaning agent when renewing pastures should be avoided, therefore, antlers and debris must be allocated specific destinations. For those who enjoy camping, you should be extremely careful with campfires, not discarding leftovers from campfires with embers, as well as not throwing leftover cigarettes during the dry season and, also, not releasing balloons, not only because it is dangerous, which can cause a fire without proportion and control, but because it is a crime, according to Federal Law No. 9,605, of February 12, 1998 (Brazil 1998). However, in the event of a fire emergency, you must quickly inform the Municipal Fire Brigade and the Fire Department by calling the number "193" (http://www.mucajai.rr.gov.br/). Protected areas generally have information signs warning about precautions, however, in addition to these, during the dry period, extra care is taken for the environment, reminding the commitment of each resident and those who visit this space.

## 6.6 Vegetation

The maintenance of endemic trees, shrubs and grasses is considered a priority for ecological services to be effective and guarantee the climatic balance and the ecological environment of forest species (de Oliveira 2023) with diverse colours of green in the crowns, such as the large *Ceiba pentandra*, *Manikarahuberi*, important plants in the region, favouring Amazonian geotourism, and other species, e.g., bryophytes, bioindicators of environmental preservation (de Souza Saldanha et al. 2021).

## 7 Final Remarks

In order to address geotourism in the context of sustainable development, it is necessary to further involve the visiting public with the Geoscience, as Geology can offer more assertive bases for geosites, with Physical Geography being the area that investigates natural characteristics, and Geomorphology and Geophysics investigating physical properties and dynamic processes. From another perspective, the need for greater harmony between the environment and society appeals to the committed citizen aware that the world is formulating strategies to integrate the 2030 Agenda and the SDGs, with a level of urgency of interdependence and integration. In the local context, citizens include transdisciplinarity, i.e., seeing holistically and, thus, making better decisions in the care movements surrounding existing paths and trails used by geotourists. This relationship must emerge spontaneously from each resident in a loving relationship with nature. Thus, this chapter concludes with an appeal to geotourists and all who are fans of this noble practice, of places that await visits, where you will find an incredible paradise with well-defined seasons, dry and rainy, with altitudes that can reach 1,000 m. The air in the morning in the mountainous territory forms a mist that gradually reveals and presents the landscape of hills and valleys framed by verdant, unique Amazonian vegetation cover, a splendid geography and a reliquary for the future, and this magic of the landscape cannot be ignored. This fight belongs to all scientists, citizens, society, managers, and the State, in order to guarantee care and witness the existence of a sustainable environment, within CS and Geotouristic approaches.

## References

Alvarenga LJ (2023) O lugar do Direito na proteção do património geológico-paisagístico: uma perspetiva luso-brasileira. [The place of law in the protection of geological-landscape heritage: a Portuguese-Brazilian perspective]. https://hdl.handle.net/1822/84693

Barnes PW, Robson TM, Zepp RG, Bornman JF, Jansen MAK, Ossola R, Paul ND (2023) Interactive effects of changes in UV radiation and climate on terrestrial ecosystems, biogeochemical cycles, and feedbacks to the climate system. Photochem & Photobiol Sci 1–43. https://doi.org/10.1007/s43630-023-00376

Bartuś T, Mastej W (2023) Morphodiversity as a tool in Geoconservation: a case study in a mountain area (Pieniny Mts, Poland). Sustainability 15(14):11357. https://doi.org/10.3390/su151411357

Batchelder M, Swinney M, O'Hara T, Goddard A, Lewis E, Cox J, Fowler HJ (2023) Experiences from a school-university partnership climate and sustainability education project in England: the value of citizen science and practical STEM approaches. Sustainability 15(12):9401. https://doi.org/10.3390/su151294401

Beserra Neta CL, Tavares Junior SS (2008) Geomorfologia do Estado de Roraima por Imagens de Sensoriamento Remoto [Geomorphology of the State of Roraima by Remote Sensing Images]. (Org) Silva PR, Oliveira R da Silva. Boa Vista (ed) Roraima 20 Anos. UFRR. ISBN-13 978–8560215331

Borges RR, Dinis MAP, Barros N (2023) Linking low family income to waste recycling in a Brazilian public university. In: Handbook of sustainability science in the future: policies, technologies, and education by 2050, 1–14. http://hdl.handle.net/10284/11240

Brazilian Citizen Science Network [Rede Brasileira de Ciência Cidadã]. In https://translate.google.com.br/?sl=pt&tl=en&text=Rede%20Brasileira%20de%20Ci%C3%AAncia%20Cidad%C3%A3%20&op=translate

Brazilian, IBGE—Instituto Brasileiro de Geografia e Estatística (2018) Regiões de Influência das Cidades [RegionsofInfluenceofCities]. Rio de Janeiro. ISBN 978–65–87201–04–7

Bussard J, Reynard E (2023) Conservation of world heritage glacial landscapes in a changing climate: The Swiss Alps Jungfrau-Aletsch case. Int J Geoheritage Park. https://doi.org/j.ijgeo.2023.06.003

Butkevičienė E, Skarlatidou A, Balázs B, Duží B, Massetti L, Tsampoulatidis I, Tauginienė L (2021) Citizen science case studies and their impacts on social innovation. Sci Citiz Sci 309–329. https://doi.org/10.1007/978-3-030-58278-4-16

Cai Y, Han J, Wu F, He G (2023) Promoting Geotourism in Dunhuang UNESCO Global Geopark. Geoheritage 15(1):22. https://doi.org/10.1007s12371-023-00796-y

Carvalho A, Ferreira V, Matos AR (2021) Ontologias do Antropoceno: Crise climática, respostas sociopolíticas e tecnologias emergentes. [Ontologies of the Anthropocene: Climate crisis, sociopolitical responses and emerging Technologies]. In: Forum Sociológico. Série II (No 38, pp 5–13). CESNOVA. https://doi.org/10.4000/sociologico.9693

De Castro LFP (2023) Campesinato: modo de vida e sujeito coletivo de direito à luz da Declaração das Nações Unidas sobre os Direitos dos Camponeses e Outras Pessoas que Trabalham nas Áreas Rurais, [The Peasantry as a CollectiveSubjectof Law: AnOverviewofthe Legal elementsofthe-UnitedNationsDeclarationontherightsofPeasants]. Rev Bras Sociol Do Direito 10(2):105–133. https://revista.abrasd.com.br/index.php/rbs/article/view/686

Coratza P, Vandelli V, Ghinoi A (2023) Increasing geoheritage awareness through non-formal learning. Sustainability 15(1):868. https://doi.org/10.3390/su15010863

Correa, Sulamita Santos, et al (2023) Natural carbon fixation and advances in synthetic engineering for redesigning and creating new fixation pathways. J Adv Res 47:75–92. https://doi.org/10.1016/j.jare,2022.07.011

Coutinho JDS (2023) Efeitos das mudanças no uso da terra sobre os atributos funcionais de plantas em florestas neotropicais. [Effectsofland use changesonthefunctionalattributesofplants in neotropical forests].(Doctoraldissertation, UFRA/Campus Belém). https://repositorio.ufra.edu.br/jspuiu/handle/123456789/2031

Covey KR, Megonigal JP (2019) Methane production and emissions in trees and forests. New Phytologist 222(1):35–51. https://doi.org/10.111nph.15624

Dax T, Tamme O (2023) Attractive landscape features as drivers for sustainable mountain tourism experiences. Tour Hosp 4(3):374–389. https://doi.org/10.3390/tourhosp4030023

Drüke M, Sakschewski B, von Bloh W, Billing M, Lucht W, Thonicke K (2023) Fire may prevent future Amazon Forest recovery after large-scale deforestation. Commun Earth & Environ 4(1):248

Elegbede IO, Goldin J, Obalola M, Matti-Sanni R, Olaleye-Haroun OS, Akindele OT (2023) Stakeholder governance and citizen science. In: Encyclopedia of sustainable management (pp 1–8). Cham: Springer International Publishing. https://www.researchgate.net/publication/370758001

Farias Filho LDP (2015) Análise da Ocorrência de Incêndios Florestais no Município de Mucajaí-RR, no Período de 2013 a 2015. http://dispace.mj.gov.br/handle/1/5628

Femarh, State foundation for the environment and water resources. http://www.femarh.rr.gov.br

Fortes SLK, Gonçalves JFDC, Costa KCPD, Lopes JDS, Ferreira MJ, Lima RMBD, Nina Junior ADR (2023) Growth and functional leaf traits of coppice regrowth of Bertholletiaexcelsa during an El Niño event in the central Amazon. Acta Amazon 53:9–19. https://doi.org/10.1590/1809-4392202103653

Fraisl D, See L, Sturn T, MacFeely S, Bowser A, Campbell J, Fritz S (2022) Demonstrating the potential of Picture Pile as a citizen science tool for SDG monitoring. Environ Sci & Policy 128:81–93. https://doi.org/10.1016/j.jag.2023.1031_103149

Frigerio D, Richter A, Per A, Pruse B, Vohland K (2021) Citizen science in the natural sciences. In Vohland K, Land-Zandstra A, Ceccaroni L, Lemmens R, Perelló J, Ponti M, Wagenknecht

K (eds) The science of citizen science (p 529). Springer Nature. https://doi.org/10.1007/978-3-030-58278-4

Gastauer M, Massante JC, Ramos SJ, da Silva RDSS, Boanares D, Guedes RS, Ribeiro PG (2022) Revegetation on tropical steep slopes after mining and infrastructure projects: challenges and solutions. Sustainability 14(24):17003. https://doi.org/10.390/su142417003

Giudice DS (2022) Morro do Chapéu: Patrimônio Geológico, Geoconservação, Geodiversidade e Geoturismo. [Morro do Chapéu: GeologicalHeritage, Geoconservation, GeodiversityandGeotourism]. In:Geodiversidade: envolvências e experiências. (Org)Wandervilson Alves- Anapolis-GO- SAMA-Solo, Água e Meio Ambiente. ISBN 978–65–0054636–1

Gomes VMA (2022) A atuação estratégica do Exército Brasileiro na Amazônia Ocidental em prol do desenvolvimento sustentável e da integração nacional. https://bdex.eb.mil.br/jspui/bitstream/123456789/11875/1/MO%206700%20-%20V%C3%ADtor%20Moreira%20AGUIAR%20Gomes.pdf

Guillén Martín C, Mateo Mederos E (2023) Geoconservation and Responsible Use of the Territory: Experiences from the Lanzarote and Chinijo Islands UNESCO Global Geopark. Geoconservation Res 6(1):161–177. https://doi.org/10.30487/GCR.2023.1982041.1127

Hadi AG, Abbas AS, Al-Refai RA (2023) Overview of the rising temperatures and the global warming crisis in the world and Iraq. https://www.researchgate.net/profile/Angham-Hadi/publication/374274143_Overview_of_the_Rising_Temperatures_and_the_Global_Warming_Crisis_in_the_World_and_Iraq/links/6516c3af3ab6cb4ec6a92b03/Overview-of-the-Rising-Temperatures-and-the-Global-Warming-Crisis-in-the-World-and-Iraq.pdf

Haklay M, König A, Moustard F, Aspee N (2023) Citizen science and Post-Normal Science's extended peer community: identifying overlaps by mapping typologies. Futures 150:103178. https://doi.org/10.1016/j.futures.2023.103178

Hartz F (2023) Leaking the IPCC: a question of responsibility? Wiley Interdiscip Rev: Clim Chang 14(3):e814. https://doi.org/10.1002/wcc.814

IBGE (2023) Instituto Brasileiro de Geografia e Estatística. Cidades e Estados. [Citiesand States]. https://cidades.ibge.gov.br/

Ioannou I, Galán-Martín Á, Pérez-Ramírez J, Guillén-Gosálbez G (2023) Trade-offs between sustainable development goals in carbon capture and utilisation. Energy & Environ Sci 16(1):113–124. https://doi.org/10.1039/d2eco1153k

Leng Y, Lyu Y, He M, Zhao J (2023) Geoheritage sites of Yan'an City for Geotourism in China's middle loess plateau. Geoherit 15(2):73. https://doi.org/10.1007/s12371-023-00825-w

Lima MAA (2023) Emc2 Project 'Exploring White crowberry coastal habitats'. Training Guide for Teachers. https://www.researchgate.net/profile/Alexandra-Lima-3/publication/371179592_Emc2_Project_'Exploring_White_crowberry_coastal_habitats'_Training_Guide_for_Teachers_E-Book_link_httpswwwmare-centreptsitesmarecentrefilesarticlesfiles2205202320Training20Guide20EMC2pdf/links/647878b179a722376503b7cf/Emc2-Project-Exploring-White-crowberry-coastal-habitats-Training-Guide-for-Teachers-E-Book-link-https-wwwmare-centrept-sites-marecentre-files-articles-files-2205202320Training20Guide20EMC2pdf.pdf.

Luo X, Li X, Liu W, Volk M, Liu C (2023) Comprehensive evaluation of natural landscapes before and after earthquake based on GST-FAHP-GIS analytical method: a case study of Jiuzhaigou Nature Reserve, China. Ecol Indic 154:110850. https://doi.org/10.1016/j.ecolind.2023.110850

Machate O, Schmeller DS, Schulze T, Brack W (2023) Mountain lakes as freshwater resources at risk from chemical pollution. Environ Sci Eur 35(1):3. https://doi.org/10.1186/s12302-022-00710-3

Manono G, Thiong'o SM, Wishitemi BE (2023) Perceived effects of climate change on plants and its potential impact on tourism–a perspective of the local residents of Maasai Mara National Reserve, Kenya. https://www.researchgte.net/profile/Simon-Thiongo-2/publication/370365592_Perceived_effects_of_climate_change_on_plants_and_its_potencial_impact_on_impact_on_tourism_-a_perspective-of_the_local_residents_of_Maasai_Mara_National_Reserve_Kenya/links/644c760597449a0e1a645df8/Perceived-effects-of-climate-change-onplants-and-its-potential-impact_on_tourism-_a_perspective-of-the-local-residents-of-Maasai-Mara-National-Reserve-Kenya.pdf

Masson-Delmotte V, Zhai P, Pirani A, Connors SL, Péan C, Berger S, Zhou B (2021) Climate change 2021: the physical science basis. Contrib Work Group I Sixth Assess Rep Intergov Panel Clim Chang 2. https://doi.org/10.1017/97815786

McGowan AH (2023) Life in the anthropocene. Environ: Sci Policy Sustain Dev 65(2):6–14. https://doi.org/10.1080/00139157.2023.2167456

Mosgrove S (2023) Mobilizing assets for sustainable rural tourism development in UNESCO's Frontenac arch biosphere. https://qspace.library.queensu.ca/handle/1974/31816

O'Donnell KL, Brundage LA (2023) It Turned into a Bioblitz: urban data collection for building scientific literacy and environmental connection. In: Transforming education for sustainability: discourses on justice, inclusion, and authenticity (pp 183–204). Cham: Springer International Publishing. https://doi.org/10.1007/978-3-031-13536-1_11

Oksanen P (2023) Accessibility in geotourism: case: Salpausselkä UNESCO Global Geopark. https://urn.fi/urn:nbn:fi:amk-202305057951

De Oliveira JMF (2023) Crescimento de três espécies florestais no município de Mucajaí, RR. [Growth of three forest species in the municipality of Mucajaí, RR]. http://infoteca.cnptia.embrapa.br/infoteca/handle/doc/1152360

Pallathadka A (2023) Urban flooding in vancouver, Canada. Integr J Res Arts Humlties 3(2):63–69. https://doi.org/10.5554/ijrah.3.2.11

Palmer CG, Fry A, Libala N, Ralekhetla M, Mtati N, Weaver M, Scherman PA (2022) Engaging society and building participatory governance in a rural landscape restoration context. Anthropocene 37:100320. https://doi.org/10.1016/j.ancene.2022.100320

Pelacho M, Ruiz G, Sanz F, Tarancón A, Clemente-Gallardo J (2021) Analysis of the evolution and collaboration networks of citizen science scientific publications. Scientometrics 126:225–257. https://doi.org/10.1007s11192-020-03724-x

Prefeitura Municipal de Mucajaí.PMM. http://www.mucajai.rr.gov.br

Prefeitura Municipal de Mucajaí. PMM Environmental Registry (*Cadastro Ambiental* Rural—CAR), accesse date. http://www.car.gov.br

Purdy LM, Sang Z, Beaubien E, Hamann A (2023) Validating remotely sensed land surface phenology with leaf out records from a citizen science network. Int J Appl Earth Obs Geoinf 116:103148. https://doi.org/10.1016/j.jag.2022.103148

Qiao H, Zhang J (2023) Enhancing global thinking can reduce the misconception of accumulation: a potential way to mitigate climate change. Environ Sci Pollut Res 30(20):58618–58629. https://doi.org/10.1007/s11356-023-26510-7

Rabelo TO, Diniz MTM, de Araújo IGD, de Oliveira Terto ML, Queiroz LS, Araújo PVDN, Pereira P (2023) Risk of degradation and coastal flooding hazard on Geoheritage in protected areas of the semi-arid coast of Brazil. Water 15(14):2564. https://doi.org/10.3390/w15142564

Raihan A (2023) A review of the global climate change impacts, adaptation strategies, and mitigation options in the socio-economic and environmental sectors. J Environ Sci Econ 2(3):36–58. 105 6556/jescae.v2i3.587

RBCC, Rede Brasileira de Ciência Cidadã. [Brazilian Citizen Science Network]. https://www.rbcc.ong.br/

Reinhart H, Rafida MR, Sofie T, Putra RD, Matoka MCR, Maulita NS, Baiquni M (2023) Assessment of geological diversity, geosites, and geotourism potencies at Menoreh Mountain for designation of geopark area. Int J Geoheritage Park 11(3):385–406. https://doi.org/10.1016/j.ijgeop.2023.05.005

Roraima (2020) Plano de Desenvolvimento Sustentável do Estado de Roraima 2020–2030. Governodo Estado de Roraima. [Sustainable Development Plan for the State of Roraima 2020–2030]. https://portal.rr.gov.br/images/CATALOGO_2030.pdf

Rosado-González EM, Lourenço JMM, Palacio-Prieto JL, Sá AA (2023) Collaborative mapping on sustainable development goals in Latin America UNESCO global geoparks: a methodological discussion. Int J Geoheritage Park 11(2):203–220. https://doi.org/10.1016/j.igeop.2023.02.002

Sanabria-Z J, Alfaro-Ponce B, González Peña OI, Terashima-Marín H, Ortiz-Bayliss JC (2022) Engagement and social impact in tech-based citizen science initiatives for achieving the

SDGs: a systematic literature review with a perspective on complex thinking. Sustainability 14(17):10978. https://doi.org/10.3390/su141710978
Santos AFD, Moura FRT, Seruffo MCDR, Santos WPD, Costa GB, Costa FAR (2023) The impact of meteorological changes on the quality of life regarding thermal comfort in the Amazon region. Front Clim 5:1126042. https://doi.org/10.3389/fclim.2023.1126042
Santos N (2021) Vozes Indígenas Marcam a COP26 em Glasgow. [Indigenous Voices Mark COP26 in Glasgow]. https://informasus.ufscar.br/vozes-indigenas-marcam-a-cop26-em-glasgow/
Signori-Müller C, Galbraith D, Tavares JV, Reis SM, Diniz FC, Gilpin M, Oliveira RS (2023).Tropical forest lianas have greater non-structural carbohydrate concentrations in the stem xylem than trees. Tree Physiol tpad096. https://academic.oup.com/treephys/advance-article/doi/10.1093/treephys/tpad096/7243038bygueston20August2023
Da Silva GDFN, Júnior SST, de Oliveira Evangelista RA, de Oliveira Júnior MCM (2008) Integração Digital e Análise Espacial Aplicadas ao Estudo da Suscetibilidade a Incêndios Florestais em Apiaú e Ribeiro Campos–Roraima. Floresta 38(4). https://revistas.ufpr.br/article/viewFile/13165/8910
Smaniotto A, Passani A (2023) Citizen science with and within the social sciences and the humanities. Etica & Politica 25(2):9–34. http://www2.units.it/etica/2023_2SMANIOTTOPASSANI.pdf
De Souza Saldanha L, Lima RA (2021) Briófitas em quintais urbanos no sudoeste da Amazônia. [Bryophytes in Urban Backyards in the Southwest of the Amazon]. S Amn J Basic Educ, Tech Technol 8(2):26–39. https://periodicos.ufac.br/index.php/SAJEBTT/article/view/3526
Štrba Ľ, Vravcová A, Podoláková M, Varcholová L, Kršák B (2023) Linking Geoheritage or Geosite assessment results with Geotourism potential and development: a literature review. Sustainability 15(12):9539. https://doi.org/10.3390/su15129539
Tyagi A, Godbole M, Vanak AT, Ramakrishnan U (2023) Citizen science facilitates first ever genetic detection of wolf-dog hybridization in Indian savannahs. Ecol Evol 13(5):e10100. https://doi.org/10.1002/ece3.10100
UN (2018) Declaración de lasNaciones Unidas sobre losDerechos de Los Campesinos y de Otras Personas que Trabajamenlas Zonas Rurales. [United Nations Declaration on the Rights of Peasants and Other People Working in Rural Areas]. In file:///C:/Users/f3012428/Downloads/declaracion%20onu%20derechos%20campesinos%202018–1.pdf
Veras ASS, Vidal DG, Barros N, Dinis MAP (2020) Landscape sustainability: contribution of Mucajaí-RR (Brazil) Region. In: Leal Filho W, Azul AM, Brandli L, Özuyar PG, Wall T (eds) Enciclopedia of the UN sustainable development goals. Responsible Consumption and Production and Production (pp 1–7). Switzerland: Springer InternationalPublishing. Disponível em. https://doi.org/10.1007/978-3-319-71062-4_1
Veras ASS, Vidal DG, Barros NA, Pimenta Dinis MA (2021) The davi trail in Mucajaí, Roraima, Brazil: An experience to (re) connect and protect nature. Geo J 87(5):4061–4075. https://doi.org/10.100/s10708-021-10484-5
Veras ASS, Vidal DG, Barros NA, Dinis MAP (2022) Geotourism social constraints and protection instruments from a sustainability perspective: evidence from the northernmost Brazilian state. In: Handbook of sustainability science in the future: policies, technologies and education by 2050 (pp 1–21). Cham: Springer International Publishing. https://doi.org/10.10.1007/978-3-030 68074-9_133-1
Veras ASS, Vidal DG, Barros NA, et al (2022) Thedavi trail in Mucajaí, Roraima, Brazil: an experience to (re)connect and protect nature. Geo J 87:4061–4075. https://doi.org/10.1007/s10708-021-10484-5
Volk GM, Carver D, Irish BM, Marek L, Frances A, Greene S, Bretting PK (2023) Safeguarding plant genetic resources in the United States during global climate change. Crop Sci. https://doi.org/10.1002/csc2.211003

Weiner D, Bloomer J, Conchúir RÓ, Dalton C (2022) The role of volunteers and citizen scientists in addressing declining water quality in Irish river catchments. Citiz Sci: Theory Pract 7(1). https://doi.org/10.5334/cstp.447

Zhanabayev D, Dzhanaleeva K, Ramazanova N, Keukenov Y, Mendybayeva G, Makhanova N (2023) Morphological characteristics of East Kazakhstan as a factor of Geotourism development. Geo J Tour Geosites 46(1):174–183.https://doi.org/10.30892/gtg.46119-1013

**Ana Sibelonia Saldanha Veras** Ph.D. student in the Postgraduate Program in Ecology and Environmental Health at Fernando Pessoa University (UFP) Porto, PT. Master in Dynamics of the Amazon Landscape from the Federal University of Roraima (UFRR), bachelor in geography from the Institute of Geosciences (UFRR). She worked as a civil servant and head of the special projects division at SEPLAN-Department of Tourism and is currently requested by the Court of Justice of the State of Roraima, Brazil. She has published scientific texts on the subject of geotourism.

**Waqas Ahmed** is currently working as Assistant Professor at NUST Business School. Researcher studied many disciplines and fields namely Supply Chain Management, Operations Research and Optmizations, Sustainable inventory Management, Mathematical Modeling, Manufacturing System, Production System Analysis, Production Planning and Control, R & D Management. His Ph.D. in Industry in year 2019 in Industrial and management Engineering from Hanyang University, South Korea and presented for best Ph.D. Thesis Award. His dissertation is positioned in the context of next-generation energy and covers the topics of sustainable supply chain management under the umbrella of economic, environmental and social strategies.

**Nelson Azevedo Barros** has a Ph.D. in Environmental Sciences, MSc in Thermal Engineering and a degree (5 years) in Environmental Engineering. Has a long experience in air quality research including Indoor air quality (monitoring, numerical modelling, impact on human health and management). Is Associate Professor at UFP, coordinator of the Ph.D. program in Ecology and Environmental Health, coordinator of the degree in Quality Management, Environment and Safety, and member of the Research Unit I3ID.

**Maria Alzira Pimenta Dinis** Habilitation, Ph.D., MS.c. (https://www.cienciavitae.pt/4710-147D-FDAF) is Associate Professor at the Faculty of Science and Technology (http://fct.ufp.pt/en/), University Fernando Pessoa (http://www.ufp.pt/), Porto, Portugal. Graduated in Metallurgical Engineering (1992) and Master Science in Materials Engineering (1995) at the Faculty of Engineering (http://www.fe.up.pt), University of Porto (https://sigarra.up.pt/up/en/), Portugal. Ph.D. in Earth Sciences (2010) and Habilitation in Ecology and Environmental Health (2023) at the Faculty of Science and Technology.

# Reflections on Traditional Agriculture as a Socio-ecological System for the Management of Natural Resources

**Vázquez-Martin Ángel Eduardo, Aguilar-Rivera Noé, and Avalos-de la Cruz Dora Angelica**

**Abstract** Currently, different studies have shown that traditional agriculture represents a valuable source for the preservation and conservation of natural resources on the planet. Despite this, agrobiodiversity represents a biological resource with a high degree of ecological vulnerability. Therefore, possibly the study of the elements that make up the processes of agrobiodiversity management within traditional agriculture could establish the paradigms for its study in a sustainable way. The above will allow us to recognize how over time the biocultural knowledge of farmers has influenced the management of their resources and their processes. Therefore, we carried out a bibliometric analysis on the main studies that recognize agrobiodiversity management within traditional agriculture in the countries that are the center of origin of cultivated plant species. Our findings showed how the main processes within traditional agriculture intervene in its management—seed selection, pest control, storage, distribution, trade-. Furthermore, we identify how the modern management of agrobiodiversity is a complex socio-ecological system that affects different areas of study such as sociology, anthropology, ethnobiology, ethnobotany, bioeconomy, gastronomy, and tourism.

**Keywords** Biocultural knowledge · Ecological vulnerability · Biological resource

V.-M. Ángel Eduardo · A.de la Cruz Dora Angelica
Colegio de Postgraduados, Campus Córdoba, Veracruz, México
e-mail: vazquez.angel@colpos.mx

A.de la Cruz Dora Angelica
e-mail: davalos@colpos.mx

A.-R. Noé (✉)
Facultad de Ciencias Biológicas y Agropecuarias, Universidad Veracruzana, Veracruz, México
e-mail: naguilar@uv.mx



W. Leal Filho et al. (eds.), *Composing Worlds: Humanities, Health and Wellbeing in the XXI Century Towards a More Sustainable World*, World Sustainability Series,
https://doi.org/10.1007/978-3-031-87108-5_16

## 1 Introduction

Various scientific studies agree with the high ecological vulnerability of global agrobiodiversity (Tamariz and Baumann 2021; Matthies et al. 2023; Porcuna-Ferrer et al. 2023; Chiffoleau et al. 2024). Historically, around the world subsistence, and expansion of agrobiodiversity has been ensured by farmers over the last 4,000 to 7,000 years (Fuller et al. 2011; Smith and Zeder 2013). Furthermore, humans since the industrial period—1800´s- have drastically transformed the biodiversity of agriculture and food systems (Foley et al. 2013). Therefore, the global agricultural landscape has witnessed significant transformations since the advent of the Anthropocene era (Tsing et al. 2019). Currently, the international intensive—conventional—agricultural model with the greatest economic and ecological impact is based on the principles of the green revolution. This productive model maximizes the use of mechanical, and agrochemical inputs—regular, and deep plows, mineral fertilizers, and pesticides-(Altieri 2009; Gleissman 2015). Consequently, agricultural landscapes have been drastically deteriorated due to anthropogenic impact. Nowadays, the agricultural landscapes cover 40% of the earth's surface and include a diversity of crops from conventional and traditional agriculture. Therefore, the impact on these landscapes' ranges from the loss of semi-natural covers, the reduction of the diversity of cultigens, and the increase in the size of agricultural fields (Robinson and Sutherland 2002; Benton et al. 2003). The anthropogenic effect on agriculture—homogenization and intensification of land use—have caused severe environmental problems and the loss of local biodiversity in agroecosystems (Tilman et al. 2002). Another problem facing modern agriculture is the gradual increase in temperature, which has increased the vulnerability of the functioning of agroecosystems (Martin et al. 2019). Those could be affected by the increase in pest outbreaks, and adverse climatic phenomena. Accordingly, the resilience of agroecosystems to such phenomena will depend on their capacity for adaptability over the years. For example, in a study by Isbell et al. showed that the grassland plant diversity increases the resilience of agricultural productivity to extreme climate (Isbell et al. 2015).

The power of global agricultural corporations within the agri-food sector has increased their influence over global markets, and their political lobby (Clapp and Purugganan 2020). As a resulted, the autonomy of small producers has currently been drastically reduced (Hendrickson et al. 2020; Burchfield et al. 2022). Despite this, the worldwide public policy in the international agricultural sector has promoted socioeconomic inequality, and injustice (Graddy-Lovelace 2016; Fagundes et al. 2019). In this context, possibly the promotion of alternative, and diversified systems within the international agri-food system could become a geopolitical strategy in the face of socio-political, and economic inequality in the agricultural sector and climate change (Spangler et al. 2022). Historically, traditional system has demonstrated their potential to provide various solutions to the modern anthropic impact of agriculture through their scientific study (Toledo et al. 2003; Altieri 2004). Because these systems produce diverse ecosystem services—such as biodiversity conservation, and carbon sequestration-(Donald and Evans 2006). In addition, they provide cultural services

**Fig. 1** Small producers around the world provide agrobiodiversity within traditional agriculture. Photographs taken in Orizaba, Mexico

that contribute to the well-being and health of people (Fig. 1) (van Berkel and Verburg 2014; Gołębiewska and Pajewski 2018). As a systematic characteristic, traditional agriculture promotes a reduction in agricultural intensity combined with greater heterogeneity of the crop mosaic (Duflot et al. 2022; Oliveira et al. 2023). Moreover, the multidisciplinary study of agrobiodiversity could support research on food security, the improvement of cultivated lands, water scarcity, agrobiodiversity, the reduction of the use of pesticides, which could indirectly benefit the poor population of the rural areas of the planet (Pretty et al. 2003).

Historically, the traditional agriculture is a socio-ecological system that has demonstrated its resilience to the main modern anthropogenic impacts—loss of biodiversity, climate change, overexploitation of natural resources. Today, multidisciplinary scientific interventions are required to safeguard the agrobiodiversity, combat poverty, food insecurity in agricultural communities around the world. Therefore, scientific interventions are necessary that provide in-depth information about the ethnobiological knowledge of small farmers. Therefore, a frame study about the traditional agriculture management is required. Consequently, we have considered answering the question about how through ethnobiological knowledge there is sustained management of natural resources. Firstly, we identified the implications of ethnobiological knowledge within traditional agriculture. Secondly, we analyzed the elements that make up ethnobiological knowledge. Finally documenting the main characteristics of traditional agriculture in the main countries, centers of plant domestication on the planet suggested by the Russian botanist Nicolai Vavilov (Harlan 1971). Finally, our findings suggested that ethnobiological knowledge within traditional agriculture could be studied as a socio-ecological model for sustainable biodiversity management in a multidisciplinary manner.

## 2 Implications of Ethnobiological Knowledge Within Traditional Agriculture

Currently, traditional agriculture provides at least a third of the food required by the global agri-food system (Lowder et al. 2021). This system provides food to approximately 475 million households (Rapsomanikis 2015). However, the livelihoods, food security, and ethnobiological knowledge of small-scale farmers are threatened by the acceleration of socio-ecological changes on a global scale-biodiversity loss, climate change, resource overexploitation-(Kodirekkala 2017; Khoury et al. 2022; Antonelli 2023). This situation has caused small producers to be the most vulnerable population to these socio-ecological changes (IPCC 2022). Consequently, the ability of this population to adapt to the transformation of its environment will be crucial for its survival (Caviedes et al. 2023). For example, FAO have estimated that by 2030 more than 120 million jobs will be lost in food production, which will directly impact traditional agriculture (FAO et al. 2023).

Around the planet, recent socio-ecological changes are shaping new ways of adapting to livelihoods in agricultural communities, which directly influences their daily decisions (Adger et al. 2013; Reyes-García et al. 2024). In this context, local communities, and indigenous peoples are responsible for practicing traditional agriculture. In relation to the importance of the population of indigenous peoples, in a study by Redvers and Blondin, estimated that 6% of the world population are indigenous peoples, who are distributed in 90 countries (Redvers and Blondin 2020). This population is diversified in 5,000 different cultures. In a study by Wendimu et al. determined that in terms of economic resources and social services these indigenous communities constitute 15% of the severely impoverished population (Wendimu et al. 2021). In addition, they are the ones who historically possess a system of complex knowledge about their interaction with natural resources—ethnobiological knowledge-(Caviedes et al. 2024). The traditional farmers ethnobiological knowledge as the basis of socio-ecological systems encourages reciprocal contributions at different scales (Payne et al. 2020). Today, natural resource management through the study of local and indigenous knowledge could provide alternative approaches, and viewpoints to Western science and conventional agriculture (Melash et al. 2023). However, the tangible and intangible resources that are used within traditional agriculture are threatened. The main cause of risk is the economic marginalization of poor farmers, and the competitive disadvantage of traditional crops compared to conventional agriculture.

## 3 Ethnobiological Knowledge Elements

Studies has suggested that the main cause of the overexploitation of natural resources is the disconnection that exists between society and the environment (Van Vliet et al. 2012; Dorninger et al. 2017). Historically, the nature and human connection has been close, and multifaceted with cognitive connections, material connections, and experiential connections (Ives et al. 2018). Today, this relationship continues to evolve spatially and temporally. This interconnection is based on the interrelation between multiple material and immaterial factors-climate, geographical location, natural resources, worldview, culture-(Fonte 2008; Horlings et al. 2016). Studies have identified that certain factors influence human behavior around the use of biological resources—politics, culture, geographic location, and tourism-(Sun et al. 2013; Yuan et al. 2014a, b; Albuquerque et al. 2015; Zhang et al. 2016; Ma et al. 2017). For example, traditional culture in each place on the planet influences the behavior surrounding the use of natural resources.

There are different meanings of the term related to ancient knowledge although. Traditional knowledge, indigenous knowledge, traditional ecological knowledge, ethnobiological knowledge—ontologically we have documented the concepts that best describe it. Therefore, we have used the epistemology approach provided by ethnobiology. Currently, this science studies the relationship between human beings, their biome, and the ecosystem. As a science, its study focuses on the study of species through the documentation of cultural uses and knowledge related to flora, fauna, and fungi (Ellen 2018). As a transdisciplinary discipline, it is supported by ethnomycology, ethnobotany, ethnozoology, and ethnoecology. Ives et al. provided a modern approach to the study of ethnobiological knowledge, which is based on five categories of connections (a) experiential, (b) material, (c) cognitive, (4) emotional, and (5) philosophical connections (Fig. 2). Furthermore, Ives et al. considered these interconnections operate along a spectrum ranging from internal connections—for example, worldviews, emotions—to external connections with nature—for example, material interaction, physical appropriation-(Ives et al. 2018). This epistemological process of knowledge construction promotes regional self-sufficiency and strengthens the local ties of the population (Ives et al. 2018). Traditional knowledge study in agriculture has proven extremely important as a strategy for sustainable development to combat rural poverty and conserve natural resources (Jiao et al. 2024). Therefore, the traditional agricultural knowledge has been built on socio-ecological processes throughout human history (Brush 2005). Our review objective based on this background was to identify the current implications of traditional knowledge in six countries, centers of origin of domesticated plants. Vavilov identified 8 agricultural regions—Southern Mexico and Central America, Peru and Bolivia, China, India, Ethiopia, Central Asia, Near East, Mediterranean—which are centers of origin of cultivated plants-(Hawkes et al. 2000). Now, we showed a literature summary with the state of the traditional knowledge in Mexico, Peru, Ethiopia, Turkey, India, and China.

## *3.1 Mexican Traditional Agriculture*

The traditional agriculture characterizes Mexican agricultural systems, as the case of many emerging economies in Latin America (Nicholson et al. 2019). Subsistence farming is a traditional form of small-scale agriculture in Mexico. Its main

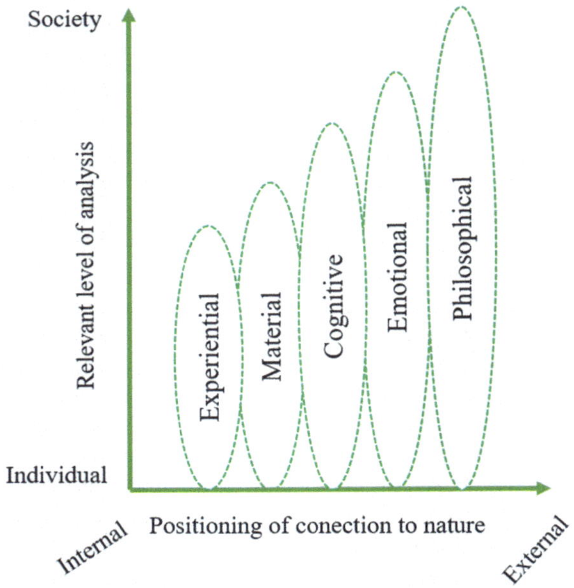


**Fig. 2** Conceptualization of different types of human-nature connections. Ives et al. presented as independent categories are interacting along a spectrum from people´s inner to outer worlds

characteristic is the producer uses all or a significant proportion of his production for home consumption. In this country the Subsistence agriculture based on corn "*Zea Mays*" has predominated. This system is called "milpa", which is characterized by the possession of small plots (<5 ha), with self-consumption of crops, dry farming practices, communal or ejidal lands -a legal mixed form between private and communal-, and high crop diversity (Zizumno and Colunga 2010). Eakin et al. estimated the productivity of corn from traditional agriculture of small producers has increased significantly in the last three decades. In addition, in this study determined the milpa system contributes more than half of the country's total corn production (Eakin et al. 2018). Furthermore, Ibarrola-Rivas and Granados-Ramírez (2017) estimated by 2050, it will be possible to feed the entire Mexican population with a basic and healthy diet, with the equivalent total surface area used by traditional agriculture for rainfed crops. Around the world, studies have shown the traditional agriculture is resilient despite the low income of rural households (Pelletier et al. 2016; Sibhatu and Qaim 2017). Traditional agriculture has a positive impact on rural food security because it provides direct access to affordable food from agrobiodiversity (Altieri et al. 2012; Bermeo and Couturier 2017).

### 3.2 *Peruvian Traditional Agriculture*

The ancient Peruvian terrace farming practices have survived to the present day. These practices for millennia have sustained populations during extreme climate change and political regime shifts (Williams et al 2005). In this country, the agriculture is characterized by food production in valleys and slopes at elevations between 3300 and 4200 m. Primarily, this system is organized to produce native tubers including at least seven species of "potatoes" -*Solanum ajanhuiri, S. curtilobum, S. phureja, S. tuberosum, S. chaucha, S. juzepzuckii, S. stenotomum*-(Huamán and Spooner 2002), olluco -*Ullucus tuberosus*- with 50 to 70 varieties, "oca" -*Oxalistuberosa*- with at least 50 technically described varieties, and "mashua" -*Tropaeolum tuberosum*- with nearly 100 botanically described varieties (Brack 2006). The terrace system characterizes one of the most difficult agroecosystems for the practice of plant cultivation in the world. This is due to the frosts, elevations, mountain slopes and soil types, which characterize the high Andean environments (Zimmerer 2010; Velazquez et al. 2011). Currently, agricultural terraces are re-emerging in Peru southern, as a system for the revitalization of hydraulic systems (Londono et al. 2017). In the current national food offering, there is a diversity of regional foods that could be explored by the food industry. As is the case of Peruvian agricultural foods such as quinoa -*Chenopodium quinoa*-, maca -*Lepidium meyenii*-, and purple corn -*Zea Mays*-, which are recognized as superfoods (Salvador et al. 2023).

## *3.3 Turkish Traditional Agriculture*

Turkey is located between two of the regions called by Vavilov as centers of origin of cultivated plants—the Mediterranean and the Middle East-(Harlan 1995). Studies have suggested in this country emerged the progenitor species of the barley -*Hordeum vulgare*-, oats -*Avena sativa,* and *Avena byzantina*-, chickpea -*Cicer arietinum*-, lentil -*Lens culinaris*-, apple (*Malus domestica*), and wheat (*Triticum spp*) (Bennett et al. 1998). Today, the cultivation of traditional wheat varieties in large Turkish agricultural areas is of crucial importance for the national food sector (Aksoy and Öz 2020). In a study by Tolunay et al. indicate the main prevailing Turkish traditional agricultural systems are family gardens and agroforestry systems. These authors provide the following classification based on their findings, (a) agroforestry systems: combination of trees and crops, (b) silvopastoral systems: combinations of trees, grasses and/or animals, (c) agrosilvopastoral systems: combinations of grasses, crops, and/or animals and trees (Tolunay et al. 2007). Traditional Turkish agricultural practices are a crucial socioeconomic component for the subsistence of the population in rural areas of the country (Oztürk et al. 2018). Studies recognized that in large economically marginalized regions the work of small producers is vital for the conservation of agrobiodiversity (Bardsley and Thomas 2005).

### 3.4 Ethiopian Traditional Agriculture

The traditional Ethiopian agriculture dominates the national production system, with limited use of modern inputs (Baye 2017). In Ethiopia, this agricultural system is strongly oriented to the production of cereals as a source of livelihood and ways of life for large sectors of its population (Hailu et al. 2015). In this country 80 ethnic groups reside with its own agricultural practices (Mekonnen et al. 2021). Therefore, small producers have the capacity to produce their own food and in certain parts of the country the barter system persists as a form of economic transaction (Tareke and Beyene 2019). Today, some ancient agricultural practices have remained such as crop rotation, terracing, manual irrigation, use of the plow, abandonment of fallow land (Baye 2017). These ancient practices and methods have varied from location to location due to differences in soil, settlement topography, and climate. For example, the free grazing on communal pastures is mostly traditionally practiced by small producers (Yisehak et al. 2013). Additionally, the agroforestry is also widely practiced and includes various components, such as tubers, vegetables, fruits, and trees grown, and managed to provide shade for coffee (Zerihun et al. 2011). In a study by Mossisa et al. demonstrated the agricultural acculturation processes allow the creation of knowledge for weed control, storage methods for agricultural inputs, selection of land for crops, and improvement in soil fertility (Mosissa et al. 2017).

### 3.5 Indian Traditional Agriculture

In India, some varieties of native tropical fruits such as melon -*Cucumis melo*-, and mango -*Mangifera indica*- have been recognized (Gulati 2006). In the last 9000 years ago, in this country have been cultivated species such as barley -*Hordeum vulgare*-, jujube -*Ziziphus jujuba*- and wheat -*Triticum* spp.-(Dev 2006). There is archaeological evidence about the rice -*Oryza sativa*- domestication by the Indus Valley population (Dev 2006). Around 4500 BC the irrigation system was implemented by the Indian civilization. The more sophisticated settlements were created using sewage and drainage (Rodda and Ubertini 2004).

Approximately 86% of Indian agricultural households are characterized by economic poverty. The average property size of these households is less than two hectares (Shah 2018). In these communities, agricultural practices are considered traditional and are part of ancestral knowledge. Examples of these traditional practices in traditional agriculture are, Darjeeling system in the Himalayas, rice-fish farming in Apatani, Zabo system, Seethampheta in Andhra Pradesh (Khumbongmayum et al. 2004, 2005). In northwest India, the use of bamboos, and stones is relevant, which are meticulously used to build various structures for irrigating crops, storing water, laying pipes, and controlling soil erosion (Kumar and Madhukar 2019). Therefore, the Indian traditional knowledge could be a sustainable solution to the management of natural resources (Tynsong et al. 2020). However, the Indian agricultural population is vulnerable to the economic, and climatological impacts that affect the planet (Joshi 2015; Bryan et al. 2017).

### *3.6 Chinese Traditional Agriculture*

The traditional Chinese knowledge within the traditional agriculture is ancient (Li et al. 2004). In a Chinese book written about 1,500 years ago titled "Important Technology of the Qi Dynasty", the author classified common food ingredients into more than 200 types of cereals, 100 types of vegetables and 100 types of animal meat products (Jie 2001). Current, Chinese smallholder farmers grow different types of grains depending on climatic, and geographical conditions. The main cereals grown are corn -*Zea Mays*-, rice -*Oryza sativa*-, wheat -*Triticum* spp.-, and there are some minor cereals such as hulled oat -*Avena sativa*-, adlay -*Coix lacryma-jobi*-, and buckwhea -*Fagopyrum esculentum*-(Li 1996). The farmers have preserved their traditional food systems within agriculture, such as rice-fish farming system, rice terrace system, traditional rice farming system, and Shexian dryland stone terrace system (Yuan et al. 2014a, b; Jiao et al. 2021). In recent decades, generational change has given evidence of the loss of traditional knowledge. For example, in a study within the population of the Mengku community of Yunnan province, China, the traditional knowledge about edible wild plant resources has disappeared with their demise from the elderly population (Ma et al. 2020).

## 4 Traditional Agriculture Characteristics as an Ecological Social System

As we have documented in this review, the agricultural landscape is closely related to traditional agriculture in the countries centers of origin of domestication of cultivated plants. This agricultural system around the world is characterized by an adaptive management process, in which the function and structure of the agricultural landscape transformation have strong human-environmental interactions at a spatiotemporal scale (Peng et al. 2015). This adaptive management in social ecological systems is based on understanding the changing interactions between humans and nature (Birgé et al. 2016). Therefore, the adaptive management is relevant for synergistic functions in traditional agriculture (Liu and Li 2017). Consequently, the Socioecological interactions are the result of people working toward collective and individual goals (Hanna and Jentoft 1996). The unpredictability, and uncertainty within human socio-ecological interactions are studied as individual, and collective behaviors (Berkes and Folke 2000). Small producers have obtained through this flow of information deep knowledge about the dynamics, distribution, behavioral changes, and intricate interactions between biotic, and abiotic resources (Lyver et al. 2018). Therefore, this process is essential to develop systemic knowledge within traditional agriculture. In consequence, the traditional knowledge can be complemented by science with multidisciplinary approaches which allow the co-evolutionary understanding of nature, and humanity (Molnar and Babai 2021). As we have documented around the world, there are agricultural practices based on traditional knowledge such as pest control, polyculture management, small animal husbandry, agroforestry, multiple cropping patterns, soil fertilization and tillage practices, uses of crop species. Wild plants seed/variety selection, storage, preservation, and cooking of food. Despite the negative effects of externalities around traditional agriculture -politics, economics, climate change- the promotion of conditions to sustain these livelihoods is a priority for international agricultural public policy (Pelletier et al. 2016; Antonelli 2023). Through sustainability science it is possible to study the problems that arise from the interaction between social, ecological, and sociopolitical systems (Fischer et al. 2015).

## 5 Conclusions

Our review presents a modern diagnosis of the current implications of the state of ethnobiological knowledge in traditional agriculture. This could serve as a conceptual framework for the design of strategies for the protection of agrobiodiversity through the protection of ancestral ethnobiological knowledge. Today, the different traditional agricultural systems are part of the world's agricultural heritage, which support the food security, and family economy of millions of people. These systems provide important ecosystem services, without which climatological alterations would increase. We have documented how this knowledge system is resilient and adaptable to the historical, political, economic, ecological, and climatological conditions of each population, and territory. Furthermore, our recommendations suggest the current state of protection of traditional knowledge and agrobiodiversity is fragile. Due to the countless externalities that affect, and limit human actions, the management of agrobiodiversity is a complex system to study. Therefore, it is necessary to promote public policy instruments to promote a solid system of traditional agricultural knowledge. International public policy could include other mechanisms for the inclusion of small farmers who depend on traditional agriculture. International initiatives such as FAO's Globally Important Agricultural Heritage Systems are not enough. The study of agrobiodiversity management as a sociological system, we have documented how sustainability science provides a holistic and multidisciplinary approach. This approach would allow scientific, and academic intervention, mainly from disciplines such as ecology, botany, zoology, mycology, anthropology, ethnobiology, bioeconomic, sociology, political sciences, agricultural sciences, gastronomy, and tourism.

## References


Adger WN, Barnett J, Brown K, Marshall N, O'Brien K (2013) Cultural dimensions of climate change impacts and adaptation. Nat Clim Chang 3:112–117

Aksoy Z, Öz Ö (2020) Protection of traditional agricultural knowledge and rethinking agricultural research from farmers' perspective: a case from Turkey. J Rural Stud 80

Albuquerque U, Ramos M, Lucena R (2015) What drives the knowledge and local uses of timber resources in human-altered landscapes in the semiarid region of northeast Brazil? Int J Sustain Dev World Ecol 22(6):545–559

Altieri MA (2009) Agroecology, small farms, and food sovereignty. Mon Rev 61(3):102–113

Altieri MA, Funes-Monzote FR, Petersen P (2012) Agroecologically efficient agricultural systems for smallholder farmers: contributions to food sovereignty. Agron Sustain Dev

Altieri MA (2004) Globally Important Ingenious Agricultural Heritage Systems (GIAHS): extent, significance, and implications for development. In: GIAHS 2nd international workshop. United Nations Food and Agriculture Organisation

Antonelli A (2023) Indigenous knowledge is key to sustainable food systems. Nature 613:239–242

Bardsley D, Thomas I (2005) Valuing local wheat landraces for agrobiodiversity conservation in Northeast Turkey. Agr Ecosyst Environ 106(4):407–412

Baye TG (2017) Poverty, peasantry and agriculture in Ethiopia. Ann Agrar Sci 15(3):420–430

Bennett SJ, Maxted N, Sabanci CO (1998) The ecogeography and collection of grain, forage and pasture legumes in south-west Turkey. Genet Resour Crop Evol 45:253–262

Benton T, Vickery J, Wilson J (2003) Farmland biodiversity: is habitat heterogeneity the key? Trends Ecol Evol 18:182–188

Berkes F, Folke C (2000) Linking social and ecological systems for resilience and sustainability. In: linking social and ecological systems: management practices and social mechanisms for building resilience. Cambridge University Press, Cambridge

Birgé HE, Allen CR, Garmestani AS, Pope KL (2016) Adaptive management for ecosystem services. J Environ Manage 183:343–352

Brack A, Bravo F (2006) Peru-millenary legacy lima: Universidad San Martín de Porres

Brush SB (2005) Protecting traditional agricultural knowledge. Wash Univ J Law Policy 17:59

Bryan E, Theis S, Choufani J, De Pinto A, Meinzen-Dick RS, Ringler C (2017) Conceptual framework: gender, climate change, and nutrition integration initiative. GCAN Policy Note 7. IFPRI, Washington, D.C. http://ebrary.ifpri.org/cdm/singleitem/collection/p15738coll2/id/132300

Burchfield EK, Schumacher BL, Spangler K, Rissing A (2022) The state of US farm operator livelihoods. Front Sustain Food Syst 5:795901

Caviedes J, Ibarra JT, Calvet-Mir L, Junqueira AB (2023) "Listen to us": small-scale farmers' understandings of social-ecological changes and their drivers in important agricultural heritage systems. Reg Environ Change 23(4):158

Caviedes J, Ibarra JT, Calvet-Mir L, Álvarez-Fernández S, Junqueira AB (2024) Indigenous and local knowledge on social-ecological changes is positively associated with livelihood resilience in a globally important agricultural heritage system. Agric Syst 216:103885

Chiffoleau Y, Dourian T, Enderli G, Mattioni D, Akermann G, Loconto A et al (2024) Reversing the trend of agrobiodiversity decline by co-developing food chains with consumers: a European survey for change. Sustain Prod Consum 46:343–354

Clapp J, Purugganan J (2020) Contextualizing corporate control in the agrifood and extractive sectors. Globalizations 17(7):1265–1275

Dev SM (2006) Agricultural wages and labor since 1950. Encyclopedia of India 1:17–20

Donald P, Evans A (2006) Habitat connectivity and matrix restoration: the wider implications of agri-environment schemes. J Appl Ecol 43:209–218

Dorninger C, Abson D, Fischer J, von Wehrden H (2017) Assessing sustainable biophysical human-nature connectedness at regional scales. Environ Res Lett 12(5):055001

Duflot R, San-Cristobal M, Andrieu E, Choisis JP, Esquerré D, Ladet S, Vialatte A (2022) Farming intensity indirectly reduces crop yield through negative effects on agrobiodiversity and key ecological functions. Agric Ecosyst Environ 326:107810
Eakin H, Sweeney S, Lerner AM, Appendini K, Perales H, Steigerwald DG, Dewes CF, Davenport F et al (2018) Agricultural change and resilience: agricultural policy, climate trends and market integration in the Mexican maize system. Anthropocene 23:43–52
Ellen R (2018) Ethnobiology, pp 1–4. https://doi.org/10.1002/9781118924396.wbiea1706
Fagundes C, Picciano L,Tillman W, Mleczko J, Schwier S, Graddy-Lovelace G et al (2019) Ecological costs of discrimination: racism, red cedar and resilience in farm bill conservation policy in Oklahoma. Renew Agric Food Syst 35:1–15
FAO, IFAD, UNICEF, WFP, WHO (2023) The state of food security and nutrition in the world 2023, Rome
Fischer J, Gardner TA, Bennett EM et al (2015) Advancing sustainability through mainstreaming a social–ecological systems perspective. Curr Opin Environ Sustain 14:144–149
Foley SF, Gronenborn D, Andreae MO, Kadereit JW, Esper J, Scholz D, Pöschl U, Jacob DE, Schöne BR, Schreg R, Vött A (2013) The Palaeoanthropocene-The beginnings of anthropogenic environmental change. Anthropocene 3:83–88
Fonte M (2008) Knowledge, food and place. A way of producing, a way of knowing. Sociol Rural 48:200–222
Fuller DQ, Van Etten J, Manning K, Castillo C, Kingwell-Banham E, Weisskopf A, Qin L, Sato YI, Hijmans RJ (2011) The contribution of rice agriculture and livestock pastoralism to prehistoric methane levels: an archaeological assessment. Holocene 21:743–759
Gleissman S (2015) Agroecology: the ecology of sustainable food systems, 3rd edn. CRC Press, Boca Raton, FL, 809
Golebiewska B, Pajewski T (2018) Positive and negative externalities of management illustrated by the case of agricultural production. J Agribus Rural Dev 48:113–120
Graddy-Lovelace G (2016) The coloniality of US agricultural policy: articulating agrarian (in)justice. J Peasant Stud 44:1–22
Gulati A (2006) Agricultural growth and diversification since 1991. Encyclopedia of India 1:14–17
Hailu B, Maeda E, Heiskanen J, Pellikka P (2015) Reconstructing pre-agricultural expansion vegetation cover of Ethiopia. Appl Geogr 65:357–365
Hanna S, Jentoft S (1996) Human use of the natural environment: an overview of social and economic dimensions. In: Hanna S, Folke C, Maler K-G (eds) Rights to nature. Ecological, economic, cultural, and political principles of institutions for the environment. Island Press, Washington D.C., pp 35–55
Harlan J (1971) Agricultural origins: centers and noncenters. Science, New York, N.Y., 174:468–74
Harlan JR (1995) The living fields: our agricultural heritage. Cambridge University Press
Hawkes JG, Maxted N, Ford-Lloyd BV (2000) The ex situ conservation of plant genetic resources. https://doi.org/10.1007/978-94-bbb011-4136-9
Hendrickson M, Howard P, Miller E, Constance D (2020) The food system: concentration and its impacts a special report to the family farm action alliance 26
Horlings L (2016) Connecting people to place: Sustainable place-shaping practices as transformative power. Curr Opin Environ Sustain 20:32–40
Huamán Z, Spooner D (2002) Reclassification of landrace populations of cultivated potatoes (Solanum sect. Petota). Am J Bot 89:947–965
Ibarrola-Rivas MJ, Granados-Ramírez R (2017) Diversity of Mexican diets and agricultural systems and their impact on the land requirements for food. Land Use Policy 66:235–240
IPCC (2022) Climate change 2022: impacts, adaptation, and vulnerability. In: contribution of working group II to the sixth assessment report of the intergovernmental panel on climate change, Cambridge
Isbell F, Craven D, Connolly J, Loreau M, Schmid B, Beierkuhnlein C, Eisenhauer N (2015) Biodiversity increases the resistance of ecosystem productivity to climate extremes. Nature 526(7574):574–577

Ives CD, Abson DJ, Von Wehrden H, Dorninger C, Klaniecki K, Fischer J (2018) Reconnecting with nature for sustainability. Sustain Sci 13:1389–1397

Jiao W, Cui W, Min Q, Zhang Y (2021) A review of research on agricultural heritage systems and their conservation. Resour Sci 43(4):823–837

Jiao W, Yang X, Li Y (2024) Traditional knowledge's impact on soil and water conservation in mountain agricultural systems: a case study of Shexian dryland stone terraced system, China. Ecol Indic 159:111742

Jie SX (2001) Qi Min Yao Shu (important tecnology of Qui dynasty). Juangshu Ancient Book Press, Nangjing, p 284

Joshi PK (2015) Has Indian agriculture become crowded and risky? Status, implications and the way forward. Indian J Agric Econ 70(1)

Khoury CK, Brush S, Costich DE, Curry HA, Haan S, Engels JMM, Guarino L et al (2022) Crop genetic erosion: understanding and responding to loss of crop diversity. New Phytol 233:84–118

Khumbongmayum AD, Khan ML, Tripathi RS (2004) Sacred groves of Manipur—Ideal centres for biodiversity conservation. Curr Sci 87(4):430–433

Khumbongmayum AD, Khan ML, Tripathi RS (2005) Sacred groves of Manipur, northeast India: biodiversity value, status and strategies for their conservation. Biodivers Conserv 14:1541–1582

Kodirekkala KR (2017) Internal and external factors affecting loss of traditional knowledge: evidence from a horticultural society in South India. J Anthropol Res 73(1):22–42

Kumar A, Madhukar AK (2019) Management of traditional water system and their conservation in North Eastern region through local traditional wisdom. AIP Conf Proc 2142:210001, AIP Publishing

Li SJ (1996) Chinese food culture. China Social Science Press, Beijing, p 197

Li LT, Yin LJ, Saito M (2004) Function of traditional foods and food culture in China. Jpn Agric Res Q JARQ 38(4):213–220

Liu Y, Li Y (2017) Revitalize the world's countryside. Nature 548(7667):275–277

Londono A, Williams P, Hart M (2017) A change in landscape: lessons learned from abandonment of ancient Wari agricultural terraces in Southern Peru. J Environ Manag 202:532–542

Lowder SK, Sanchez MV, Bertini R (2021) Which farms feed the world and has farmland become more concentrated? World Dev 142:105455

Lyver PO, Richardson SJ, Gormley AM, Timoti P, Jones CJ, Tahi BL (2018) Complementarity of indigenous and western scientific approaches for monitoring forest state. Ecol Appl 28(7):1909–1923

Ma N, Min QW, Yuan Z, Li WH, Yang QC (2017) The wild edible plants of four main ethnic groups in the Lahu-Va-Blang-Dai autonomous county of Shuangjiang in Yunnan Province. Resour Sci 39:1406–1416

Ma N, He S, Min Q (2020) Edible biological resource use in an agricultural heritage system and its driving forces: a case of the shuangjiang mengku ancient tea and culture system. Sustainability 12(18):7791

Martin EA, Feit B, Requier F, Friberg H, Jonsson M (2019) Assessing the resilience of biodiversity-driven functions in agroecosystems under environmental change. Adv Ecol Res 60:59–123

Matthies AE, Fayet CM, O'Connor LM, Verburg PH (2023) Mapping agrobiodiversity in Europe: different indicators, different priority areas. Ecol Ind 154:110744

Mekonnen Z, Kidemu M, Abebe H, Semere et al (2021) Traditional knowledge and institutions for sustainable climate change adaptation in Ethiopia 3:100080

Melash AA, Bogale AA, Migbaru AT, Chakilu GG, Percze A et al (2023) Indigenous agricultural knowledge: a neglected human based resource for sustainable crop protection and production. Heliyon 9:e12978

Molnar Z, Babai D (2021) Inviting ecologists to delve deeper into traditional ecological knowledge. Trends Ecol Evol 36(8):679–690

Mosissa R, Jimma W, Bekele R (2017) Knowledge management strategy for indigenous knowledge on land Use and agricultural development in western Ethiopia. Univers J Agric Res 5(1):18–26

Nicholson CF, Stephens EC, Jones AD, Kopainsky B, Parsons D, Garrett J (2019) Setting priorities to address the research gaps between agricultural systems analysis and food security outcomes in low- and middle-income countries. In: CCAFS working paper no. 255. CGIAR research program on climate change, agriculture and food security (CCAFS), Wageningen, the Netherlands. https://www.ccafs.cgiar.org

Oliveira W, Cruz-Neto O, Tabarelli M, Galetto L, Peres C, Lopes A (2023) Markedly declining reproductive functional diversity of food plants in the world's largest tropical country despite rapid cropland expansion. Agr Ecosyst Environ 357:108673

Oztürk M, Jongerden J, Hilton A (2018) The re(production) of the new peasantry in Turkey. J Rural Stud 61:244–254

Payne D, Snethlage M, Geschke J, Spehn EM, Fischer M (2020) Nature and people in the Andes, East African mountains, European alps, and Hindu Kush himalaya: current research and future directions. Mt Res Dev 40(2):A1

Pelletier B, Hickey GM, Bothi KL, Mude A (2016) Linking rural livelihood resilience and food security: an international challenge. Food Secur 8:469–476

Peng J, Liu Y, Wu J, Lv H, Hu X (2015) Linking ecosystem services and landscape patterns to assess urban ecosystem health: a case study in Shenzhen City, China. Landsc Urban Plan 143:56–68

Porcuna-Ferrer A, Labeyrie V, Alvarez-Fernandez S, Calvet-Mir L, Faye NF, Ouadah S, Reyes-García V (2023) Crop biocultural traits shape seed networks: implications for social-ecological resilience in south eastern Senegal. Agric Syst 211:103750

Pretty JN, Morison JIL, Hine RE (2003) Reducing food poverty by increasing agricultural sustainability in developing countries. Agr Ecosyst Environ 95:217–234

The economic lives of smallholder farmers: An analysis based on household data from nine countries. Food Agric Organ UN 1–39, Rome

Redvers N, Blondin B (2020) Traditional Indigenous medicine in north America: a scoping review. PLoS ONE 15(8):e0237531

Reyes-García V, Garcia del Amo D, Alvarez-Fernandez S, Benyei P, Calvet-Mir L, Junqueira A et al (2024) Indigenous Peoples and local communities report ongoing and widespread climate change impacts on local social-ecological systems. Commun Earth Environ 5(1):29

Robinson R, Sutherland W (2002) Post-war changes in arable farming and biodiversity in great Britain. J Appl Ecol 39:157–176

Rodda JC, Ubertini L (2004) The basis of civilisation-water science? International association of hydrological science 2004, IAHS Press, Centre of Ecology and Hydrology, Wallingford, Oxfordshire, OX 10 8BB, UK

Salvador R, Furlan L, Martinez-Villaluenga C, Dala-Paula B, Clerici M (2023) From ancient crop to modern superfood: exploring the history, diversity, characteristics, technological applications, and culinary uses of Peruvian fava beans. Food Res Int 173:113394

Shah D (2018) Climate change and Indian agricultural sector: adaptation, mitigation and impact assessment

Sibhatu KT, Qaim M (2017) Rural food security, subsistence agriculture, and seasonality. PLoS ONE 12(10):e0186406

Smith BD, Zeder MA (2013) The onset of the anthropocene. Anthropocene 4:8–13

Spangler K, Schumacher B, Bean B, Burchfield E (2022) Path dependencies in US agriculture: regional factors of diversification. Agr Ecosyst Environ 333:107957

Sun Y, De la Cruz MJ, Min Q, Liu M, Zhang L (2013) Conserving agricultural heritage systems through tourism: exploration of two mountainous communities in China. J Mt Sci 10:962–975

Tamariz G, Baumann M (2021) Agrobiodiversity change in violent conflict and post-conflict landscapes. Geoforum 128:217–222

Tareke A, Beyene A (2019) Climate-smart agricultural practices and welfare of rural smallholders in Ethiopia: does planting method matter? Land Use Policy 85:387–396

Tilman D, Cassman KG, Matson PA, Naylor R, Polasky S (2002) Agricultural sustainability and intensive production practices. Nature 418:671–677

Toledo V, Ortiz B, Cortés L, Moguel P, Ordoñez MJ (2003) The multiple use of tropical forests by indigenous peoples in mexico: a case of adaptive management. Conserv Ecol 7(3)
Tolunay A, Alkan H, Korkmaz M, Bilgin SF (2007) Classification of traditional agroforestry practices in Turkey. Int J Nat Eng Sci 1(3)
Tsing A, Mathews A, Bubandt N (2019) Patchy anthropocene: landscape structure, multispecies history, and the retooling of anthropology: an introduction to Supplement 20. Curr Anthropol 60(S20):S186–S197
Tynsong H, Dkhar M, Tiwari BK (2020) Review: traditional ecological knowledge of tribal communities of north east India. Biodiversitas 21(7):3209–3224
Van Berkel D, Verburg P (2014) Spatial quantification and valuation of cultural ecosystem services in an agricultural landscape. Ecol Ind 37:163–174
Van Vliet N, Mertz O, Heinimann A, Langanke T, Pascual U, Schmook B, Adams C, Schmidt-Vogt D, Messerli P et al (2012) Trends, drivers and impacts of changes in swidden cultivation in tropical forest-agriculture frontiers: a global assessment. Glob Environ Chang 22(2):418–429
Velásquez-Milla D, Casas A, Torres-Guevara J et al (2011) Ecological and socio-cultural factors influencing in situ conservation of crop diversity by traditional Andean households in Peru. J Ethnobiology Ethnomedicine 7:40
Wendimu A, Tekalign W, Asfaw B (2021) A survey of traditional medicinal plants used to treat common human and livestock ailments from Diguna Fango district, Wolaita, southern Ethiopia. Nord J Bot 39(5)
Williams PR, Keefer DK, Nash DJ, Sims KY, Dayton C, Moseley M (2005) Envisioning the invisible: reconstructing the taphonomy of ancient Andean agricultural landscapes through GIS slope stability models. In: The reconstruction of archaeological landscapes through digital technologies: proceedings of the 2nd Italy-United States Workshop, Archaeopress, Oxford, pp 3–60
Yisehak K, Taye T, Aynalem H (2013) Characteristics and determinants of livestock production in Jimma zone/southwestern Ethiopia. Afr J Basic Appl Sci 5(2):69–81
Yuan Z, Lun F, He L, Cao Z, Min Q, Bai Y, Liu M, Cheng S, Li W, Fuller AM (2014a) Exploring the state of retention of traditional ecological knowledge, TEK in a Hani rice terrace village, southwest China. Sustainability 6(7):4497–4513
Yuan Z, Lun F, He L, Cao Z, Min Q, Bai Y, Liu M et al (2014b) Exploring the state of retention of traditional ecological knowledge (tek) in a Hani rice terrace village, southwest China. Sustainability 6:4497–4513
Zerihun K, Weyessa G, Adugna D (2011) Understanding homegarden in household food security strategy, case study around Jimma, southwestern Ethiopia. Res J Appl Sci 6(1):38–43
Zhang D, Min QW, He L, Yuna Z (2016) Agrobiodiversity features, conservation and utilization of China's globally important agricultural heritage systems. Chin J Eco-Agric 24:451–459
Zimmerer K (2010) Biological diversity in agriculture and global change. Annu Rev Environ Resour 35:137–166
Zizumbo D, Colunga P (2010) Origin of agriculture and plant domestication in west Mesoamerica. Genet Resour Crop Evol 57:813–825

# Material and Technical-Constructive Culture Contours and the Challenges of Sustainability

**Marluci Menezes and Maria do Rosário Veiga**

**Abstract** The responsibility of the construction sector and urban buildings for $CO_2$ emissions challenges the material and technical-constructive culture accumulated throughout history. This, in parallel, influences a broad social field, as this sector interactively encompasses a diversity of actors (experts, designers, decision-makers, investors, entrepreneurs, users, etc.). Eco-efficient materials, waste reuse, recycling, rehabilitation, usage of materials with low embodied energy together with more energy efficient buildings is the path to environmental sustainability. The chapter focuses on the potential and challenges faced by the need of (re)composing a material and technical-constructive culture regarding the use of eco-efficient materials, specifically in wall coverings. The coatings perform protective, aesthetic-decorative and finishing functions. Hence, their acuity in safeguarding the comfort, health, and well-being of the building users. Experimental research in civil engineering has innovated and contributed to the development of eco-efficient materials in wall coverings, which can potentially be integrated into a circular economy perspective. Nevertheless, what are the obstacles and drivers for the (re)composition of a material and technical-constructive culture related to the broad social field affected by this sector? The chapter reflects on this issue and is supported by a literature consultation while exploring a first set of semi-structured and in-depth interviews with experts working in Portugal.

**Keywords** Responsible construction · Eco-efficient materials · Stakeholder involvement · Sustainability · Possible futures

---

M. Menezes (✉) · M. do Rosário Veiga
National Laboratory for Civil Engineering (LNEC), Lisbon, Portugal
e-mail: marluci@lnec.pt

M. do Rosário Veiga
e-mail: rveiga@lnec.pt



W. Leal Filho et al. (eds.), *Composing Worlds: Humanities, Health and Wellbeing in the XXI Century Towards a More Sustainable World*, World Sustainability Series,
https://doi.org/10.1007/978-3-031-87108-5_17

# 1 Introduction

To achieve climate neutrality by 2050 is the goal set by the European Union, considering a reduction in emissions by 55% before 2030 (EU 2019). This long-term vision reflects a growing social concern for the environment and a more resilient society that tends to adapt to the multiplicity of inevitable impacts of climate change. It is essential to face these challenges and reduce impacts at various levels with greater eco-efficiency (Szyja 2015; Caiado et al. 2017).

The construction sector, in addition to its fundamental role for society, contributes to a significant emission of greenhouse gases, which has a consistent impact on sustainability (Kazemi et al. 2023). Buildings account for a third of emissions in Europe. It is essential to address these emissions by enhancing energy efficiency and reducing energy consumption. The construction sector's environmental impact is considerable and affects many other aspects as well. The sector's impact is profound, generating significant waste—33% within EU—and high greenhouse gas emission. The cement industry alone is responsible for around 7% of global CO2 emissions associated with human activity. Buildings consume about 35% of global energy resources during construction and operation with 40% of this energy attributed to European buildings stock (Salvia et al. 2021; Sáez & Osmani 2019; IEA 2020; Laaroussi et al. 2020).

The notion of responsible construction stands as a decisive element in the pursuit of sustainability within the sector. The transition encompasses interconnected environmental, social and economic dimensions, prioritizing the entire life cycle of the building matters. By fostering the development of more resilient buildings, it aims simultaneously enhance the well-being, comfort, and health of people while safeguarding the judicious use of natural resources. Furthermore, the correlation between environmental actions and the economic dimension of the construction industry should be regarded not merely as an expense, but as a means of promoting social justice and creating value across environmental, economic and social realms (Oikanrinen 2022).

For the benefit of sustainability in construction, experimental research in civil engineering (and related areas) has innovated and contributed to transforming challenges into opportunities (Pacheco-Torgal 2014). Here, the challenge is the eco-efficiency of construction materials. The principles of these materials involve minimizing embodied energy; reduce impacts on nature, and minimizing waste and pollutants derived from the production process of goods and services. Further considering that the use of eco-efficient materials is multiplied by reusing and recycling waste into new construction materials—with an added value of generating performance improvements in some areas (Farinha et al. 2018; Ranesi et al. 2024a, b); reduction of emissions and even capture of greenhouse gases through construction elements (Bastos et al. 2024; Ranesi 2024a). Moreover, adopting concepts and methodologies to ensure healthy and comfortable conditions can be achieved with ecological solutions that reduce energy consumption. Observing the inclusion of these aspects in assessing of the life cycle of materials and constructions, allows a more objective quantification of the gains obtained. This path assumes that gains in sustainability occur without significant losses in the performance of construction materials (Farinha et al. 2021).

The evolution towards the sustainability of the construction sector needs not only the advances in knowledge of materials and their production, which are carried out through scientific-technological research, but it also requires a strong motivation of all the players and their involvement. Besides, these factors must be aligned. In fact, the research in materials and solutions will be effective only if real implementation occurs and this depends on how the users face and accept the new solutions. Both the construction professionals, such as designers and constructers, and the end users, such as inhabitants of the built environment, strongly influence the final decisions about the construction technologies to use in practice. Therefore, to know the perspective of the different players is an important step to find solutions with better acceptance by the users and thus increase their implementation capacity, respecting people, nature and the planet.

The challenge of environmental sustainability in the construction sector instigates, in parallel, the material and technical-constructive culture accumulated throughout history. This involves social values, beliefs, practices, and behaviours. Nevertheless, in the (re)composition of a material and technical-constructive culture related to the broad social field affected by the sector, what is the perception that stakeholders have of the obstacles and drivers of environmental sustainability, particularly regarding eco-efficient materials?

## 2 Study Framework, Research Objectives and Methodology

### *2.1 Objectives*

The Research Program of the National Laboratory for Civil Engineering (LNEC) called "RESUME—Buildings for the Future: Resilience and Environmental Sustainability" aims to contribute to increasing the environmental sustainability of construction, without loss of performance, encompassing different activities and scientific projects. It also aims to promote increased comfort, durability, and resilience of buildings. Among the justifications for carrying out RESUME, the incipient sensitivity (or lack of knowledge of the potential) of the construction industry and society in general towards the adoption of passive, more eco-efficient and sustainable construction solutions stand out. Therefore, one of the specific objectives of the program is to understand the perspective of diverse audiences on the outlined knowledge themes. By doing so, it seeks to contribute to the dissemination, communication, and awareness initiatives pertaining to environmental sustainability within the sector in a timely and appropriate manner.

Looking at eco-efficient materials, the reuse, and recycling of waste into new construction materials, particularly wall coverings, the increase in experimental research in these areas is evident. Taking advantage of the framework, LNEC researchers provides a range of potential opportunities for the development of eco-efficient technologies and materials. However, there are various obstacles to the effective implementation of these procedures and the use of these materials by the sector, as well as by society in a broader sense.

To advance one of the objectives of the RESUME Program, it was deemed appropriate to initiate a campaign aimed at gathering and organizing opinions, experiences, and perceptions on these topics from a diverse range of stakeholders.

This chapter presents an exploratory study carried out with sector experts in Portugal. The aim is to base the first essay on the challenges facing material and technical-constructive culture related to the eco-efficiency of materials and sustainability in construction. More specifically, we look at eco-efficient coating materials, considering that they perform protective, aesthetic-decorative and finishing functions, respectively influencing the comfort, health, and well-being of building users.

## *2.2 Theoretical Framework*

The construction sector is essential for the broader safeguarding of environmental sustainability. It affects the health, comfort, and well-being of humans, along with the quality of the environment, the use of resources and the production building process. However, alongside the sector's growing interest in sustainability issues, the relationship between innovation in the construction sector, environment, governance and the social dimension requires more knowledge and performance (Cruz et al. 2023). For Romanova (2016), the adoption of the principles of mutual adaptation between society and nature is essential for environmental sustainability in the sector. Approaches within a circular economy logic (Andreucci et al. 2021; Valencia et al. 2023), imply an increase in ecologically motivated behaviours, together with a rise in ecological awareness. Noting further that considering the achievement of carbon neutrality targets solely based on technical-scientific parameters can prove to be a difficult task to accomplish. It is necessary to involve objectives related to changing behaviour and mentality, the specificity of sociocultural contexts and the increase in a more responsive social understanding of energy efficiency goals (Mustika et al. 2021; Lorente & Diego 2021; Lencastre et al. 2023; Vidal et al. 2024).

The construction industry has been criticized for primarily pursuing profit. In their analysis of literature on sustainability reports in this sector, Kazemi et al. (2023) identified gaps in knowledge and research related to the incorporation of various sustainability approaches, as well as the measurement, monitoring, evaluation and reporting of sustainability practices. The article highlights the insufficient focus on social sustainability in many large-scale studies on environmental sustainability in the construction industry. Kazemi et al. (2023) focus in more detail on the business and organizational component of environmental sustainability reports on the construction industry. However, it is crucial to retain the idea that social sustainability encompasses a diversity of social values, beliefs, activities and performances, driven by a diversity of stakeholders. In general, stakeholders are seen as having the ability to affect decision-making, shaping political strategies that prioritizes environmental sustainability, or influencing others trough persuasion (Hosseini-Teshnizi 2015; Oikarinen 2022).

For Hosseini-Teshnizi (2015), the stakeholders that constitute the main players in the different phases of a building's life cycle are internal (considered as strategic) and external (considered as normative), plus a third group, made up of internal and external in simultaneously. Strategic stakeholders are those who are directly involved in the project; normative ones are those who have the power to influence (or can be influenced) by the project, although they do not act directly on the project. As outlined by author, these different stakeholders embody a range of values that are interdependent and interconnected, prompting curiosity about the distinctions between these values and their mutual influences.

## 2.3 *Methodology*

The study is a small exploratory research employing a qualitative approach, which involves conducting semi-structured and detailed interviews with experts in Portugal. The aim was to investigate the views on eco-efficient construction materials, specifically in wall coverings, highlighting the challenges and opportunities for a more sustainable transition. The study focus on the perspective of the stakeholders who were interviewed. The goal was to pinpoint key ideas that could lead to strategies for sustainability transition in the construction sector. Table 1 displays the analysis dimensions and the questions for interview.

The study was conducted through interviews with five experienced individuals in design, project, management and applied research—two architects and three civil engineers from our knowledge network, between February and March 2024. Because it was challenging to have face-to-face interviews with chosen experts, the decision was made to send the interview questions via email in order to standardize the information gathering process. The interviews send back their answers by email. Throughout their professional performance, the experts interviewed worked with eco-efficient coating building materials (see Table 2).

The interest in interviewing architects stems from their influence in selecting construction materials—from the design phase—and designing buildings (Kubba 2012). The selection of materials related to visibility and appearance, in addition to the functionality of buildings is important (Oikarinen 2022), which special impact on covering wall materials. The choice of civil engineering specialists derives from their general knowledge of the various aspects linked to the construction and performance of buildings. They are experts in construction technology and physics with

**Table 1** Analysis dimensions and questions for interview

| Use of eco-efficient materials | Impact of eco-efficient materials | Initiatives to increase the eco-efficient and sustainable technical-constructive culture |
|---|---|---|
| How do you perceive—in broad and specific terms (for Portugal) the use of eco-efficient construction materials for the current and future of the construction sector?<br>Are you used or planning to use eco-efficient construction materials, particularly as coating materials?<br>What is your opinion on the potential, obstacles and challenges in the short and medium term in the construction sector, especially in relation to the confrontation between the existing material and construction culture? A culture that makes use of eco-efficient materials? | [on the health, comfort, and well-being of building users]<br>How could a new cultural approach in the construction sector contribute to a healthier, safer and more comfortable environment?<br>[on the overall cost of construction]<br>Do you consider that changes in the approach to the use of eco-efficient construction materials could have an impact on overall construction costs?<br>[on the durability and maintenance of buildings]<br>What impact do you think the use of eco-efficient materials tends to have on the maintenance needs and durability of buildings? | Do you believe we should shift our construction practices towards greater sustainability and eco-efficient? Why?<br>In relation the various stakeholders involved in the construction sector, what do you believe is essential to do in order to increase the use of eco-efficient construction materials? |

**Table 2** Identification of interviewed experts

| Title | Expertise | Reference |
|---|---|---|
| Architect | Architectural Designer in Portugal | A-1 |
| Architect | Architectural Designer in Portugal | A-2 |
| Engineer | Researcher at a Portuguese public scientific research institution, specialist in sustainable construction materials, products, and technologies | E-3 |
| Engineer | Researcher at a Portuguese public scientific research institution, specialist in sustainable construction materials, products, and technologies | E-4 |
| Engineer | Professor at the Portuguese Public University, where he is also a researcher, specialist in traditional, bio-constructive and eco-efficient materials, products, and technologies | E-5 |

in-depth knowledge of the composition, technology, and performance of construction materials.

The responses obtained were subject to content analysis. To give examples, the most representative comments as expressed by the interviewees are cited. The analysis results will be presented in the following section. After that, we will conduct a SWOT analysis to identify key ideas related to sustainability in the construction sector and the use of eco-efficient materials based on the responses obtained.

# 3 Expert Insights: Results and Discussion

## *3.1 Use of Eco-Efficient Materials*

All interviewees share the idea that environmental sustainability is crucial for the sector, as also observed by Hosseini-Teshnizi (2015) and Oikarinen (2022) regarding the stakeholders linked to the construction sector in which the authors worked. Likewise, all our interviewees consider that society is increasingly more aware of the importance of sustainability. They also share the opinion that eco-efficient materials play a key role in safeguarding sustainability.

Although there is an increase in demand for eco-efficient materials in the country because of "increasingly demanding legislation in terms of energy efficiency" (A-2), most interviewees consider that currently in the sector, the use of these materials is still little expressive. Noting that this is due to three main aspects that influence each other: the low supply of eco-efficient materials on the market, little market demand and the high price of existing products. One of the interviewed architects (A-1) highlights, however, three other aspects: the lack of variety of products on the market, especially finishing materials and colours, the lack of information about these materials and the lack of skilled labour-work to apply them. This highlights not only concerns with the visual aspects and appearance of the building (Kubba 2012), but also with practical aspects regarding the need for knowledge for its effective use and application.

The relationship between sustainability, nature-based solutions, bio-construction, environmentally friendly materials and eco-efficient materials with the tourism sector (Kelly et al. 2007; Tyrväinen et al. 2014; Ekhaese & Ndimako 2023) is particularly evident in the statement below, even though it is noted that, in Portugal, the investor is mainly foreign: "There has been a growing concern in the application and investment on this type of materials, especially in the tourism sector (rural tourism/nature tourism/ ecotourism)" (A-2).

The engineers interviewed, although they share the opinion that socio-environmental awareness and sustainability are increasing, add the fruitful interest in the theme of environmental and constructive sustainability in academia. However, they suggest that this awareness is not as evident among most entrepreneurs in the sector, for example:

> Academia is very concerned about sustainability in construction. Numerous studies encompass the development of new materials with the introduction of waste from the construction industry as well as waste from other industries, such as textile fibre waste. Despite an increasing body of research, in practice the corporate embrace of sustainability is a relatively recent development. It is imperative to implement these extensive studies in order to

> commercialize innovative, sustainable materials. For instance, consider industrial mortars incorporating recycled aggregates (E-4).

In a study on eco-drivers applied to ecological building design, Vakili-Ardebili and Boussabaine (2007) already had insights that designers reported, although sensitive to more sustainable building design options, real practices did not necessarily correspond to this perception. Regarding the effective adoption of strategies that increase more environmentally friendly construction practices, in a study on the motivations and barriers to implementing these strategies among stakeholders, Hosseini-Teshnizi (2015) observes that even though they attribute importance to these strategies, this is not reflected in practice.

In addition to the negative environmental impacts of the construction sector being widely known, as well as the importance of sustainability being increasingly propagated:

> There is a lot of talk, but in situ there are other concerns. There is greenwash of some materials and products (E-5).

The greenwashing behaviours adopted by the sector to obtain environmental legitimacy and value their products (Bowen & Aragon-Correa 2014; He et al. 2020; Nemes et al. 2022; Bernini & La Rosa 2024), also contribute to the still weak awareness of the construction business sector regarding environmental sustainability—especially the construction and real estate industries.

The transition to environmental sustainability in the construction sector, particularly using eco-efficient materials, is not, however, a simple and direct consequence, as much will have to be accomplished in the short to medium term. For example, at a legislative level, awareness-raising, economic advantages, greater coordination between scientific-technological research, academia and industries, and increasing the scale of production and supply of these materials.

> There has been an increasing demand in recent years for these materials, so I want to believe that in the future, they will be increasingly common and accessible on the market, and soon there will be greater acceptance on the part of the customer (A-2).
>
> This change will have to come from a change in legislation or some commercial advantage that companies may have. Raising awareness of the use of these products among construction owners or architects could also influence the market (E-3).
>
> It is clear that companies need to invest time and money in developing more sustainable products. Therefore, universities and research laboratories (e.g. LNEC) can work in partnership with the construction industries towards a future in which what is done in the laboratory is effectively seen on an industrial scale. My professional experience is based on sustainability in the construction sector, covering the field of waste management, from the characterization of the final destination to waste recycling and the development of new materials through the incorporation of waste. Specifically, the objective of these studies is to produce coating mortars with recycled aggregates at an industrial level. Thus, applying what has been tested and validated in the laboratory to larger commercialization scales. I believe that using these materials will no longer be a choice in the future, but rather a common method, as natural resources are limited (E-4).
>
> There must be measures that promote/encourage the design and execution of constructions with eco-efficient materials, starting with the training of architects and engineers. But also lifelong training, including "benefits" when these materials are effectively applied (E-5).

The idea that it is essential to reduce our environmental footprint is, in broad terms, transversal to the opinions of those interviewed, with eco-efficient materials being one of the elements that contribute to achieving this objective. Opinions also point to the importance of this attitude as a creation of environmental, social and economic value.

> Eco-efficient construction materials offer a variety of benefits, such as reducing energy consumption, reducing environmental impacts, creating more pleasant and natural spaces, and improving people's quality of life. Finally, it should be noted that the use of this type of material... ends up increasing the value of the property to which they are applied (A-2).

The path towards more responsible and sustainable construction benefits not only from socio-environmental awareness and public opinion, but also from the increased scientific and technological interest of researchers and professionals in the sector. A set of dynamics can be observed that influence each other. These echo the idea that it is important to understand the construction process in an integrated way—encompassing the environmental, economic and social dimensions, in which the role of the circular economy is central (Coley & Lemon 2009; Szyja 2015; Mustika et al. 2021; Lorente & Diego 2021; Andreucci et al. 2021; Valencia et al. 2023; Vasić et al. 2023).

> Request from society, with increasingly more ecological concerns. Greater knowledge of younger civil engineers and architects in the area. Greater availability of effectively eco-efficient materials on the market and at competitive costs (E-5).
>
> The advantages of using an environmentally efficient material are clear, as thinking about environmental viability is extremely indispensable for the planet. Using sustainable materials has a lower carbon footprint. The circular economy begins to effectively make sense, at a level where the useful life cycles of all materials are thought of in an integrated way (E-4).

In the short term, however, many obstacles arise. For example: high price (especially considered in terms of initial investment), mentality of contractors, lack of experience and specialized labour, lack of knowledge and adequate training, resistance to change, lack of raw materials and demand for eco-efficient materials, lack of advertising about these materials, insufficiency of standards and legislation.

> An obstacle... is resistance to change and lack of knowledge on the part of some construction professionals, who are used to using traditional methods and materials (A-2).
>
> Initially, I believe that part of the construction industry has a certain objection to using eco-efficient products due to the finishing of the material, the associated costs, and the unknown market. The target audience (buyers) also presents some resistance to changes. Therefore, the consumption of eco-efficient materials may encounter some obstacles (E-4).
>
> Association with less durable materials and products. Effective lack of knowledge of civil engineers, architects, licensees. Due to lack of scale effect, some materials, and products are more expensive than other competitors (E-5).
>
> (...) little publicity in this area, from the companies that represent these materials, which leads the designer, due to lack of knowledge, not to apply them in projects (A-2).

Obstacles present economic, technological, communicative, and social challenges, reflecting necessary paradigm shifts:

Paradigm change, better prices, greater durability of materials, fire resistance. Pay greater attention to waste management, use less wasteful materials (A-1).

Reach the theoretical knowledge already developed to companies (E-3).

I believe that the biggest challenge found is the supply chain of raw materials to manufacture an eco-efficient product. For a sustainable product, for example industrial mortars with recycled aggregates, to enter the market, the availability of this waste needs to be large enough to support industrial production. In this case, construction and demolition waste operators need to be aligned with mortar factories so that this sustainable product raw material is always available in great quantity and quality for future production (E-4).

Legislation that helps promote use, for example giving some benefit in terms of licensing, area that can be built, tax reductions on this type of products, etc. Expansion of technical, industrial and scientific knowledge in the area (E-5).

## *3.2 Impacts of Eco-Efficient Materials*

### 3.2.1 Impacts on the Health, Comfort and Well-Being of Building Users

The use of eco-efficient materials must safeguard the health, safety, comfort, and well-being of people, thus complying with requirements regarding air quality, ventilation and natural light, humidity, use of non-toxic materials, among other aspects (Lima et al. 2016; Ranesi 2021).

To be efficient, materials must meet health requirements to be applied, namely they must be subject to toxicity tests. Passing these tests, in my opinion, they will have no influence on the health of building users. Regarding comfort, it will depend on the material itself, if they are thermal mortars they will probably have an influence on comfort, otherwise it will not affect users. In relation to well-being, it will depend on the sensitivity of users to the use of eco-efficient materials. This can contribute to the use of materials with less environmental impact, which indirectly contributes to the health of populations (E-3).

Nevertheless, with health and well-being requirements being safeguarded, the challenge lies in shifting mentalities, paradigms, and cultures to align with the increasing global awareness of environmental issues and the necessary shift towards sustainability.

I think it will have a greater impact on users' awareness than on comfort itself. Since the current situation in society is beginning to think about sustainability in a less individualized way (E-4).

### 3.2.2 Impacts on the Overall Cost of Construction

All interviewees share the opinion that the use of eco-efficient materials has a negative impact on overall construction costs, although they admit that this may depend on the type of material.

However, they recognize that this impact is mainly in the short term, as in long term, the use of these materials contributes to sustainability by reducing energy consumption and waste.

> Although, in the short term, it may have an impact on the costs of purchasing this type of construction materials (higher costs given the investment in innovative technologies and processes), in the long term, the adoption of more eco-efficient construction materials can bring economic and environmental significant benefits. It can, for example, result in lower energy consumption, or a saving of natural resources, or a reduction in waste, or a reduction in maintenance throughout the life of the building (A-2).

> I believe that the use of more eco-efficient construction materials will have a negative impact on overall construction costs. Continuing with the example given: for the production of industrial mortars with recycled aggregates, an investment by the factory is necessary to install a new silo, a new conveyor belt, and costs associated with the processing of recycled aggregates, as they cannot be used "as is". Therefore, the increase in operating costs for the manufacture of this new product is undeniable. However, what will be more important? Protect the environment or pay more in the near future? The question remains for the companies themselves (E-4).

This initial negative impact on construction costs could be minimized if, for example, "CO2 rates are applied" (E-3) or increased "the scale of use" in parallel with expanding knowledge, regarding performance and relevance of its usefulness (E-5).

### 3.2.3 Impacts on the Durability and Maintenance of Buildings

The issue of durability and maintenance of buildings is still a challenge for scientific-technological knowledge (Parracha et al. 2022), being the one in which the interviewees presented more distinct opinions, although not necessarily divergent. For the architects interviewed, issues of impact on the durability and maintenance of buildings are associated with the type of material, the location where it is applied (whether indoor or outdoor) and the existence of greater or lesser human activity in its proximity.

The engineers interviewed reflect perceptions of durability and efficiency that are complementary to each other, although distinct. These opinions range from the consideration that in order to have eco-efficiency, durability must be guaranteed from the outset, to the need to deepen scientific-technological knowledge regarding the durability of these materials. This includes safeguarding the quality of their design, the type of use, and their application.

> To be considered eco-efficient, the durability of these materials must be taken into account, in comparison with materials usually produced. In terms of maintenance, I think there are no additional concerns due to the use of these new materials (E-3).

> Another factor that makes the use of these eco-efficient materials more limited is their durability. In relation to the example I presented of mortars with recycled aggregates, we saw that durability is an issue that needs further investigation. Both on how to improve this durability and on introducing maintenance strategies to mitigate such effects. Recycled aggregates can affect mortars in a way that makes them more porous and fragile. However, other approaches to treating recycled aggregate before its incorporation can also be considered (E-4).

> It will depend on the type of use/application (correct/incorrect) of many of these materials and products. If properly designed and applied, I think there will be no impact compared to other materials and products (E-5).

### 3.3 *Initiatives to Increase the Eco-Efficient and Sustainable Technical-Constructive Culture*

All opinions highlight the importance of introducing changes to the dominant technical-constructive culture, since: "Sustainability must be associated with everything that involves human activity" (A-1). It is the life cycle of the environment and the way we consume it that is important to take care of, as this is in everyone's interest:

> We should change the way we build and should move towards more sustainable and eco-efficient building, promoting energy efficiency (thus reducing energy consumption), helping not only the environment but also, in the long term, resulting in cost savings. There is also the health and well-being factor, as sustainable buildings often use natural, non-toxic materials, thus improving air quality. The more sustainable and eco-efficient construction is essential to protect the environment and promote people's health and well-being! (A-2).
>
> With this change in consciousness, and with the growing concern in the environmental aspect, I believe that the use of new materials can make the construction sector more interested in this aspect, which leads to thinking more generally in a healthier way in terms of consumption (E-4).
>
> It is necessary for the entire supply chain to be activated so that the production of these new materials can be carried out continuously. The target audience (consumers) must be informed and in agreement with new perspectives. The implementation of this more sustainable culture depends largely on the end consumer. Companies produce to sell, so if there are consumers for this new "green" product, companies will begin to be interested in this production (E-4).

The assertiveness with which all interviewees listed a variety of initiatives to unblock obstacles and increase the use of eco-efficient materials reveals the urgent need to change the current situation. This also reveals the applicability of these initiatives in the current situation. In this sense, Hosseini-Teshnizi (2015) observed that the effective adoption of strategies regarding materials with environmental significance by the construction sector is more related to stakeholder motivations or the lack of these, than to obstacles. The author also highlights that the adoption of strategies regarding materials with environmental significance has less weight in the planning phase, compared to the other phases analysed, although when adopted there is a greater relationship with environmental motivations. It also states that the limitations related to the areas of action, time, control, and responsibility on the part of stakeholders are reflected in a lack of interest in implementing these strategies. The set of suggestions presented were systematized into four major interrelated topics: (1) Knowledge and technical training: Coordination between the construction industry, universities, research centres and technical orders; (2) Knowledge, social empowerment and dissemination; (3) Research, knowledge, technical training and academic curriculum; (4) Legislation, regulation and incentives (see Fig. 1).

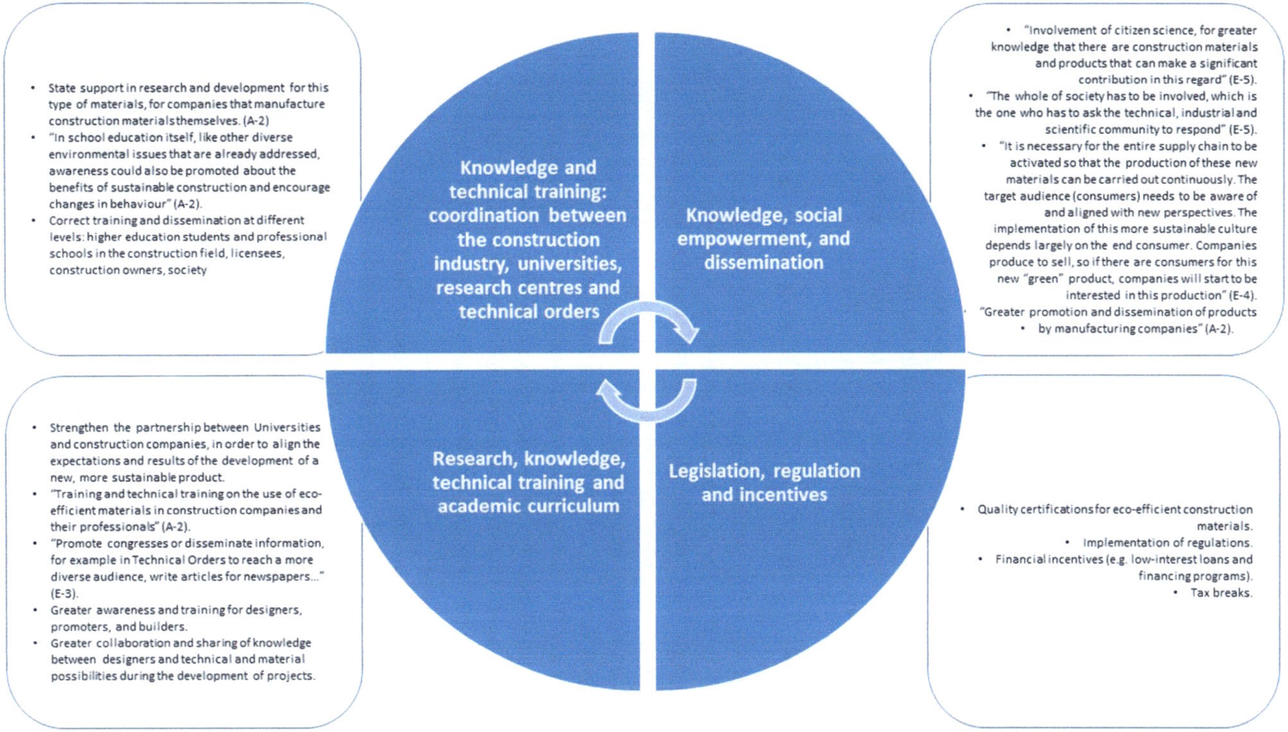


**Fig. 1** Key ideas to increase the use of eco-efficient construction materials

## 4 Lessons Learned

The responses provided emphasize the significance and advantages of adopting more responsible construction practices, whit eco-efficient materials playing a key part. It also points that sooner or later the construction sector will have to embrace constructive eco-efficiency as the way forward towards sustainability. This is due to regulatory goals and requirements, but also due to an increase in society's environmental awareness, thus influencing this necessary change. Despite the path forward, in this sense, much progress has been made in terms of scientific-technological knowledge, legislation, goals and regulations, and the manifestation of more responsible practices towards the environment and respective construction processes. Nevertheless, there are still obstacles to overcome.

In alignment with the literature reviewed, the statements reveal particularities of the construction sector in Portugal. This enable the identification a set of key ideas the internal and external aspects affecting the relationship between current material and technical-constructive culture for a more eco-efficient and sustainable approach (see Table 3).

Identifying these, key ideas through a SWOT analysis enable us to contemplate:

- The identified strengths are important for raising awareness among the broad social field covered by the construction sector about the importance of using eco-efficient materials. This specifically targets the construction industry, which plays a significant role in creating environmental, economic, and social value, as well as driving technological innovation.
- The weaknesses are aspects that need improvement. On the one hand, weakness demonstrates the strong influence of economic and market factors for better implementation and use of construction materials. On the other hand, stakeholders emphasize the need to investigate certain aspects linked to eco-efficient materials, and the significance of training and raising awareness among technicians working directly in the sector.
- Regarding opportunities, the benefits of using these materials to contribute to sustainability by related to environmental, economic, and social aspects. Simultaneously, this reinforces and expands the role of scientific, academic, and technical innovation, as well as emphasizes the increasing importance of socio-environmental values in our consumption of the Earth's natural resources.
- The threats that were underlined are the areas that need to be adjusted or changed. Once again, economic and market issues emerge as key factors in increasing the supply and use of these materials. However, these aspects emphasize the importance of implementing initiatives that contribute to changing practices and expanding knowledge about these materials. As well as their use and application.

**Table 3** Identification of key ideas about the context (internal and external) related to the use of eco-efficient construction materials from SWOT analysis

| Internal Factors | |
|---|---|
| **Strengths +** | **Weaknesses –** |
| Contributes to the sustainability of construction: reduced energy consumption, reduced environmental impacts<br>The use of construction materials with a lower environmental impact indirectly contributes to the comfort, health, and well-being of building users<br>Increased importance is given to the circular economy, with the useful life cycles of all materials thought of in an integrated way<br>Growing concern about the application and investment in these materials<br>Academic concern with sustainability in construction<br>Greater knowledge/interest on the part of younger civil engineers and architects<br>The use of eco-efficient materials adds value to the properties where they are applied | Materials with little use and commercialization<br>Weak demand/supply in the construction market<br>High cost<br>Due to the scale effect, certain eco-efficient materials are more expensive than their competitors<br>Lack of knowledge of long-term material performance<br>Less durability<br>Incipient interest in constructive sustainability on the part of the business/industrial sector linked to construction<br>Its application is dependent on the promoter's acceptance<br>Effective lack of knowledge on the part of civil engineers, architects, and licensors<br>Insufficient standardization and regulation |
| **External Factors** | |
| **Opportunities +** | **Threats –** |
| Society's growing concern with environmental, ecological issues and resource consumption<br>Materials with a lower ecological footprint<br>Eco-efficiency is the future of the construction sector<br>Eco-efficient materials will be more common, accessible and accepted<br>Significant environmental, social and economic benefits (e.g.: lower energy consumption, saving natural resources, reducing waste, reducing maintenance throughout the life of buildings)<br>Increased scientific and technological research on sustainability in the construction sector | Resistance to change and lack of knowledge on the part of some construction professionals<br>Lack of experience in its use/application<br>Lack of adequate training on the use/application of eco-efficient materials by professionals<br>Need for specialized labour<br>Incipient dissemination/advertising of these materials by the companies that represent them<br>The reluctance of the construction industry and its clientele (buyers/users) to use eco-efficient materials due to the material's finish, associated cost and lack of knowledge of the market<br>Need for larger-scale production of constituent raw materials<br>Greenwash Effect |

## 5 Conclusions

The research led to the identification of a series of key ideas about the challenges and benefits of utilizing eco-efficient materials in the construction sector, as mentioned by experts. The identified key ideas are useful as a roadmap for future research and enhancing strategies that promote sustainability.

The analysis revealed that the diverse interviewees hold different understandings of eco-efficient materials: some view them as natural resources, while others see them as residues or materials with low energy consumption in production. It is crucial to enhance understanding of the exact meaning of eco-efficient material. Here, there is an interest to broaden the research to involve additional and varied stakeholders, allowing for the gathering—and subsequent analysis—of a greater diversity of perspectives. A stronger dedication to sustainability in the construction sector, including norms, classifications, and laws (among other identified aspects), is also associated with a more precise delineation of eco-efficient building materials definition.

The study still provides evidence around the importance of changing mentalities and behaviours, but also in training and educational attributes in favour of sustainability. This emphasizes the significance of cultural transition, whether it is related to technical-constructive culture, formative culture, consumer culture, etc.

The importance of investing in scientific-technological knowledge and effectively linking it with construction industries/companies, designers, and universities was emphasized. Nevertheless, the experts interviewed demonstrate that achieving a sustainable shift in the construction sector goes beyond what scientific-technological expertise alone can offer. The key ideas identified indicate strategic directions to follow, emphasizing the importance of work that effectively connects environmental, economic, and social aspects. Concurrently, it demonstrates the significance of extending the debate on eco-efficiency and sustainability in the construction sector to incorporate public opinion and adherence. New solutions with better environmental performance are being developed, but their effective implementation needs the adhesion of people that use the built environment. These players should therefore be included in the process of change. The aim of getting an ecofriendly built environment should be aligned with peoples' needs and expectations, in parallel with safeguarding possible futures. This can be achieved by involving the social sciences more fully, as has been the case with other study themes related to the challenges posed by climate change and the essential green-ecological transition.

**Acknowledgements** This study is part of LNEC's Research and Innovation Strategy for 2024-2027 under the RESUME Program. We wish to express our gratitude to the experts who participated in the study for their cooperative efforts.

## References

Andreucci M, Marvuglia A, Baltov M, Hansen P (eds) (2021) Rethinking sustainability towards a regenerative economy (Vol 15). Springer International Publishing. https://doi.org/10.1007/978-3-030-71819-0

Bastos D, Farinha CB, Pederneiras CM, Veiga MR, Bogas A, Gomes R, Santos SA (2024) Pathway to carbon neutrality in the cement industry: CO2 uptake by recycled aggregates from construction and demolition waste. Appl Sci 14(12):5224. https://doi.org/10.3390/app14125224

Bernini F, La Rosa F (2024) Research in the greenwashing field: concepts, theories, and potential impacts on economic and social value. J Manage Governance 28:405–444. https://doi.org/10.1007/s10997-023-09686-5

Bowen F, Aragon-Correa JA (2014) Greenwashing in corporate environmentalism research and practice: the importance of what we say and do. Organ Environ 27(2):107–112. https://doi.org/10.1177/1086026614537078

Caiado RGG, Dias RF, Mattos LV, Quelhas LG, Filho WL (2017) Towards sustainable development through the perspective of eco-efficiency—a systematic literature review. J Clean Prod 165:890–904. https://doi.org/10.1016/j.jclepro.2017.07.166

Coley FJS, Lemon M (2009) Exploring the design and perceived benefit of sustainable solutions: a review. J Eng des 20(6):543–554. https://doi.org/10.1080/09544820802001276

Cruz PL, Ávila LV, Dinis P, MA, Baggio DK (2023) Environmental, social and governance (ESG) and innovation in the construction sector: systematic literature review. Rev Adm Da UFSM 16(4). https://doi.org/10.5902/1983465975315

Ekhaese EN, Ndimako OO (2023) Eco-friendly construction materials and health benefits in the design of an all-inclusive health resorts Nigeria. Front Built Environ 9:1011759. https://doi.org/10.3389/fbuil.2023.1011759

Farinha C, Brito J, Veiga MR, Fenández JM, Jiménez JR, Esquinas AR (2018) Wastes as aggregates, binders or additions in mortars: selecting their role based on characterization. Materials 11(3):453. https://doi.org/10.3390/ma11030453

Farinha CB, Brito J, Veiga MR (2021) Eco-efficient rendering mortars. WP Woodhead Publishing Series in Civil and Structural Engineering, Elsevier, Use of Recycled Materials

He Q, Wang Z, Wang G, Zuo J, Wu G, Liu B (2020) To be green or not to be: how environmental regulations shape contractor greenwashing behaviors in construction projects. Sustain Cities Soc 63:102462. https://doi.org/10.1016/j.scs.2020.102462

Hosseini-Teshnizi ZS (2015) The influence of stakeholders' motivations and barriers on construction material use effectiveness. Master thesis on Advanced Studies in Architecture, The University of British Columbia, Vancouver. Retrieved July 20 2024, from http://hdl.handle.net/2429/54498

IEA (2020) Global energy review 2019. International Energy Agency Report. Retrieved June 15 2024, from https://www.iea.org/reports/global-energy-review-2019

Kazemi MZ, Elamer AA, Theodosopoulos G, Khatib SFA (2023) Reinvigorating research on sustainability reporting in the construction industry: a systematic review and future research agenda. J Bus Res 167:114145. https://doi.org/10.1016/j.jbusres.2023.114145

Kelly J, Haider W, Williams PW, Englund K (2007) Stated preferences of tourists for eco-efficient destination planning options. Tour Manage 28(2):377–390. https://doi.org/10.1016/j.tourman.2006.04.015

Kubba S (2012) Handbook of green building design and construction: LEED, BREEAM, and green globes. Butterworth-Heinemann, Boston

Laaroussi Y, Bahrar M, Zavrl E, El M, Stritih U (2020) New qualitative approach based on data analysis of European building stock and retrofit market. Sustain Cities Soc 63:102452. https://doi.org/10.1016/j.scs.2020.102452

Lencastre MPA, Vidal DG, Barroso EP, Campelo A, Toldy T, Magalhães S, AG, Estrada R (2023) Climate change and health: essay on the limits of naturalism. In Leal Filho W, Vidal DG, Dinis MAP (eds) Climate change and health hazards: Addressing hazards to human and environmental health from a changing climate (pp 25–39). Springer. https://doi.org/10.1007/978-3-031-26592-1_2

Lima J, Faria P (2016) Eco-efficient earthen plasters. The influence of the addition of natural fibers. In: Fangueiro R, Rana S (eds) Natural fibres: advances in science and technology towards industrial applications (pp 315–327). RILEM Bookseries, Vol 12. Springer: Dordrecht. https://doi.org/10.1007/978-94-017-7515-1_24

Lorente JI, Diego R (eds) (2021) Natulazes en fuga. Ecocrítica(s) de la ciudad en transformación. Anthropos, Barcelona

Mustika MNW, Sueca NP, Dwijendra NKA, Putra AD (2021) Sustainable Socio-cultural Aspect within Green Building User Behavior in Bali, Indonesia. IOP Conf Ser: Earth Environ Sci 933(1):012021. https://doi.org/10.1088/1755-1315/933/1/012021

Nemes N, Scanlan SJ, Smith P, Smith T, Aronczyk M, Hill S, Lewis SL, Montgomery AW, Tubiello FN, Stabinsky D (2022) An integrated framework to assess greenwashing. Sustainability 14(8):4431. https://doi.org/10.3390/su14084431

Oikarinen M (2022) Internal stakeholders' perceptions of sustainable construction in the field of industrial construction. Master's Thesis on Marketing, Department of Marketing Hanken School of Economics, Helsinki. Retrieved June 20 2024, from https://helda.helsinki.fi/server/api/core/bitstreams/d29cfa7d-1bc4-429a-8254-c8a451df7e98/content

Pacheco-Torgal F (2014) Eco-efficient construction and building materials research under the EU Framework Programme Horizon 2020. Constr Build Mater 51:151–162. https://doi.org/10.1016/j.conbuildmat.2013.10.058

Parracha JL, Borsoi G, Veiga MR, Flores-Colen I, Nunes L, Viegas CA, Moreira LM, Dionísio A, Gomes MG, Faria P (2022) Durability assessment of external thermal insulation composite systems in urban and maritime environments. Sci Total Environ 849:157828. https://doi.org/10.1016/j.scitotenv.2022.157828

Ranesi A, Faria P, Veiga MR, Gall ET (2024b) Eco-efficient coatings for healthy indoors: ozone deposition velocities, primary and secondary emissions. Build Environ 254:111306. https://doi.org/10.1016/j.buildenv.2024.111306

Ranesi A, Faria P, Veiga MR (2021) Eco-efficiency of plasters for rehabilitation and new buildings. In Horszczaruk E, Brzozowski P (eds) Proceedings of the 2nd international conference on sustainable, environmentally friendly construction materials—ICSEFCM. Poland: West Pomeranian University of Technology, Szczecin Publishing House, 91–96. Retrieved July 20 2024, from https://novaresearch.unl.pt/en/publications/eco-efficiency-of-plasters-for-rehabilitation-and-new-buildings

Ranesi A, Faria P, Freire T, Gonçalves M, Veiga MR (2024) Eco-efficient plastering mortars for improved indoor comfort. The influence of A. dealbata bark addition. Constr Build Mater 421:135572. https://doi.org/10.1016/j.conbuildmat.2024.135572

Romanova E (2016) Research of ecological motivation of the heads and specialists getting an education in the field of production and application of construction materials. MATEC Web Conf 73:07010. https://doi.org/10.1051/matecconf/20167307010

Sáez PV, Osmani M (2019) A diagnosis of construction and demolition waste generation and recovery practice in the European Union. J Clean Prod 241:118400. https://doi.org/10.1016/j.jclepro.2019.118400

Salvia M et al (2021) Will climate mitigation ambitions lead to carbon neutrality? An analysis of the local-level plans of 327 cities in the EU. Renew Sustain Energy Rev 135:110253. https://doi.org/10.1016/j.rser.2020.110253

Szyja P (2015) Eco-efficiency in creating the ecological safety. Wroclaw Sch Bank Res J 15(7):877–886. Retrieved June 10 2024, from https://ojs.wsb.wroclaw.pl/index.php/WSBRJ/article/view/120/179

Tyrväinen L, Uusitalo M, Silvennoinen H, Hasu E (2014) Towards sustainable growth in nature-based tourism destinations: clients' views of land use options in Finnish Lapland. Landsc Urban Plan 122:1–15. https://doi.org/10.1016/j.landurbplan.2013.10.003

Vakili-Ardebili A, Boussabaine AH (2007) Design eco-drivers. J Archit 12(3):315–332. https://doi.org/10.1080/13602360701470117

Valencia M, Bocken N, Loaiza C, De Jaeger S (2023) The social contribution of the circular economy. J Clean Prod 408:137082. https://doi.org/10.1016/j.jclepro.2023.137082

Vasić MV, Goel G, Dubale M, Živković S, Trivunić M, Pezo M, Pezo L (2023) Socio-economic analysis of the construction and building materials' usage—ecological awareness in the case of Serbia. Sustainability 15(5):4080. https://doi.org/10.3390/su15054080

Vidal DG, Lencastre MPA, Lopes HS, Magalhães S, Estrada R (2024) Compor mundos: explorando a metamorfose do mundo na era das alterações climáticas e as suas implicações para a saúde. Rev Lusófona Estud Cult 11(1):e024005. https://doi.org/10.21814/rlec.5405

**Marluci Menezes** holds a master's degree and a doctorate in Anthropology by NOVA University, Lisbon, and is a geographer. She is a Senior Researcher at the National Laboratory for Civil Engineering (LNEC), Lisbon. At LNEC, she has been involved in several national and international competitive research projects exploring the connection between space, society, and culture. Several articles are connected to the research she carried out. She has a post-doctorate in Urbanism from Lusófona University, where she also works as an invited professor in the Masters and Doctorate in Urban Planning.

**Maria do Rosário Veiga** is a Civil Engineer by Technical Institute of the University of Lisbon and has a Ph.D. from the University of Porto. She is Coordinating Researcher at LNEC, where She is Head of the Building Coatings and Insulation Research Unity. At LNEC, she develops consultancy work for external entities and participates in national and international competitive research projects, having coordinated four of them. She specializes in coating materials and performance requirements and develops studies on the design of sustainable, durable and compatible mortars and wall coverings. She has been a member of European groups for the preparation of Guides for European Technical Approval since 1991 and participates of the RILEM Committees. She supervised several doctoral and master's theses. She is a member of the Board of the Doctoral Program financed by FCT—Eco-Construction and Rehabilitation (EcoCoRe). She is author/coauthor of more than 100 articles in indexed scientific journals and more than 20 books/chapters of books.

# Enhancing Environmental Performance Indicators Through Stochastic Multi-Attribute Analysis: A Novel Approach Applied to the Ecosystem Vitality

**Alexandre Morais Nunes, Paulo Caldas, Guilherme Vara, and Diogo Cunha Ferreira**

**Abstract** The measurement and monitoring of environmental performance are critical components in assessing the sustainability of human activities. Traditional environmental performance indicators (EPIs) have faced challenges in accurately capturing the complex and dynamic interactions within ecosystems, often resulting in oversimplified assessments and inadequate policy responses. In this book chapter, we propose a novel approach to address these limitations by integrating Stochastic Multi-Attribute Analysis (SMAA) into the construction of EPIs. SMAA offers several distinct advantages over traditional methods. Firstly, it embraces uncertainty inherent in environmental data, providing a more robust and reliable assessment of environmental performance. By considering uncertainties in input data, SMAA reduces the risk of biased conclusions and enhances the credibility of EPIs. Secondly, SMAA enables the incorporation of multiple criteria and attributes, allowing for a comprehensive evaluation of environmental performance across various dimensions such as air quality, water resources, biodiversity, and ecosystem services. This multidimensional perspective provides a more holistic understanding of environmental impacts,

A. M. Nunes (✉)
Center for Administration and Public Policies, Institute of Social and Political Sciences, University of Lisbon, Lisbon, Portugal
e-mail: anunes@iscsp.ulisboa.pt

Rua Almerindo Lessa—Polo Universitário Alto da Ajuda, Lisbon, Portugal

P. Caldas
Business and Economic School, Instituto Superior de Gestão, Lisbon, Portugal
e-mail: paulo.caldas@isg.pt

P. Caldas · G. Vara
CEG-IST, Instituto Superior Técnico, Universidade de Lisboa, Lisbon, Portugal
e-mail: guilherme.vara@tecnico.ulisboa.pt

D. C. Ferreira
Instituto Superior Técnico, University of Lisbon, Lisbon, Portugal
e-mail: diogo.cunha.ferreira@tecnico.ulisboa.pt



W. Leal Filho et al. (eds.), *Composing Worlds: Humanities, Health and Wellbeing in the XXI Century Towards a More Sustainable World*, World Sustainability Series,
https://doi.org/10.1007/978-3-031-87108-5_18

facilitating informed decision-making and policy formulation. Moreover, SMAA facilitates the integration of stakeholder preferences and values into the assessment process, promoting transparency, accountability, and inclusivity in environmental governance. By engaging stakeholders in the identification and weighting of evaluation criteria, SMAA enhances the legitimacy and acceptability of EPIs, fostering consensus-building and collaborative action towards sustainability goals. Furthermore, SMAA offers flexibility in handling diverse types of data, including qualitative and quantitative information, spatial and temporal data, and expert judgments. This flexibility enables the adaptation of EPIs to different contexts and scales, catering to the specific needs and priorities of stakeholders. Additionally, SMAA provides a systematic framework for sensitivity analysis, allowing for the exploration of the robustness of EPIs to changes in input parameters and assumptions. Through a case study application, we demonstrate the utility and effectiveness of SMAA in improving the construction of EPIs. By addressing the limitations of traditional approaches, SMAA contributes to advancing the field of environmental assessment and management, offering a valuable tool for promoting sustainable development and preserving the integrity of ecosystems for future generations.

## 1 Introduction

Public policies encompass a collection of governmental measures, choices, and endeavours that seek to examine matters across various domains, including but not limited to health, the environment, education, and economic development. The areas covered include social welfare, economic development, finance, education, health, and environment, among others (Hill and Hill 2016). According to Howlett et al. (2009), the government plays a fundamental role in public policy by making choices between several courses of action. This includes deciding to intervene in certain areas or to not take any action, which is a deliberate choice rather than a lack of action.

Public policies are intricate processes that occur at various levels of activity and decision-making, including the local, regional, national, and international levels. These measures might manifest as laws, rules, programs, and initiatives that are specifically crafted to shape or manage parts of public life. Their primary objective is to cultivate the welfare of citizens, tackle social issues, and advance the collective benefit (Hill and Hill 2016). These policies are deemed crucial for establishing a comprehensive framework that governs the conduct of the government, citizens, and all other public and private entities. They involve various stakeholders, including government officials, legislators, voters, public administration, and interest groups, with the objective of addressing public issues and facilitating the equitable distribution of power and resources (Nunes and Catarino 2024).

Public policies have a vital impact on determining environmental outcomes in the context of the environment. The reason for this is that well-executed environmental policies can alleviate pollution, safeguard natural resources, and encourage sustainable practices. Conversely, inadequately formulated policies can result in detrimental impacts on the environment (Lim 2021). To achieve optimal outcomes and ensure favourable policies, it is imperative to meticulously plan and execute each stage, including scheduling, formulation, decision-making, and implementation. This comprehensive approach will facilitate a favourable review and satisfy the desired expectations (Nunes and Catarino 2024). Hence, the formulation of public policy necessitates a well-structured and consistent approach, ensuring the active participation of all stakeholders, including the government, citizens, and other interest groups.

An essential aspect of environmental issues is ensuring the sustainable incorporation of human activities into the global ecosystem to protect the integrity of ecosystems for future generations (Böhme et al. 2022).

Regarding assessment, conventional environmental performance indicators (EPIs) are employed. These indicators rely on data pertaining to the global state of sustainability, encompassing categories such as biodiversity, air pollution, agriculture, water resources, levels of noxious gases in the atmosphere, sanitation, and access to clean drinking water, among others (Block et al. 2024).

In this book chapter, we propose a novel approach to address these limitations by integrating Stochastic Multi-Attribute Analysis (SMAA) into the construction of EPIs. SMAA offers several distinct advantages over traditional methods. Firstly, it embraces uncertainty inherent in environmental data, providing a more robust and reliable assessment of environmental performance. By considering uncertainties in input data, SMAA reduces the risk of biased conclusions and enhances the credibility of EPIs. Secondly, SMAA enables the incorporation of multiple criteria and attributes, allowing for a comprehensive evaluation of environmental performance across various dimensions such as air quality, water resources, biodiversity, and ecosystem services. This multidimensional perspective provides a more holistic understanding of environmental impacts, facilitating informed decision-making and policy formulation. Moreover, SMAA facilitates the integration of stakeholder preferences and values into the assessment process, promoting transparency, accountability, and inclusivity in environmental governance. By engaging stakeholders in the identification and weighting of evaluation criteria, SMAA enhances the legitimacy and acceptability of EPIs, fostering consensus-building and collaborative action towards sustainability goals.

Contemporary socio-ecological challenges, such as climate change, biodiversity loss, and resource depletion, necessitate a comprehensive approach to sustainability assessment. These challenges are multifaceted, impacting various aspects of the environment and society simultaneously. To address these complexities, composite indicators are constructed by aggregating multiple environmental performance indicators into a single, cohesive measure. This approach allows for a more holistic evaluation of environmental sustainability, capturing the interplay between different factors. For instance, indicators such as carbon emissions, water usage, and waste generation can be combined to provide an overall sustainability score. The construction of these composite indicators involves selecting relevant base indicators, normalizing the data, and applying appropriate weighting methods to reflect their relative importance.

## 2 Public Policies and Environment

Public policies encompass the decisions and actions taken by a government, or the lack thereof, to address specific issues across various domains such as social well-being, economic development, finance, education, health, and environment. These policies reflect the government's initiatives and aim to analyse and tackle these issues. Public policies can be classified into various categories based on different criteria and associated characteristics (refer to Table 1).

Adapted from: Nunes and Catarino (2024)

Public policies are instrumental in determining environmental outcomes, as they are deliberate choices made by the government with the explicit aim of reducing the impact of pollution, safeguarding natural resources, and encouraging sustainable behaviours. Nevertheless, governmental environmental policies may not always achieve their intended objectives. This occurs when inadequately formulated policies can have counterproductive effects and be detrimental to the environment and the overall well-being of populations. Considering the information shown in Table 1, environmental policies can be categorised as regulatory, market-based (restrictive), or development-oriented (incentive-based).

Regarding regulatory policies, these aim to regulate and control various activities to provide guarantees. Examples include:

**Table 1** Public policies definition

| Criterion/ dimension | Definition |
|---|---|
| Aim | Public policies can be proactive, aiming to face emerging challenges, or reactive, responding to existing problems |
| Coverage | Public policies can take the form of economic policies (address issues related to economic development, fiscal and monetary policies, taxation, trade and employment); social policies (focus on social issues (health, education, well-being, housing, poverty), or environmental policies (aim to manage and preserve natural resources, combat pollution and promote sustainable practices) |
| Scope | Public policies are macropolicies (address broad issues at the national level), mesopolitics (target specific sectors or industries) or micropolicies (address local issues, specific communities or groups) |
| Approach | Public policies can appear as regulatory policies (aim to regulate and control various activities to provide guarantees), redistributive policies (aim to reduce economic and social inequalities through mechanisms such as progressive taxation and social welfare programs), development policies (focus on economic growth, innovation and infrastructure development), or restrictive market-based policies |
| Length | Public policies can be permanent (aim to be lasting and consistent in the long term) or temporary (aim to address specific issues or crises with a fixed term) |
| Targets | Public policies can take the form of universal policies (apply to the entire population without discrimination) or targeted policies (focus on the needs of certain groups) |

- the Clean Air Act (CAA) authorized the U.S. Environmental Protection Agency (EPA) to establish National Ambient Air Quality Standards (NAAQS) to protect public health and the environment, is a comprehensive Federal law that regulates all sources of greenhouse gas emissions (U.S. Department of the interior, Bureau of Ocean Energy Management 2024);
- the Clean Water Act (CWA) that establishes the basic structure for regulating discharges of pollutants into the waters of the United States and regulating quality standards for surface waters, leading to the water quality improvement in the United States (Environmental Protection Agency (EPA) 2024); and
- the Endangered Species Act that:
  - establishes protections for fish, wildlife, and plants that are listed as threatened or endangered;
  - provides for adding species to and removing them from the list of threatened and endangered species, and for preparing and implementing plans for their recovery;
  - provides for interagency cooperation to avoid take of listed species and for issuing permits for otherwise prohibited activities;
  - provides for cooperation with States, including authorization of financial assistance; and
  - implements the provisions of the Convention on International Trade in Endangered Species of Wild Flora and Fauna (CITES) (U.S. Fish and Wildlife Service 2024).

Environmental regulations that have a more stringent approach are designed to consider the impact on prices and markets. Cap-and-Trade Programs implement laws that set specific limitations on the emission of pollutants. These limits are enforced by requiring enterprises to acquire or sell permits. Two examples of successful environmental programs are the Acid Rain Program in the United States and the Emission Trading System (EU ETS) in Europe. The Acid Rain Program, implemented since 1990, has effectively reduced emissions of sulphur dioxide (SO2) and nitrogen oxides (NOx), which are the main contributors to acid rain, from the energy sector. The EU ETS, on the other hand, is a crucial component of the European Union's climate change policy and serves as a cost-effective method for reducing the carbon market is the first of its kind globally and continues to be the largest under the European Commission until 2024. Another important aspect to consider in this category is the "Carbon Pricing" element, which involves the implementation of higher carbon taxes and cap-and-trade systems. These policies provide a monetary cost for carbon emissions, motivating companies and consumers to decrease their impact on the environment by emitting less carbon dioxide (Kojima and Asakawa 2021).

Environmental policies focused on development prioritise incentives, which manifest in two forms: subsidies to encourage the use of renewable energy sources like wind, solar, and bioenergy, and energy efficiency credits that reward companies and individuals investing in energy-efficient equipment, appliances, buildings, and practices (Hoinaru 2023).

Examples include the designation of protected areas such as national parks and marine protected areas. These places provide additional protection to ecosystems by promoting biodiversity and implementing sustainable forestry techniques, which ensure the long-term management of forests in a sustainable manner. It is noteworthy to mention the commendable initiative taken by numerous nations to advocate for comprehensive agreements that extend beyond their own territories, such as the Paris Agreement. This worldwide treaty on climate change aims to restrict global warming to a level much below 2 degrees Celsius over pre-industrial levels. The Montreal Protocol is an international pact created to gradually eliminate the manufacturing and use of compounds that harm the ozone layer. This protocol is supported by several industries (Cracknell and Varotsos 2012; Savaşan 2019).

The issue of the impact of public environmental policies on health is closely associated with the issue of health in all policies that although only recently discussed, has been perpetuated for several decades, namely since the very declaration that created the World Health Organization (WHO):

> "*Governments have a responsibility for the health of their peoples which can be fulfilled only by the provision of adequate health and social measures*" (World Health Organization 1946).

According to Nunes and Ferreira (2022), health in all policies "*represents an inter-sectoral approach's clear strategy, based on evidence, such that actions and policies of initiative from sectors outside health have repercussions on health and equity in access to it*". As stated by the same authors, these effects can have either bad or good consequences. Hence, enhancing the standard of living and public health metrics necessitates a united and synchronised effort from the entire government, encompassing health-related institutions and, in the context of the environment, the entire industry responsible for emissions, waste disposal, and the release of pollutants into water sources. This idea seeks to enhance the correlation between health and other sectors, such as the economy, environment, agriculture, and industry. It emphasises that the improvement of health outcomes is not solely reliant on individual actions, but also on communal factors (Greaves and Bialystok 2011; Puska 2007).

## 3 Sustainability of Human Activities and Impact in Environment

Authors who analyse the influence of human activity on the environment prioritise the sustainable integration of human activities into the global ecosystem. Human activity is responsible for significant degradation of the ecosystem, resulting in detrimental effects on the population, including premature mortality, social instability, immigration, and the risk of wars that jeopardise favourable living conditions (Mont and Power 2010). Scientific advancements and research have established the following facts: (i) elevated levels of carbon dioxide emissions have detrimental effects on the atmosphere and oceans; (ii) nitrogen oxide dissolved in water leads to global warming and ice melting, and when released through combustion, it can contribute to heart, lung, and brain diseases such as atherosclerosis and thrombosis; (iii) the transmission of nuclear energy to the oceans and the convective heat from combustion result in temperature.

Lorge (2020) argues that anthropogenic heat poses a significant ecological challenge and emphasises the need for immediate technical interventions to mitigate its effects. These interventions are crucial for transitioning out of the "thermal era" and advancing into the "electroprotonic era". According to the author, electroprotonics is a sustainable form of energy that can be generated through photoenzymatic systems. This method is cost-effective and produces green and renewable hydrogen with minimal transportation expenses. However, there are challenges and unfair competition from carbon-based products, which are driven by significant economic interests. Lorge (2020) suggests a technical solution to complement the transition to green hydrogen, which involves minimising the use of biomimicry by utilising enzymes such as redox enzymes to generate energy from water and air using sunlight. Additionally, Lorge proposes a socioeconomic solution that involves stricter enforcement of market economy laws, including imposing costs on companies that do not prioritise sustainability.

The literature widely agrees that sustainable lifestyles are crucial for achieving sustainability, by promoting a social and ecological shift towards sustainability. Nevertheless, the aforementioned literature emphasises the pressing necessity to advocate significant transformations in order to attain sustainability and foster life on Earth (IGES 2019; Akenji and Chen 2016; Gilby et al. 2019).

There is an increasing agreement regarding the connection between sustainability and sustainable lifestyles, as well as their impact on the attitudes of both the general people and politicians (Gilby et al., 2019). Nevertheless, the process of modifying behaviour to embrace healthier lifestyles necessitates a reevaluation of food production, energy consumption, and transportation. It also needs political determination to establish effective plans and achieve widespread agreement with the global industry (Böhme et al. 2022). The matter at hand involves the presence of dominant social paradigms, which differ from country to country and can either facilitate or hinder sustainability. These paradigms have an impact on individuals in terms of their values, motivation, attitudes, and beliefs, as well as on society in economic, business, institutional terms, and in their perception of the modern world (Fischer and Riechers 2019; Gilby et al. 2019; Wahl 2016).

Raworth (2018) argues that environmental impacts cannot be fully separated from economic progress, and therefore, these effects should be incorporated into countries' development measures. Thus, policies that foster sustainability must have the capacity to bring about significant transformations, given that human existence is closely linked to the overall health of the planet (Gilby et al. 2019).

## 4 Traditional Environmental Performance Indicators (EPIs)

The Yale Centre for Environmental Law & Policy and the Centre for International Earth Science Information Network Earth Institute at Columbia University have been active in recent years. The Environmental Performance Index (EPI), supported by The McCall MacBain Foundation, has been published. It offers a data-driven overview of the global level of sustainability.

The original EPI utilises 58 performance indicators across 11 thematic categories to assess and evaluate 180 nations based on their performance in the areas of climate change, environmental health, and ecosystem vitality. These indicators offer a comprehensive assessment of the extent to which countries have achieved set environmental policy objectives at a national level. Table 2 displays the factors addressed by this group, the policies encompassed, classifications, and their corresponding performance metrics.

Block et al (2024) argue that utilising these indicators enables us to identify issues, establish objectives, track patterns, comprehend outcomes, and pinpoint optimal political strategies. This is particularly valuable for policymakers, as it emphasises the importance of engaging in discussions with various interest groups, including citizens. Furthermore, it highlights the fact that the highest-income nations do not necessarily exhibit the most favourable global environmental performance.

Conducting an examination of environmental indicators is crucial and provides valuable recommendations for countries striving to transition to a more sustainable future.

## 5 The measurement and Monitoring of Environmental Performance: Novel Approach to Address These Limitations by Integrating Stochastic Multi-Attribute Analysis (SMAA)

Ensuring the sustainable development of nations worldwide as well as monitoring their environmental performance has been the mission of many entities abroad, including (but not limited to) scholars, regulators, industry, and national/international agencies. It is imperative to conduct an environmental performance assessment of systems abroad to check which systems (countries, continents, regions, etc.) are performing well and which can act as benchmarks of the ones performing poorly. That way, it should be possible to identify the best practices to adapt and adopt by the latter.

To do so, it is mandatory a benchmarking exercise that usually resorts in a set of key performance indicators (KPIs), often aggregating them into composite indicators. This kind of indicators has gained huge popularity in many research fields, capturing the attention of policymakers. EPI is precisely one of the most widely spread tool

**Table 2** Indicators used in the EPI formulation

| Policy objectives | Categories | Indicators |
|---|---|---|
| Climate Changes (30%) | Climate Change Mitigation (30%) | CO2 (7,5%)<br>CO2 to allocated carbon budget (0,5%)<br>CH4 (3%)<br>F-gases (2%)<br>N2O (1%)<br>Black Carbon (1,5%)<br>LILCF—net C flux (1%)<br>GHG trend adjusted by emissions intensity (6%)<br>GHG trend adjusted by emissions per capita (6%)<br>GHG in 2050 (1%)<br>GHG Budget Use (0,5%) |
| Ecosystem Vitality (45%) | Biodiversity & Habitat (25%) | Marine KBA protection (3%)<br>Marine habitat protection (3%)<br>MPA stringency (0,5%)<br>Protected area representativeness index (3%)<br>Species protection índex (4%)<br>Biome protection (2,5%)<br>KBA protection (2,5%)<br>PA effectiveness (0,5%)<br>Cropland in PAs (0,5%)<br>Red list índex (3%)<br>Species Habitat Índex (2%)<br>Bioclimatic resilience (0,5%) |
| | Forest (5%) | Primary forest loss (1,5%)<br>Intact forest landscapes loss (1,5%)<br>Lasting tree cover loss (1,5%)<br>Net change (0,5%)<br>Forest Landscape integrity (0,25%) |
| | Fisheries (2%) | Fish stock status (0,3%)<br>Fish catch discardes (0,4%)<br>Domestic bottom trawling (0,6%)<br>Global bottom trawling (0,6%)<br>Marine trophic índex (0,1%) |
| | Air pollution (6%) | SO2 trend (2,5%)<br>NOx Brand (2,5%)<br>02 exposure croplands (0,5%)<br>03 exposure KBAs (0,5%) |
| | Agriculture (3%) | Yield gap (1,2%)<br>Pesticida pollution (0,5%)<br>Phosphorus surplus (0,1%)<br>SNMI (1,2%) |

(continued)

**Table 2** (continued)

| Policy objectives | Categories | Indicators |
|---|---|---|
| | Water Resources (5%) | Wastewater reuse (0,5%)<br>Wastewater collection (2%)<br>Wastewater collection (2%)<br>Wastewater generated (0,5%) |
| Environmental Health (25%) | Air Quality (17%) | Ambientara-me PM2.5 from human sources (6,5%)<br>Household solidariedade fuels (6,5%)<br>Ozone (1,5%)<br>Nitrogen dióxido (1%)<br>SO2 (0,5%)<br>CO (0,5%)<br>VOCs (0,5%) |
| | Sanitation & Drinking Water (5%) | Unsafe water (3%)<br>Unsafe sanitation (2%) |
| | Heavy Metals (2%) | Lead (2%) |
| | Waste Management (1%) | Waste generation per capita (0,4%)<br>Waste (0,2%)<br>Waste recovery (0,4%) |

*Source* Environmental Performance Index 2024

using KPIs for such an exercise. While being a valuable tool on a global scale, the interpretation of the EPI's results may somehow be skewed, because of multiple reasons, identified and described below:

1. *Data Availability and Quality*: The EPI heavily relies on the availability and quality of data from various sources. In some cases, data might be incomplete, outdated, or inconsistent across countries, which can lead to inaccuracies and bias in the assessment. Certain countries, particularly developing nations, may have limited data available for certain indicators due to inadequate monitoring systems or resource constraints. This can result in incomplete assessments that do not fully capture a country's environmental performance.
2. *Indicator selection*: The process of selecting indicators for the EPI involves trade-offs and subjective decisions. Some important environmental aspects may be left out, while others might receive disproportionate attention. This can lead to a skewed representation of environmental issues.
3. *Weighting of indicators*: The EPI assigns different weights to indicators based on expert judgments. These weightings can impact the overall index score and ranking of countries. However, the subjectivity in weighting can raise questions about the fairness and objectivity of the assessment.
4. *Country comparability, cultural and regional variation, and lack of contextual information*: The EPI ranks countries based on their overall environmental performance. However, comparing countries with vastly different socio-economic,

geographic, and political contexts can be challenging and might not provide a complete picture of their efforts. The EPI does not fully account for cultural, regional, and local variations in environmental priorities and approaches. What works well in one country might not be directly applicable to another due to differences in values, governance, and circumstances.

5. *Temporal dynamics*: Environmental conditions and policies can change rapidly over time. The periodic updates of the EPI might not capture these dynamic changes accurately, leading to a lag in reflecting the current state of environmental performance.
6. *The aggregation method and the weighting scheme used*: the research team who developed this tool considered a straightforward additive model with linear value functions and questionable weights that have changed over time and are equal for all entities. However, in the words of Greco et al. (2019),

> [...] *the differential weighting scheme between units (e.g. countries) is potentially a desirable property for policymakers, because each unit chooses its own weights in such a way as to maximise its performance. Thus, any potential conflicts, for example the chosen weights not favouring any unit, are in fact dismissed.*

In other words, weights should be entity-specific to account for different policy goals for different areas. Likewise, weights should be, as much as possible, endogenously calculated. As such, the EPI model is compensatory, compensating the poor performance in some dimensions with better results in others.

7. *Deterministic nature of the index*, which does not allow for statistical inference of the outcomes.

There are other alternatives to the EPI, including the Carbon Footprint (by the Global Carbon Project and organizations like the Carbon Trust), the Human Development Index (by the United Nations Development Programme), the Environmental Impact Assessment and Life Cycle Assessment, and the discontinued Environmental Sustainability Index (by the World Economic Forum). Some consider similar indicators for their construction (differing in the aggregation method or the weighting procedure), others are not solely environmental performance assessments but highlight the interconnections between human well-being and environmental sustainability, and finally others do not consider any aggregation of indicators. Still, several of those drawbacks remain unsolved. This kind of tools is very valuable for policymakers, researchers, and the public to understand and compare environmental performance across countries. Their multidimensional assessment helps highlight areas for improvement and encourages proactive approaches to environmental sustainability. However, one should tailor a single or composite indicator capable of ranking entities in terms of their performance about environmental protection, never forgetting the three pillars of sustainability: environment, society, and economics.

This work aims at suggesting an enhanced EPI, although focusing for now on a single policy: Ecosystem Vitality (EV), which is the one with the highest weight in the construction of the original EPI, thus suggesting this as the most important policy of the three considered. The new composite EPI-EV uses the same indicators as the original tool concerning EV (to make them comparable), but the former is featured by:

- Computation: we developed a heuristic based on simulation and the cross efficiency-Data Envelopment Analysis (CE-DEA) model of Sexton et al. (1986). Simulation is based on SMAA and the so-called Hit & Run algorithm (Ferreira et al. 2018) to generate values for replacing data gaps and the cases of imperfect knowledge of data. CE-DEA improves upon classic DEA by integrating peer evaluations, resulting in a more precise and resilient measurement of efficiency. To address the problem of many optimal solutions, a cross-efficiency score is calculated for each country. This score considers the country's own performance as well as evaluations from other countries. This strategy enhances equity and uniformity by considering the performance in relation to the entire group of countries, rather than solely focusing on the individual unit's most advantageous viewpoint. As a result, CE-DEA offers a more detailed and thorough examination, which becomes especially valuable in determining rankings and comparing performance in competitive settings.
- Simplicity of interpretation: each score ranges from 0 (bottom level of performance) to 1 (upper level), or even 0 to 100, and the larger the gap between the country's score and the unit, the lower its performance in terms of EV.
- Mitigation of the imperfect knowledge of data: The Hit & Run (SMAA) runs every time a data observation is not perfectly known, replacing it with suitable candidates—for details, please refer to Ferreira et al. (2023). Given the stochastic nature of Hit & Run and its merging with CE-DEA, we must run CE-DEA for every iteration of the former algorithm (which should occur thousands of times), ending up with as many estimates of EPI-EV per country as the number of iterations (in this case, we used 1000 iterations). It gives us a statistical distribution of EPI-EV scores (rather than a deterministic score), which allows us to perform statistical inference over them as well as to retrieve confidence intervals.
- Endogeneity of weights computation: since weights assigned to each indicator composing the EPI-EV are optimized based on data themselves, they are not biased by subjective judgments of experts—rather, they are objectively estimated by the linear programming model.
- Benevolent or aggressive approach: the CE-DEA can be featured by the benevolent or the aggressive approaches, which stand for the minimization or maximization of the impact made by outstanding weights. For instance, if a country gets a weight in a specific indicator that highly benefits or penalizes its overall performance, this impact can be mitigated or exacerbated. To account for the two approaches, we run them both and estimate the average score. This way, the compensatory nature of traditional benchmarking methods likely vanishes.

Figure 1 portrays the histograms of performance scores for ecosystem vitality obtained for EPI-EV (a) and SMAA-CE-DEA (b). These figures highlight that each model skews the results differently. EPI-EV has a clear right-skewed distribution, where the average (44.25) is larger than the median. The maximum score was 73.90, achieved by Austria. Only a handful countries obtained a score larger than or equal to

70: Austria, United Arab Emirates, Luxembourg, and Slovenia. Meanwhile, countries like Germany, Slovakia, Romania, Malta, Latvia, and Croatia, all in Europe, also performed well, making these the top 10 lists of countries in EPI-EV. These countries excel in ecosystem vitality as a result of a synergistic blend of efficient environmental regulations, robust governance, and favourable geographical factors. A number of European nations have well-established environmental rules that give priority to maintaining high standards of air and water quality, protecting biodiversity, and managing resources in a sustainable manner. EU member states are obligated to adhere to strict environmental regulations, including the EU Water Framework Directive and the Natura 2000 network, which protect ecosystems and encourage biodiversity. Moreover, these nations reap the advantages of comparatively elevated levels of economic advancement, which facilitate investments in environmentally friendly technology and infrastructure, such as sustainable energy and waste disposal systems. Furthermore, there is generally a high level of public knowledge and education regarding environmental issues, which results in increased adoption of sustainable behaviours and greater support for conservation initiatives. Geographic factors also contribute to this phenomenon; several of these nations possess varied natural terrains that they place great importance on conserving, such as Austria's mountainous areas or Croatia's coastal sections along the Adriatic Sea. Non-European countries like the UAE can enhance their ecosystem vitality by making substantial expenditures in sustainability initiatives, such as water desalination and renewable energy projects, even in the face of environmental constraints like desert temperatures. Their outstanding performance in ecosystem vitality is primarily supported by a combination of policy, investment, public participation, and natural heritage.

When looking at the histogram of SMAA-CE-DEA, one can notice a skew to the left, meaning that the model gives the benefit of the doubt when computing (optimizing) weights, thus accounting for the different policies each country develops or adopts. The average score of SMAA-CE-DEA is 67.30, and the maximum is 96.10 (Slovenia). The top 10 list is composed of Slovenia, Croatia, Austria, Norway, Iceland, Hungary, Canada, Czech Rep., Bahamas, and Lithuania. Norway and Iceland benefit from vast, unspoiled natural landscapes and have strong policies for preserving biodiversity and maintaining clean energy production through hydropower and geothermal resources. Similarly, Slovenia and Croatia have invested heavily in protecting their rich biodiversity and pristine natural areas, especially in the context of eco-tourism, which is a key economic driver. Meanwhile, Canada's strong environmental performance is supported by its vast natural reserves and a robust framework for environmental protection. The inclusion of the Bahamas, a small island nation, underscores the importance of coastal and marine ecosystem management, which is critical for its economy and environmental health. These countries are known for their proactive environmental stewardship and have the resources, policies, and public support needed to maintain high levels of ecosystem vitality. Their commitment to sustainability reflects in their strong performance on such indices.

It is interesting to note that, in both approaches, no country excels in all dimensions considered in ecosystem vitality, meaning that there is still room for improvement, even for the most efficient ones. However, given the benefit of the doubt nature of

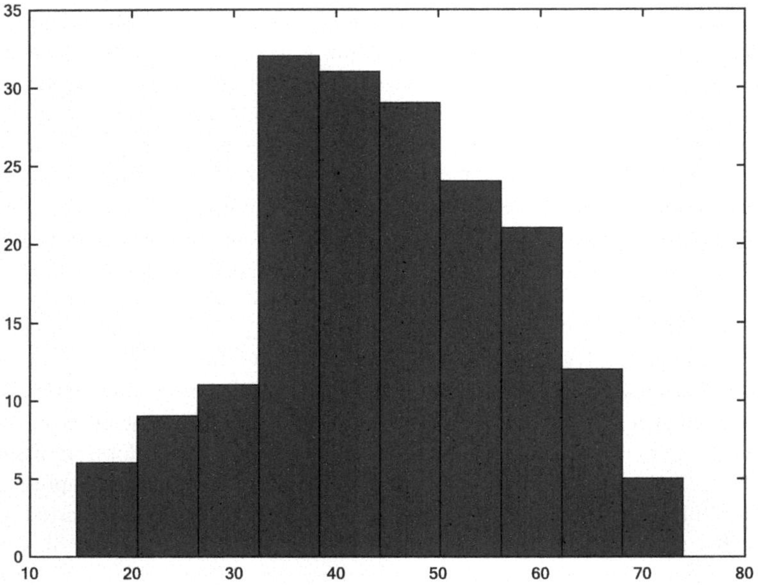


(a) EPI-EV

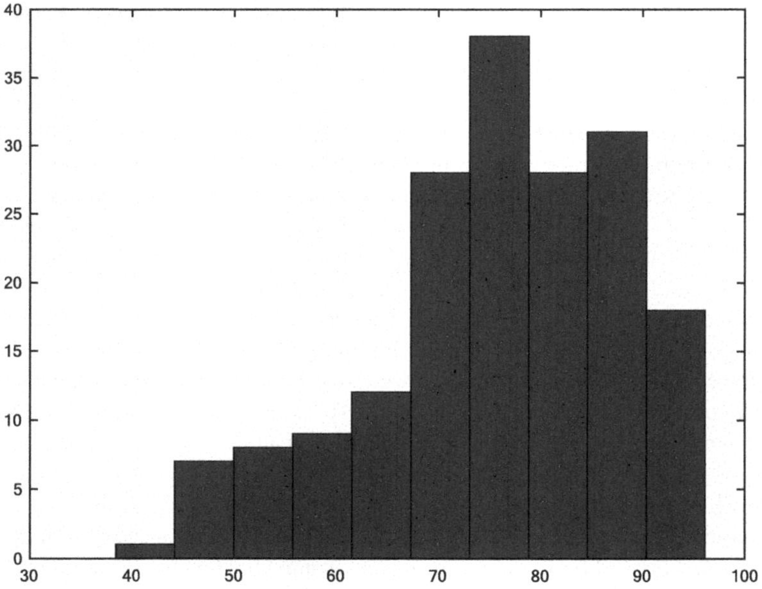


(b) SMAA-CE-DEA

**Fig. 1** Histogram of EPI-EV **a** and SMAA-CE-DEA **b** performance scores

SMAA-CE-DEA, the identified gap to excellence is smaller than the one observed for EPI-EV.

Using the available data for the EPI-EV, we constructed the map displayed in Fig. 2**a**. However, when using the raw data of the same indicators as per the original index but with SMAA-CE-DEA instead, the map in Fig. 2**b** was obtained. In these figures, three colors are used: red for low performance countries, green for the best ones, and yellow that represents the average. There are seemingly differences between both maps, especially for some countries in Middle East and Africa, which are heavily penalized by the weights chosen for the EPI-EV, thus placing them at the bottom of the ranking, while their performances scores are not that low when weights are optimized via CE-DEA. It becomes evident, however, that both approaches agree that the worst performers are in Asia (especially South Asia) and Central America.

When looking at rankings, both approaches seem to lead to meaningful differences. On average, the difference between ranking positions is about 30.6 (in a sample of 180 countries). Only in a handful of cases (19 to be precise, about 10% of the sample), that difference was 5 or less positions in the ranking: Albania, Armenia, Austria, Benin, Belarus, Dem. Rep. Congo, Dominica, Ecuador, Estonia, UK, Georgia, Guinea-Bissau, Liberia, Moldova, Nigeria, Romania, Slovenia, Eswatini, and South Africa. There is no apparent reason for such low difference other than the fact that the weights used to construct EPI-EV closely match the ones determined by the SMAA-CE-DEA model. However, those weights are only valid for these countries, while not being adequate for assessing the ecosystem viability performance of the other 161 countries.

As mentioned earlier, the original EPI-EV is purely deterministic, as the standard CE-DEA model is. Nonetheless, when introducing SMAA, results become stochastic and permit statistical inference. Figure 3 exhibits the histogram of Portugal's performance scores for ecosystem viability, and one can easily see that, beyond the bell-shape curve that could fit the bar plot, the performance score of this country cannot be smaller than 69 nor larger than 75. In fact, the 95% confidence interval for this country's score is [69.94; 74.48], which is not a wide gap. It states that this country is not performing well in terms of its ecosystem viability, exhibiting an average room for improvement of 27.64% in the related indicators—as such, Portugal occupies the 119th position in the ranking. Interestingly, when comparing Portugal to Spain, two neighbor countries, the latter's 95% confidence interval is [77.92; 83.04]. As this lower bound is higher than Portugal's upper bound, it is thus evident that Spain outperforms Portugal. Mathematically, since both histograms can be well-fitted by a Gaussian curve, the probability of Spain's performance score being larger than the Portugal's one can be easily computed: it is near 100%.

## 6 Conclusions

This study proposes a novel approach to address these limitations by integrating SMAA into the construction of environmental performance indicators.

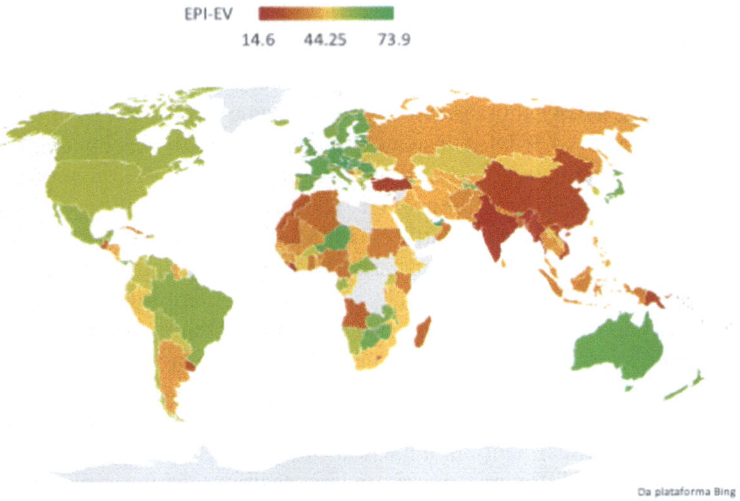


(a) EPI-EV

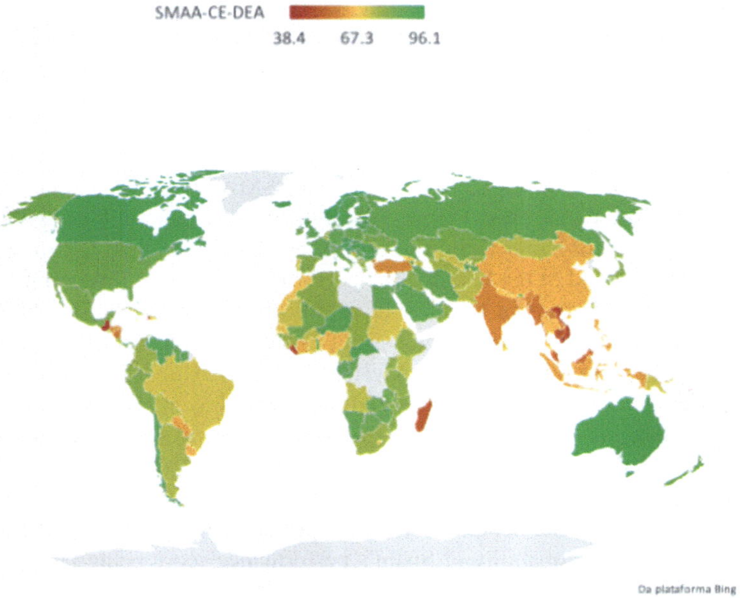


(b) SMAA-CE-DEA

**Fig. 2** Maps constructed with the EPI-EV **a** and the SMAA-CE-DEA **b** results

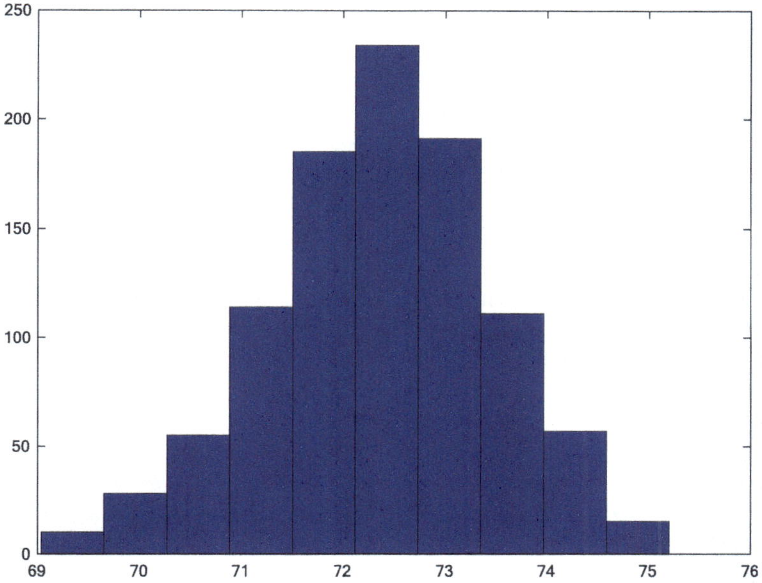


**Fig. 3** Histogram of performance estimates for Portugal

The SMAA offers flexibility in handling diverse types of data, including qualitative and quantitative information, spatial and temporal data, and expert judgments. This flexibility enables the adaptation of EPIs to different contexts and scales, catering to the specific needs and priorities of stakeholders. Additionally, SMAA provides a systematic framework for sensitivity analysis, allowing for the exploration of the robustness of EPIs to changes in input parameters and assumptions. Through a case study application, we demonstrate the utility and effectiveness of SMAA in improving the construction of EPIs. By addressing the limitations of traditional approaches, SMAA contributes to advancing the field of environmental assessment and management, offering a valuable tool for promoting sustainable development and preserving the integrity of ecosystems for future generations.

The proposed model enhances the development of effective and tangible environmental public policies by facilitating a rigorous assessment based on indicators. This assessment promotes the construction of a data base that reinforces the formulation of tailored environmental public policies for each country/region, considering the existence and significance of various variables. The development of public policies is an ongoing process involving multiple actors, including political decision-makers, public opinion, and other interested parties. However, this process is constrained by available resources, legal restrictions, and economic considerations. This will allow for better results in existing policies, allowing: (i) policies such as the Clean Air Law and the Clean Water Law to become more efficient with even more significant reductions in air and water pollution, with fewer negative effects on health, which also has increased social/economic costs; (ii) that cap-and-trade and/or carbon pricing programs are more adjusted and specific, thus allowing a greater reduction in greenhouse gas emissions (an important milestone in the fight against climate change), protecting the biodiversity and threatened species; (iii) that incentives are directed towards solving real problems and not as a source of financing based on subsidization, in order to accelerate the transition to clean energy sources and dependence on non-renewable resources; (iv) it is possible to reach international agreements that are more appropriate to the reality of each country and that, as a whole, positive results are achieved in unified cooperation against climate change.

This methodology will promote innovation and enhance the ability to meet sustainability indicators. Additionally, it will prevent the ineffective implementation of environmental policies, which can have significant economic and social consequences, such as increased costs, without yielding the desired outcomes. Simultaneously, the suggested model would consider problems of equality and justice when assigning weights to indicators. This will ensure that environmental degradation or policy costs do not disproportionately impact more disadvantaged groups.

The construction of these composite indicators involves selecting relevant base indicators, normalizing the data, and applying appropriate weighting methods to reflect their relative importance. This process not only simplifies the interpretation of complex data but also aids policymakers in identifying priority areas for intervention. By integrating diverse environmental metrics, composite indicators offer a robust tool for monitoring progress towards sustainability goals and informing strategic decision-making.

Ultimately, meticulously crafted and executed public policies grounded in scientific principles and utilising appropriate metrics will prove increasingly effective and indispensable in achieving desired outcomes, particularly in the realms of environmental preservation and sustainability. These policies will lead to notable enhancements in air and water quality, as well as the conservation of biodiversity, with a specific emphasis on transitioning to renewable energy sources. This approach can be easily replicated in several nations, making it a highly adaptable instrument that will significantly contribute to the shared goal of supporting a more sustainable environment.

## References

United States Environmental Protection Agency (EPA) (2024b). Acid rain program. https://www.epa.gov/acidrain/acid-rain-program

Akenji L, Bengtsson M (2010) Is the customer really king? stakeholder analysis for sustainable consumption and production using the example of the packaging value Chain. In: Sustainable consumption and production in the Asia-pacifc region: efective responses in a resource constrained World. Institute for Global Environmental Strategies, pp 23–46. https://www.iges.or.jp/en/pub/customer-really-king-stakeholder-analysis/en

Block S, Emerson JW, Esty DC, de Sherbinin A, Wendling ZA, et al (2024) Environmental performance index. Yale Center for Environmental Law & Policy

Böhme J, Walsh Z, Wamsler C (2022) Sustainable lifestyles: towards a relational approach. Sustain Sci 17:2063–2076
Chem Q, Braun A (2017) Protons and the hydrogen economy. MRS Energy Sustain 4:1–4
European Commission (2024) EU emissions trading system (EU ETS). https://climate.ec.europa.eu/eu-action/eu-emissions-trading-system-eu-ets_en
Cracknell AP, Varotsos CA (2012) The montreal protocol. In: Cracknell A Varotsos C (eds) Remote sensing and atmospheric ozone. Springer Praxis Books. Springer, Berlin, Heidelberg. https://doi.org/10.1007/978-3-642-10334-6_5
Environmental Performance Index (2024) EPI framework. https://epi.yale.edu/
Ferreira DC, Marques RC, Pedro MI (2018) Explanatory variables driving the technical efficiency of European seaports: an order-α approach dealing with imperfect knowledge. Transp Res Part E: Logist Transp Rev 119:41–62
Ferreira DC, Figueira JR, Greco S, Marques RC (2023) Data envelopment analysis models with imperfect knowledge of input and output values: an application to Portuguese public hospitals. Expert Syst Appl 231:120543
Fischer J, Riechers M (2019) A leverage points perspective on sustainability. People Nat 1:115–120
Gilby S, Mao C, Koide R, Watabe A, Akenji L, Timmer V (2019) Sustainable lifestyles policy and practice: challenges and way forward. Institute for Global Environmental Strategies, Havama.
Greco S, Ishizaka A, Tasiou M, Torrisi G (2019) On the methodological framework of composite indices: A review of the issues of weighting, aggregation, and robustness. Soc Indic Res 141:61–94
Hill M, Hill M, Varone F (2016) The public policy process, 7th edn. Routledge, London, United Kingdom
Hoinaru R (2023) Green economic incentives. In: Brinkmann R (ed) The palgrave handbook of global sustainability. Palgrave Macmillan, Cham. https://doi.org/10.1007/978-3-031-01949-4_115
Howlett M, Ramesh M (2009) Studying public policy: policy cycles and policy subsystems, 3rd edn. Oxford University Press, United Kingdom
Institute for Global Environmental Strategies, Alto University, and D-mat ltd (2019) 1.5-degree lifestyles: targets and options for reducing lifestyle carbon footprints. Technical report. Institute for Global Environmental Strategies, Hayama
Kojima S, Asakawa K (2021) Expectations for carbon Pricing in Japan in the global climate policy context. In: Arimura TH Matsumoto S (Eds) Carbon pricing in Japan. Economics, Law, and Institutions in Asia Pacific. Springer, Singapore. https://doi.org/10.1007/978-981-15-6964-7_1
Lim S (2021) Stability of the main policy actors, resilience of their control of the agenda, and intensified conflicts as underlying dynamics of network governance in the environmental conflict resolution process. In: Policy network ties in the dynamic process of environmental conflict resolution. Environment & Policy, vol 60. Springer, Cham. https://doi.org/10.1007/978-3-030-70855-9_3
Lorge P (2020) Sustainable integration of human activities into the global ecosystem. MRS Energy & Sustainability 7:29
Mont O, Power K (2010) The role of formal and informal forces in shaping consumption and implications for a sustainable society. Sustainability 2(7):2232–2252
Nunes AM, Ferreira DC (2022) Social inequity and health: from the environment to the access to healthcare in composite indicators, the portuguese case. In: Leal Filho W, Vidal DG, Dinis MAP, Dias RC (eds) Sustainable policies and practices in energy, environment and health research. World Sustainability Series. Springer, Cham.
Nunes A, Catarino J (2024) Methodological approaches in public policies: relevance of analysis in the context of good governance. Acad J Polit Public Adm 1(2):1–9
Raworth K (2018) doughnut economics: seven ways to think like a 21st-century economist. Chelsea Green Publishing

Savaşan Z (2019) Introduction. In: paris climate agreement: a deal for better compliance? the anthropocene: politik—Economics—Society—Science, 11. Springer, Cham. https://doi.org/10.1007/978-3-030-14313-8_1
Sexton T, Silkman R, Hogan S (1986) Data envelopment analysis: critique and extensions. New Dir Program Eval 32:73–105
U.S. Fish & Wildlife Service (2024) Endangered species act. https://www.fws.gov/law/endangered-species-act
U.S. Department of the interior, Bureau of Ocean Energy Management (2024) Air Quality Act (1967) Or The Clean Air Act (CAA). https://www.boem.gov/air-quality-act-1967-or-clean-air-act-caa
United States Environmental Protection Agency (EPA) (2024a). Summary of the clean water act. https://www.epa.gov/laws-regulations/summary-clean-water-act
Wahl DC (2016) Designing regenerative cultures. Triarchy Press, Axminster

**Alexandre Morais Nunes** Completed his Ph.D. in Health Administration (2016) at the Institute of Social and Political Sciences (ISCSP), University of Lisbon. He also holds a Master's degree in Public Administration, in Health Administration specialty (2013) from the same educational institution. He is an Associate Professor at ISCSP, where he teaches Administration, Management, Health Policies and scientific research methodologies.

**Paulo Caldas** Doctor of philosophy in economy in 2017 from the University of New England, Australia. He is director of the economics, financing and innovation unit of the portuguese industrial association chamber of commerce and industry. Associate Professor at Grupo Ensinus—Instituto Superior de Gestão and visiting Associate Professor at the University of Lisbon, Instituto Superior Técnico. He is also an adjunct associate professor at the University of New England, Australia. He has published several articles in specialized journals. He works in the area(s) of social sciences with an emphasis on economics and management.

**Guilherme Vara** Ph.D. student in Engineering and Management|Instituto Superior Técnico. Bachelor's and a master's degree in civil engineering, specializing in Urban Planning and Systems, from Instituto Superior Técnico (IST) at the University of Lisbon.

**Diogo Cunha Ferreira** Completed an integrated Master's degree in Biomedical Engineering in 2012 from the University of Lisbon, Instituto Superior Técnico and a Ph.D. in Engineering and Management in 2018 from the same educational institution. He is a Professor at Instituto Superior Técnico and works in the area(s) of Social Sciences with an emphasis on Economics and Management.

# Knowledge and Preparation of the Civil Protection Services in the Climate Change Adaptation in Mainland Portugal

**P. F. Silva, R. Dantas, H. S. Lopes, and P. Remoaldo**

**Abstract** Similar to the current global paradigm of Climate Change (CC), there was also a need in Portugal to create adaptive processes and resilience in the face of increasing extreme phenomena. The Civil Protection (CP) activity is fundamental for the prevention and control of collective risks of natural origin. With the increasing occurrence of natural hazards, it is necessary to adapt CP operational models for preparedness, adaptation and mitigation. Due to its geophysical and climatic characterization, the Portuguese mainland presents conditions conducive to the occurrence of extreme natural events, favoured by the effects of CC. Preparation for all types of climatic-atmospheric phenomena, especially episodes of extreme heat caused by CC, should be at the center of discussion and concern of local and national governments. This study aims to assess what are the main concerns and weaknesses, as well as the level of knowledge and preparation of the Municipal Civil Protection Services (MCPS) in terms of climate adaptation and responses to the quality of life of the population. For the purposes of the study, a questionnaire survey was conducted with 72 municipal CP officials in mainland Portugal to obtain primary data between March and June 2024. The *ArcGIS Survey 123* was used to collect the main opinions. The pre-test took place on March 12, 2024, with 7 people, which allowed to measure (i) the average response time; (ii) the change of the *software* to be used; (iii) the rewording of sections and questions to reduce the average response time (5 min).

P. F. Silva · H. S. Lopes · P. Remoaldo
Heritage and Territory Laboratory/IN2PAST/Department of Geography/ICS, University of Minho, Guimarães, Portugal
e-mail: franciscosilva258@gmail.com

P. Remoaldo
e-mail: premoaldo@geografia.uminho.pt

R. Dantas
Department of Geography/ICS, University of Minho, Guimarães, Portugal

H. S. Lopes (✉)
University of Barcelona, Barcelona, Spain
e-mail: helderlopes@geografia.uminho.pt

Higher School of Education of Paula Frassinetti (ESEPF), Porto, Portugal

 
W. Leal Filho et al. (eds.), *Composing Worlds: Humanities, Health and Wellbeing in the XXI Century Towards a More Sustainable World*, World Sustainability Series,
https://doi.org/10.1007/978-3-031-87108-5_19

This questionnaire survey consists of 20 questions divided into 3 sections and had the following objectives: (i) the respondent's opinion on the performance of the MCPS on the effects of CC; (ii) the implementation of territorial management instruments for adaptation, action and mitigation to CC; (iii) the degree of importance of awareness and education of MCPS. After data collection and processing, in a universe of 278 MCPS, a total of 72 responses allows us to conclude that there is still a need to reconfigure civil protection in terms of the application of operational measures and actions. The proposed methodology could be very useful for researchers in the field of civil protection and policy makers, since it contributes with adaptive measures regarding the integration in adaptation and mitigation policies.

**Keywords** Climate change · Mainland Portugal · Adaptation · Civil protection

## 1 Introduction

Climate change (CC) has been increasing the frequency and intensity of extreme events, directly affecting the resilience and response capacity of communities (Aguiar et al. 2018; Birchall et al. 2023; Huang et al. 2023). Civil Protection (CP) has evolved in response to the challenges posed by CC (Cook & Dorussen, 2021). Recent studies, such as Zandvoort et al. (2017) or Lopes et al. (2023) on adaptation pathways, highlight the importance of flexibility in the design of adaptation policies. This approach allows for the assessment of different response pathways to CC, considering the inherent uncertainty of future climate scenarios. In the Portuguese context, coastal and water resources risk management has been a priority area, with significant efforts made in the implementation of local adaptation strategies (Campos et al. 2017; Costa & Lopes, 2024). Furthermore, Aguiar et al. (2018) highlight the need to integrate local and technical knowledge in the development of effective measures to mitigate climate impacts, especially in vulnerable urban areas.

The literature also suggests that CP services face institutional and resource barriers to implementing these strategies. According to Schmidt et al. (2014), risk perceptions and planning culture influence the effectiveness of local responses to extreme events. In Portugal, the Municipal Civil Protection Services (MCPS) face a heterogeneous territory in terms of risk exposure, making coordination between different levels of government a fundamental challenge for effective climate adaptation (Amaro, 2023; Bento-Gonçalves, 2021).

This chapter explores the main concerns, weaknesses, and the level of preparedness of the MCPS in mainland Portugal, based on data collected from 72 municipal officials. The methodology adopted could significantly contribute to the development of adaptation policies and more effective mitigation strategies in the context of CC.

At present, CC is considered a global challenge with the potential to impact human health, with the exacerbation of vector-borne infectious diseases being one of the most serious risks. Parasitic and infectious diseases are one of the most sensitive causes of death in CC (Alves et al. 2023). In addition, pre-existing socio-economic conditions are a determining factor in the spread of these types of diseases, especially among the poorest communities, where access to basic health care services or to clean water supply and sanitation is low or non-existent.

This scenario requires the adaptation of CP operational models in the face of phenomena such as heatwaves, floods, and wildfires, which are becoming more frequent in mainland Portugal and can lead to the displacement of communities and disrupt health systems, making them more susceptible to disease (Githeko et al. 2000; Campbell-Lendrum et al. 2015; Alves et al. 2023). Adapting CP to CC aims to reduce collective risks and increase the resilience of populations through the implementation of prevention, preparedness and mitigation measures.

Based on these studies, the adaptation of municipal CP actors in Portugal requires a focus on the dynamics of inter-institutional cooperation, technical capacity building and community involvement, which are essential for the implementation of policies that consider local and regional specificities.

## 2 Importance of Civil Protection and Operational Challenges in Implementing Adaptive Measures to Climate Change

The increasing frequency and intensity of extreme climate events, such as heatwaves, storms, and floods, highlight the need for civil protection structures worldwide (Arnell, 2022; Hua et al. 2021; Zandvoort et al. 2017) and for Portugal to adapt. CC not only increases the vulnerability of infrastructure and populations, but also poses new challenges in terms of coordination and the implementation of adaptation measures (Salimi & Al-Ghamdi, 2020). In addition, the increase in extreme weather events, mentioned above, increases the level of risk of disease outbreaks, with a particular focus on vulnerable social groups such as the elderly and children. Awareness of these risks matter must be the target of measures that ensure the prevention and surveillance of vector-borne pests and diseases.

CP plays a central role in risk mitigation, but its success depends on the effective integration of adaptation strategies that consider both mitigation and community resilience (Aguiar et al. 2018; Haque & Doberstein, 2021; Lopes et al. 2023, 2022).

The adaptive pathways approach, as proposed by Zandvoort et al. (2017), emphasizes the importance of adaptive planning that incorporates multiple options and scenarios to mitigate the uncertainty associated with climate projections. This model allows decision-makers to plan based on adaptive tipping points, identifying critical moments when strategies can be adjusted or replaced (van Ginkel et al. 2022). Alves et al. (2023), highlights the urgent need to create new operational systems, but also systems for monitoring and identifying areas most vulnerable to these risks. In Portugal, the diverse geography and population distribution require personalized adaptation strategies based on local and regional data to identify areas most susceptible to natural disasters (Bittencourt et al. 2024; Figueiredo et al. 2020).

CP adaptation should thus be understood as an ongoing process involving the participation of various stakeholders, including local and national governments and the private sector (Anastasopoulou, 2023), to ensure the effectiveness of responses to the challenges posed by CC. Campos et al. (2017) point out that, while there are significant efforts at the municipal level, institutional barriers and limited resources continue to hinder the implementation of effective adaptation policies.

The main challenges that Municipal Civil Protection Services (MCPS) face in adapting to CC are related to institutional barriers, lack of resources, and a planning culture focused on technical solutions rather than integrated strategies (Amaro, 2023; Bento-Gonçalves, 2021). According to Schmidt et al. (2014), risk perception is influenced by planning culture and the lack of community awareness, limiting the acceptance of preventive and adaptive policies. This is particularly relevant in Portugal, where extreme climate events such as wildfires and floods have a significant impact on the most vulnerable areas (Aguiar et al. 2018; Campos et al. 2017; Schmidt et al. 2014).

The literature identifies that the fragmentation of responsibilities across different levels of government is one of the main obstacles to implementing adaptive measures (Aguiar et al. 2018). In Portugal, municipalities have limited autonomy in applying adaptive policies and are often dependent on state resources or European programs (Campos et al. 2017; Cossu, 2020; Zandvoort et al. 2017). Furthermore, inter-institutional coordination is crucial to ensuring the effectiveness of emergency responses. In this sense, the integration of adaptation policies into Municipal Emergency Plans and the strengthening of the technical and financial capacities of MCPS are essential steps to face climate risks (Campos et al. 2017).

# 3 Methods and Data

## 3.1 Case Study of Mainland Portugal

Portugal is located in south-eastern Europe, with the Portuguese mainland bordering the Atlantic Ocean to the north and west, the Mediterranean Sea to the south and Spain to the east, forming the Iberian Peninsula. Administratively, mainland Portugal has 278 municipalities, divided into 23 NUTS III regions in 2022 (Fig. 1).

Located approximately between latitudes 37° N and 42° N and longitudes 9.5° W and 6.5° W, in the transition zone between the subtropical anticyclone (Azores Anticyclone) and the subpolar low-pressure zone, the Atlantic Ocean strongly influences the Portuguese climate due to its proximity (Miranda et al. 2006). The orographic factor is decisive for the climate, with areas in the north and centre exceeding 1000 metres in altitude.

Temperatures tend to be higher in the summer months (June-September) and lower in the winter months (December-February). According to the 1961–90 climatic normal, the average annual temperature varies between 7°C in the northern and

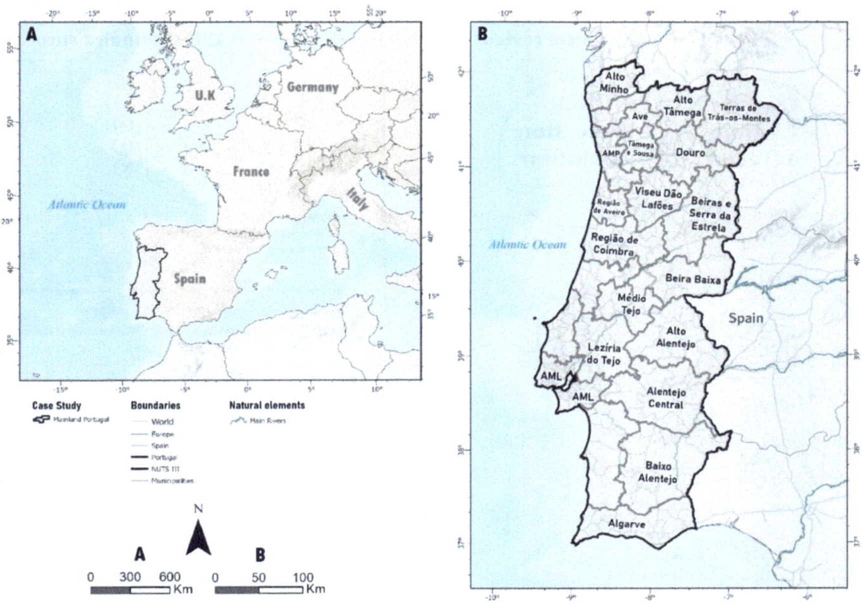


**Fig. 1** European view (**a**) and case study of mainland Portugal by NUTS III and municipalities (**b**). *Source* Authors' own elaboration

central interior highlands and about 18°C on the south coast (IPMA, 2011). Rainfall tends to be more abundant in the autumn and winter months, especially in December. Conversely, the lowest rainfall is recorded in the summer months, especially in August. Based on the 1961-90 climatological normal, the average annual rainfall is highest in Minho and Douro Litoral, and lowest in the interior of Baixo Alentejo (IPMA, 2011).

Given its geographical location and climatic context, Portugal is particularly vulnerable to the impacts of CC (Carvalho et al. 2011). Sea level rise, coastal erosion, increased air temperature, and dry seasons, among other impacts, are the most likely to affect the Portuguese territory due to its proximity to the Atlantic Ocean and the Mediterranean Sea.

## 3.2 *Procedures and Data*

This study was carried out between November 2023 and March 2024 and was divided into three distinct phases. The first phase, related to the literature review, made it possible to gain knowledge about the relationship between CP and CC and how the emergency preparedness team should approach these increasingly common risks.

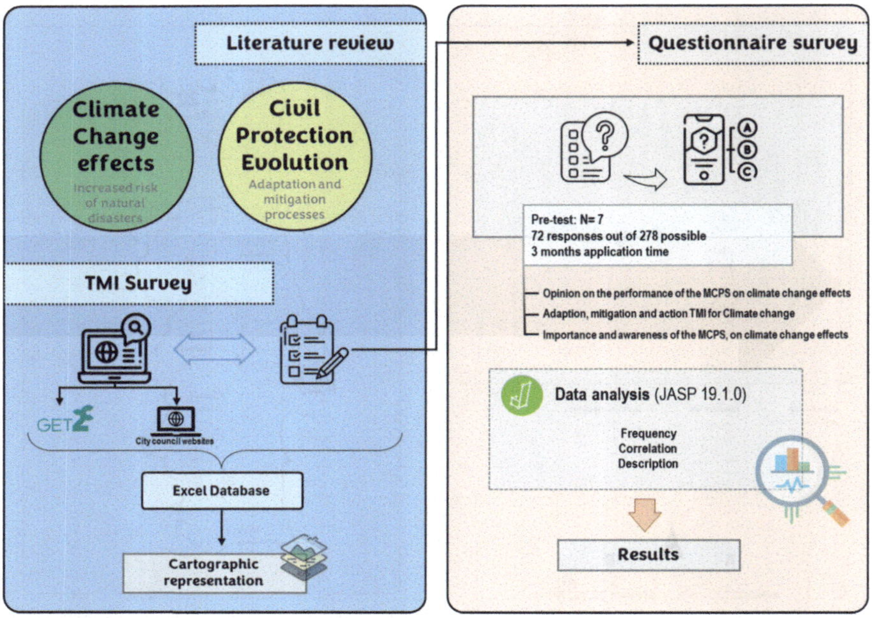


**Fig. 2** Research methodological scheme. *Source* Authors' own elaboration

The second part of the research consisted in the collection, storage and subsequent cartographic representation of the elaboration of the different Territorial Management Instruments (TMI) of adaptation, mitigation and local climate action. These data allow us to assess, geographically, the degree of preparation and the importance that each municipality attaches to risk planning issues related to the consequences of the impacts of CC.

The final phase is to obtain real information on the opinion of Municipal Civil Protection Services (MCPS) on CC-related issues. For this purpose, a questionnaire was sent to the 278 MCPS of mainland Portugal. Finally, the data was analysed using the JASP statistical software, and the results were presented. Figure 2 illustrates the methodological development of research.

### 3.2.1 Gathering Adaptation, Mitigation and Action of Territorial Management Instruments (TMI) of Climate Change

The study also looked at ways in which the different types of plans could work together. Although there is a need to explore the complementary and subsidiary relationships between these plans, it was clear that the municipal TMI must always consider information from the regional TMI. This can help to adapt to climate change in different ways. Figure 3 shows our two-stage approach to data collection.

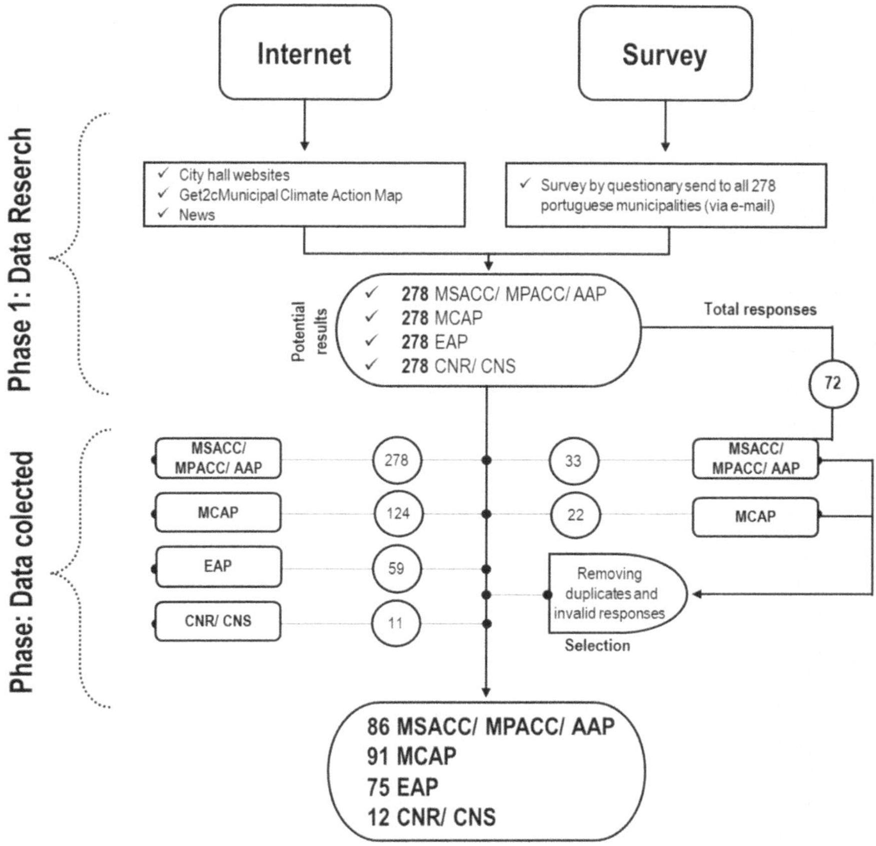


**Fig. 3** Two-stage approach to data collection of TMI. *Source* Authors' own elaboration

The first phase was to collect information from the municipalities through a questionnaire. The second approach is to search for the CC TMI online. It used four ways of finding information: (i) Municipality official websites; (ii) News about the preparation of plans; (iii) Email contact with municipal institutions; (iv) GET2C Municipal Climate Action Map. This longer method was more accurate and reliable. It found flaws and gaps on the GET2C website and in the responses received from the questionnaire.

### 3.2.2 Data Treatment

The data was stored on an Excel spreadsheet (see Appendix 1). We marked the municipalities that have CC TMI, are preparing for it and don't have it. We checked the information from the questionnaires against the GET2C website, the municipality websites and news to make sure it was correct and up to date. Despite our best efforts, it's possible that some plans were not marked.

To meet the goals and show the TMI data on a map, it was used with ArcGIS Pro. The tool was used to check which municipalities have 1 to 4 of the mentioned TMI. This method showed how prepared each municipality is for climate change planning and identified areas that are not prepared for the risks posed by CC.

Once the data had been collected, it was stored in an Excel spreadsheet database and subsequently treated.

### *3.3 Questionary Survey to Analyse the Relationship Between Municipal Civil Protection Services (MCPS) and Climate Change (CC)*

The utilisation of questionnaire surveys as a methodology enables the comprehensive evaluation of a group of individuals' perceptions on a specific topic at a macro level (Sá et al. 2021; Santos & Henriques, 2021).

To obtain information and opinions from each of the 278 municipal institutions in mainland Portugal, a questionnaire survey was designed. This allowed for the collection of reliable results with analytical capacity, based not only on instruments developed specifically for this purpose but also on existing literature on the subject.

The survey was based on previous studies, including those by Naess et al. (2005), Groven et al. (2012), Velosa (2015) and Campos et al. (2017). Additionally, new questions were included based on findings from the literature review that were deemed relevant for the Portuguese civil protection system and its dynamics.

The questionnaire survey (see Appendix 2) was designed to address pertinent questions regarding the policies adopted within the context of climate change (CC) at the municipal level in mainland Portugal. It sought to gain insights from agents of the Ministry of the Sea, Ports and Climate (MSPC) on the relationship between the impacts of CC and the dynamics of action in matters of adaptation and mitigation by municipal teams.

The initial version was created in Word and the pre-test phase commenced. However, for reasons of ease of communication of answers and consequent ease of reading of the results, the same survey was conducted online via Google Forms®.

Following an initial review of the proposed questions, the different sections and subsequent questions were reformulated with the aim of facilitating the respondent's response and reducing the overall response time. Subsequently, the decision was taken to employ the Survey Software 123, which enabled the storage of data and their georeferencing by municipality in mainland Portugal. This constituted the final version.

The pre-test was conducted between November 1, 2023, and December 3, 2024, following seven iterations of the questionnaire.

### 3.3.1 Questionnaire Structure and Sample

The survey was structured in three parts to evaluate the risk perception. The first part pertains to characterisation and aims to classify and verify plans, strategies and measures, mandatory or optional, that each municipality presents in the areas of civil protection and climate change. In the civil protection area, the survey considers the potentiality and degree of operation of the SMPC and the elaboration of the Municipal Emergency and Civil Protection Plan (MECPP). In the climate change area, the survey considers the Municipal Strategy for Adaptation to Climate Change (MSACC), the Municipal Plan for Adaptation to Climate Change (MPACC) and the Municipal Climate Action Plan (MCAP), or another document relevant to the theme.

Following the administration of the questionnaire survey, it was determined that the Energy Action Plans (EAP) and the Municipal Carbon Neutrality Strategies/ Roadmaps (CNR/CNS), which were developed subsequently, were not included. It was requested that an annex be sent with the questionnaire response for all plans and instruments. The responses to this initial section then permitted the execution and comparison of the municipal (TMI) survey, which was then contrasted with the GET2C Climate Action Map and the Portuguese Environment Agency (PEA) survey.

Subsequently, it became necessary to assess the risk pressure levels of each contracted agent in each municipality, with the understanding that the instrument referenced would prevail. To facilitate the response to this second part, the principal risks were identified in accordance with the National Risk Assessment (2019), as well as the most prevalent risks in the previously revised MSACC. Nevertheless, respondents were permitted to provide supplementary information regarding any additional potential risks not previously mentioned in the form of an open question.

The third section of the questionnaire included questions on prevention, adaptation and mitigation measures, which allowed the respondents to provide open answers regarding their level of concern and the strategies employed by the municipality for environmental education and training. Finally, respondents were asked about the relationship between the Regional TMI (Intermunicipal Plans for Adaptation to Climate Change/Metropolitan Plans for Adaptation to Climate Change (IPACC and MPACC)) and the most recent Regional Climate Action Plans (RCAP), with the measures and strategies implemented at the municipal level, to address a personal curiosity.

A total of 72 responses were collected, representing 27% of the 278 possible responses (see Appendix 3).

## 4 Spatial Mapping of Territorial Municipal Instruments (TMI) to Climate Change (CC)

The results of the survey, as illustrated in Fig. 4, permit the geographical representation of the various plans presented by each municipality in mainland Portugal.

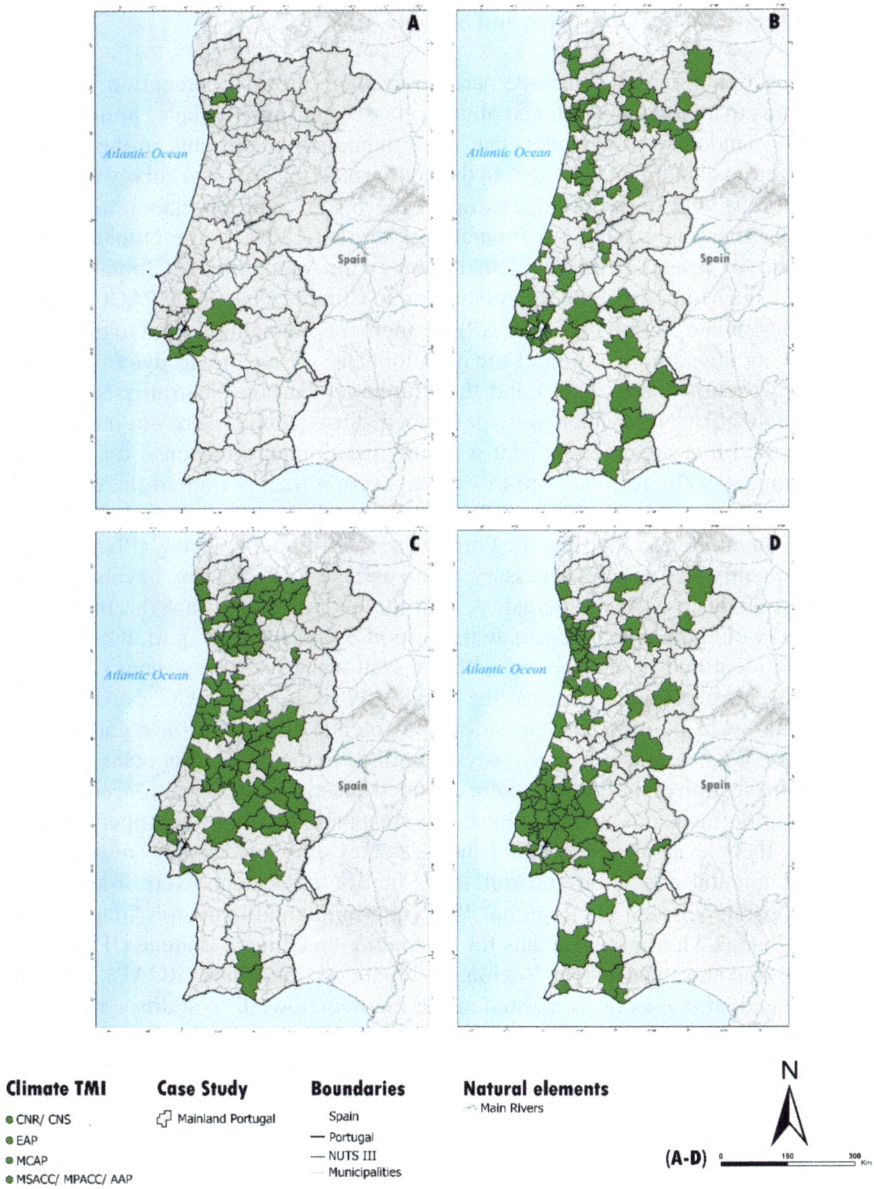


**Fig. 4** Municipal climate change territorial management instruments representation, **a**—CNR/CNS; **b**–EAP; **c**—MPAC & **d**—MSACC/MPACC/*AAP*. *Source* Authors' own elaboration

The Municipal Plan for Climate Action (MPAC) is the most representative, with 91 municipalities having presented the plan and 11 currently under preparation. The greatest representativeness of this IMT is related to its mandatory nature, which arises from the Climate Framework Law. This law demanded that all Portuguese municipalities must have a MPAC ready by the first day of February 2024 (Law No. 98/2021, of 31 December). However, despite its mandatory nature, not all municipalities have presented one.

In contrast, the Roadmap to Carbon Neutrality (CNR/CNS) is less prevalent, with 264 municipalities not presenting this TMI and only two municipalities currently in the process of developing one. This can be attributed to the similarity of this TMI with the Energy Action Plans (EAP).

At the EAP level, our findings indicate that 195 municipalities (70.1%) lack an Energy Action Plan, while only 8 municipalities are in the preparatory phase. It is also noteworthy that this TMI is applicable at the intermunicipal level (NUTS III), with only two intercity communities (CIM) currently utilising it: the CIM of Alto Minho and the Region of Coimbra. However, despite the EAPs of the CIMs in question having been presented, it was found that some municipalities had not yet submitted their respective EAPs. In the CIM of Alto Minho, of the 10 member municipalities, half (5) are not in possession of the plan and are not currently engaged in its preparation. Regarding the CIM of the Region of Coimbra, of the 19 municipalities that belong to it, 14 do not have the EAP, and it is probable that they are utilising the TMI developed by their CIM.

Regarding the MSACC/MPACC/AAP, which are the adaptation plans, it was found that 179 municipalities lack this TMI (64.4%), while 86 municipalities have developed their respective climate change adaptation plan. Figure 4 illustrates that most municipalities that have developed this TMI are situated on the coast, near metropolitan areas such as Porto and Lisbon. At the level of the CIM, an intermunicipal/metropolitan adaptation plan to climate change (IPAAC/MPACC) was identified, which was found to be very similar to the TMI. About the latter, only one CIM, the Aveiro Region, is not in possession of such a plan (see Appendix 4). Figure 5 shows the total number of TMI for each municipality. Smaller municipalities have fewer TMI related to CC. This makes them less informed and less able to adapt to the risks of CC.

Some large cities are better prepared for climate planning than smaller ones. The larger the population, the greater the risk. This forces entities to learn more about the risks in their area. This helps them respond to the public.

All 23 intermunicipal communities (CIM) must present a Regional Climate Action Plan (RCAP) by February 1, 2024. This is in addition to the intermunicipal plans for adaptation to climate change and the obligation to prepare the MPAC. This is in addition to the intermunicipal plans for adaptation to climate change, which are required to be prepared in accordance with the climate framework law, and the obligation to prepare the MPAC. The results are even worse than at the municipal level. Only eight northern CIMs have started to create a regional climate action plan (Fig. 6). This plan will have five sections.

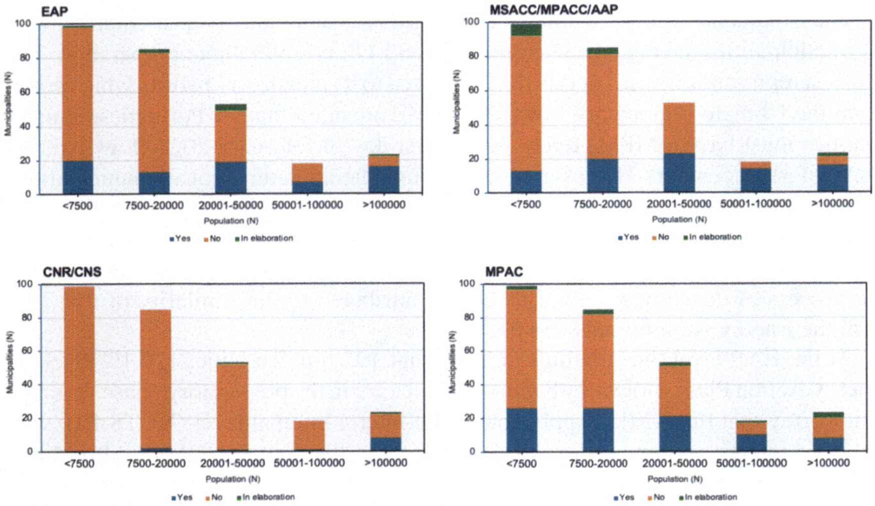


**Fig. 5** Chart of elaboration of municipal CC TMI by population clusters. *Source* Authors' own elaboration

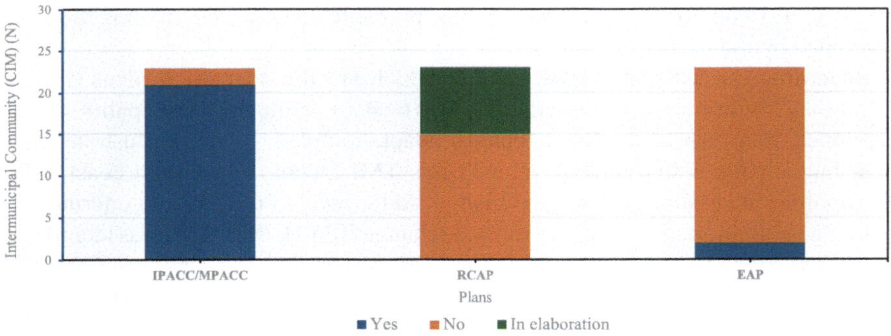


**Fig. 6** Regional and Intermunicipal CC TMI preparation chart. *Source* Authors' own elaboration

It is notable that the municipalities which only have one plan represent the largest group (116 municipalities). Conversely, the number of municipalities that have not adopted any TMI is also noteworthy, indicating a potential lack of concern regarding the impacts of climate change. It is also noteworthy that the municipalities of Guimarães, Porto, Coruche and Setúbal have demonstrated a commendable commitment to addressing climate change by presenting the four TMI plans under consideration. Conversely, the municipalities belonging to the NUTS III regions of Viseu Dão Lafões and Alentejo Litoral have not yet formulated any plans. Even though the research results do not directly indicate that municipalities that did not prepare the TMI´s are more vulnerable to the risks arising from climate change, it necessarily implies a lower capacity for adaptation and resilience, provided from the lack of clear identification of what these risks are.

While the vast majority of CIMs include an IPAAC plan, which identifies and maps the primary exposure risks and delineates vulnerable areas, municipal plans play a significant role in enhancing the level of detail and pinpointing circumstances within each territory.

### *4.1 Risk Perception and Analysis*

The responses to the questionnaire survey submitted to the municipalities of mainland Portugal were analysed to identify the principal risks faced by the respondents. The results are presented in Fig. 6.

By analysing the risks presented, a similarity between municipalities about the main risks they face was identified. This was based on the responses, which focused on the risks of extreme heat, rural fires, excessive precipitation, floods and drought. These risks were found to be more prominent in mainland Portugal. The most significant risk identified is that of extreme heat, with 63 municipalities indicating this as one of the principal risks they face. Almeida (2012) posits that the unrelenting rise in global temperatures will precipitate a surge in the frequency and intensity of extreme heat events. Conversely, most municipality' agents that responded ($n = 61$) identified rural fires as a significant risk. Despite a decline in the extent of burned areas, Portugal has witnessed a proliferation of extensive forest fires (Santos et al. 2023).

From an alternative standpoint, it was noted that only one municipal agent (that of the municipality of Vidigueira) identified frost and fog as a significant risk factor affecting the area. It is, however, crucial to underscore the fact that the municipalities in question perceive the dissemination of diseases and vectors as one of the most significant risks inherent to their respective territories. It is plausible that the effects of CC will manifest in human health through the emergence of novel diseases from other geographic regions or the resurgence of previously existing infectious diseases (Alves et al. 2023). A total of 13 municipalities identified this risk, namely: Guimarães, Mirandela, Moura, Oliveira de Frades, Ovar, Penacova, Penela, Póvoa de Lanhoso, Santa Comba Dão, Santa Maria da Feira, Santarém, Vila Pouca de Aguiar, and Viseu.

The georeferencing of the municipalities (see Fig. 7) revealed that most municipalities are situated in the northern region of the mainland. A review of the population data for the municipalities in question revealed that 8 have a population cluster between 7,501 and 20,000 inhabitants. Additionally, the municipalities of Guimarães and Santa Maria da Feira merit mention, as they have a population exceeding 100,001 inhabitants. Consequently, of the 13 municipalities, 10 have identified this risk as a principal one within their respective territories, due to the elevated population figures, which directly influence their susceptibility to disease and vector transmission (Figs. 8 and 9).

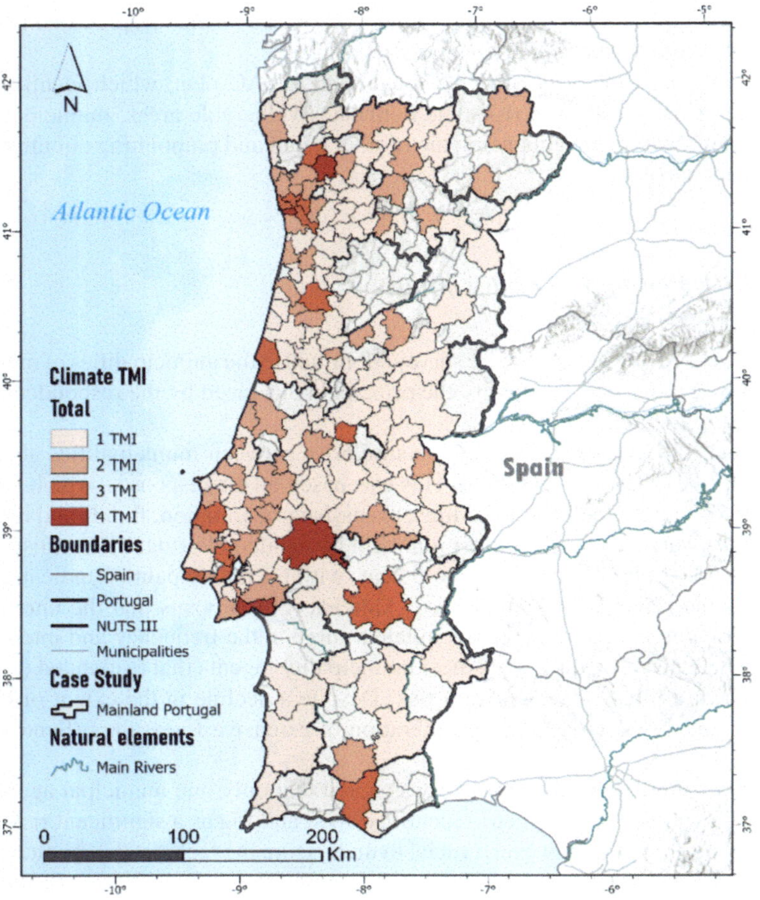


**Fig. 7** Municipal climate change territorial management instruments (TMI) (total). *Source* Authors' own elaboration

# 5 Conclusion

The growing prevalence of extreme weather events, exacerbated by the impacts of climate change, is a global phenomenon, particularly evident in mainland Portugal. This underscores the necessity for the development of adaptive strategies and pathways across diverse sectors.

It is thus imperative to establish a connection between the planning of measures to mitigate the impact of these risks and the coordination of prevention and relief teams. In Portugal, as previously stated, the National Emergency and Civil Protection Authority (NECPA) is responsible for overseeing civil protection efforts at the

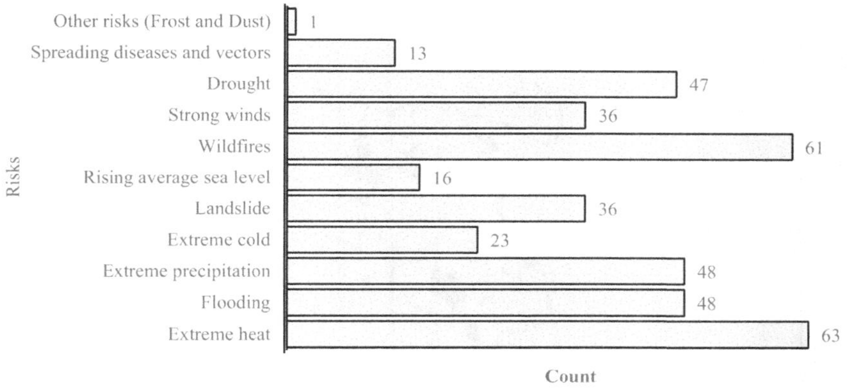


**Fig. 8** Representation of the main risks, stated by the municipalities, during the questionnaire survey. *Source* Authors' own elaboration, based on 72 responses of MPCS

national level. At the municipal level, the NECPA is represented by municipal civil protection services.

The establishment of consistent information channels enables prevention and action teams to collectively gain insight into the impacts of climate change.

It is thus imperative to reestablish and reorganise the existing MCPS in Portugal and initiate the integration of adaptation issues. In this regard, the municipal level plays a pivotal role in identifying the nuanced risks experienced in each territory, thereby enabling the formulation of bespoke adaptation measures and, when necessary, direct action in the catastrophic events.

In terms of limitations, the questionnaire survey proved to be the most challenging aspect of the investigation. Despite repeated efforts to establish contact with the relevant municipalities via email and telephone, the number of responses received has been disappointingly low. Communication with public institutions represents a significant challenge in the process of investigating municipal and regional policies in Portugal. Conversely, the number of distinct TMI sources about CC represented a challenging and time-consuming process of compilation. The availability of an updated and reliable database renders this work relevant and important in terms of spatialising the level of preparedness of each municipality and establishing the relationship between civil protection and the effects of CC.

On this subject, in Portugal, the existing bibliography is not very extensive. Studies like this have already been prepared, however, they all reveal the normal lack of knowledge and interconnection between MCPS and climate issues, motivated by different logistical obstacles. This study, through the results of the questionnaire, also reveal these consequences. We thus outline the need to modify the relationship between MCSP and current climate issues. There is an urgent need for greater inter-institutional interconnection between the different preparation teams for the various plans mentioned and the emergency services.

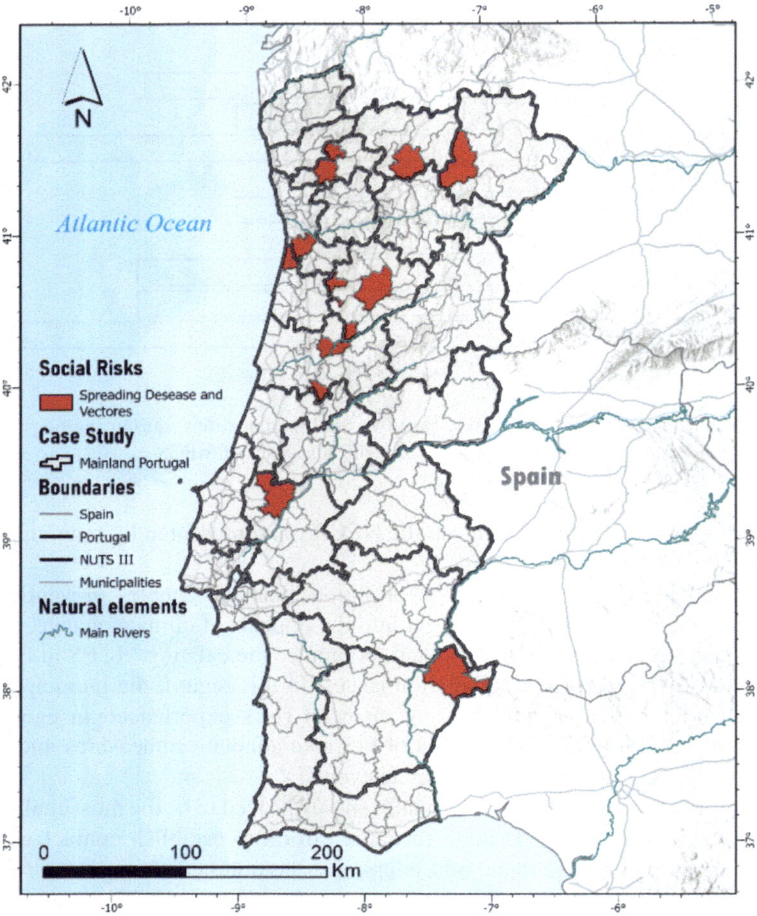


**Fig. 9** Geographical referencing of municipalities, respondents to the questionnaire survey, which are affected by the spread of diseases and vectors. *Source* Authors' own elaboration, based on 72 responses of MPCS

## Abbreviations and Acronyms

| | |
|---|---|
| AAP | Climate change adaptation plan |
| AML | Lisbon metropolitan area |
| AMP | Porto metropolitan area |
| CC | Climate change |
| CNR | Carbon neutrality roadmap |
| CNS | Carbon neutrality strategy |
| CP | Civil protection |
| EAP | Energy action plan |

IPAAC Intermunicipal climate change adaptation plan
IPMA Portuguese institute of sea and atmosphere
MCPA Municipal climate action plan
MCPS Municipal civil protection services
MECPP Municipal emergency and civil protection plan
MPACC Metropolitan climate change adaptation plan
MSACC Municipal climate change adaptation plan
NECPA National emergency and civil protection authority
NUTS Nomenclature of territorial units for statistical purposes
RCAP Regional climate action plan
TMI Territorial management instruments

## Appendix List

### Appendix 1—Database Construction, for Geographic Representation, and Analysis

| DICOFRE | Population | Municipalities | EAP | MSACC/ MPACC/ AAP | CNR/ CNS | MPAC |
|---|---|---|---|---|---|---|
| 1401 | 20,001–50,000 | Abrantes | 2 | 2 | 2 | 1 |
| 0101 | 20,001–50,000 | Águeda | 1 | 1 | 2 | 1 |
| 0901 | > 7500 | Aguiar da Beira | 2 | 2 | 2 | 3 |
| 0701 | > 7500 | Alandroal | 2 | 3 | 2 | 2 |
| 0102 | 20,001–50,000 | Albergaria-a-Velha | 2 | 2 | 2 | 1 |
| 0801 | 20,001–50,000 | Albufeira | 2 | 2 | 2 | 2 |
| 1501 | 7501–20,000 | Alcácer do Sal | 2 | 2 | 2 | 2 |
| 1402 | 7501–20,000 | Alcanena | 2 | 1 | 2 | 1 |
| 1001 | 20,001–50,000 | Alcobaça | 2 | 1 | 2 | 2 |
| 1502 | 7501–20,000 | Alcochete | 1 | 2 | 2 | 2 |
| 0802 | > 7500 | Alcoutim | 2 | 2 | 2 | 2 |
| 1101 | 20,001–50,000 | Alenquer | 3 | 1 | 2 | 2 |
| 0401 | > 7500 | Alfândega da Fé | 1 | 1 | 2 | 2 |
| 1701 | 7501–20,000 | Alijó | 2 | 2 | 2 | 2 |
| 0803 | > 7500 | Aljezur | 1 | 2 | 2 | 2 |
| 0201 | > 7500 | Aljustrel | 2 | 2 | 2 | 2 |
| 1503 | > 100,001 | Almada | 1 | 2 | 3 | 3 |
| 0902 | > 7500 | Almeida | 2 | 2 | 2 | 2 |

(continued)

(continued)

| DICOFRE | Population | Municipalities | EAP | MSACC/ MPACC/ AAP | CNR/ CNS | MPAC |
|---|---|---|---|---|---|---|
| 1403 | 7501–20,000 | Almeirim | 2 | 1 | 2 | 2 |
| 0202 | > 7500 | Almodôvar | 2 | 1 | 2 | 1 |
| 1404 | > 7500 | Alpiarça | 2 | 1 | 2 | 2 |
| 1201 | > 7500 | Alter do Chão | 2 | 2 | 2 | 2 |
| 1002 | > 7500 | Alvaiázere | 1 | 2 | 2 | 2 |
| 0203 | > 7500 | Alvito | 2 | 2 | 2 | 2 |
| 1115 | > 100,001 | Amadora | 2 | 2 | 2 | 2 |
| 1301 | 20,001–50,000 | Amarante | 2 | 2 | 2 | 1 |
| 0301 | 7501–20,000 | Amares | 2 | 2 | 2 | 2 |
| 0103 | 20,001–50,000 | Anadia | 1 | 2 | 2 | 1 |
| 1003 | 7501–20,000 | Ansião | 2 | 2 | 2 | 2 |
| 1601 | 7501–20,000 | Arcos de Valdevez | 2 | 3 | 2 | 2 |
| 0601 | 7501–20,000 | Arganil | 2 | 2 | 2 | 2 |
| 1801 | > 7500 | Armamar | 1 | 2 | 2 | 2 |
| 0104 | 7501–20,000 | Arouca | 2 | 1 | 2 | 3 |
| 0702 | > 7500 | Arraiolos | 2 | 3 | 2 | 2 |
| 1202 | > 7500 | Arronches | 2 | 2 | 2 | 1 |
| 1102 | 7501–20,000 | Arruda dos Vinhos | 2 | 1 | 2 | 2 |
| 0105 | 50,001–100,000 | Aveiro | 2 | 1 | 2 | 1 |
| 1203 | > 7500 | Avis | 1 | 2 | 2 | 1 |
| 1103 | 7501–20,000 | Azambuja | 2 | 1 | 1 | 3 |
| 1302 | 7501–20,000 | Baião | 2 | 2 | 2 | 1 |
| 0302 | > 100,001 | Barcelos | 3 | 2 | 2 | 3 |
| 0204 | > 7500 | Barrancos | 2 | 2 | 2 | 2 |
| 1504 | 50,001–100,000 | Barreiro | 1 | 1 | 2 | 2 |
| 1004 | 7501–20,000 | Batalha | 2 | 2 | 2 | 2 |
| 0205 | 20,001–50,000 | Beja | 1 | 2 | 2 | 2 |
| 0501 | > 7500 | Belmonte | 2 | 2 | 2 | 2 |
| 1405 | 20,001–50,000 | Benavente | 2 | 1 | 2 | 2 |
| 1005 | 7501–20,000 | Bombarral | 2 | 1 | 2 | 2 |
| 0703 | > 7500 | Borba | 2 | 3 | 2 | 2 |
| 1702 | > 7500 | Boticas | 2 | 2 | 2 | 1 |
| 0303 | > 100,001 | Braga | 1 | 1 | 2 | 1 |
| 0402 | 20,001–50,000 | Bragança | 1 | 1 | 2 | 2 |
| 0304 | 7501–20,000 | Cabeceiras de Basto | 2 | 2 | 2 | 1 |
| 1104 | 7501–20,000 | Cadaval | 2 | 1 | 2 | 2 |

(continued)

(continued)

| DICOFRE | Population | Municipalities | EAP | MSACC/ MPACC/ AAP | CNR/ CNS | MPAC |
|---|---|---|---|---|---|---|
| 1006 | 20,001–50,000 | Caldas da Rainha | 1 | 1 | 2 | 2 |
| 1602 | 7501–20,000 | Caminha | 2 | 2 | 2 | 1 |
| 1204 | > 7500 | Campo Maior | 2 | 2 | 2 | 2 |
| 0602 | 20,001–50,000 | Cantanhede | 2 | 2 | 2 | 1 |
| 0403 | > 7500 | Carrazeda de Ansiães | 2 | 2 | 2 | 2 |
| 1802 | 7501–20,000 | Carregal do Sal | 2 | 2 | 2 | 2 |
| 1406 | 7501–20,000 | Cartaxo | 2 | 1 | 2 | 1 |
| 1105 | > 100,001 | Cascais | 1 | 1 | 1 | 2 |
| 1007 | > 7500 | Castanheira de Pêra | 2 | 2 | 2 | 2 |
| 0502 | 20,001–50,000 | Castelo Branco | 2 | 1 | 2 | 2 |
| 0106 | 7501–20,000 | Castelo de Paiva | 2 | 2 | 2 | 1 |
| 1205 | > 7500 | Castelo de Vide | 2 | 1 | 2 | 1 |
| 1803 | 7501–20,000 | Castro Daire | 2 | 2 | 2 | 2 |
| 0804 | > 7500 | Castro Marim | 2 | 2 | 2 | 2 |
| 0206 | > 7500 | Castro Verde | 2 | 2 | 2 | 2 |
| 0903 | > 7500 | Celorico da Beira | 2 | 2 | 2 | 2 |
| 0305 | 7501–20,000 | Celorico de Basto | 2 | 2 | 2 | 1 |
| 1407 | > 7500 | Chamusca | 2 | 1 | 2 | 2 |
| 1703 | 20,001–50,000 | Chaves | 2 | 2 | 2 | 1 |
| 1804 | 7501–20,000 | Cinfães | 2 | 2 | 2 | 1 |
| 0603 | > 100,001 | Coimbra | 2 | 1 | 2 | 2 |
| 0604 | 7501–20,000 | Condeixa-a-Nova | 2 | 2 | 2 | 1 |
| 1408 | > 7500 | Constância | 2 | 2 | 2 | 1 |
| 1409 | 7501–20,000 | Coruche | 1 | 1 | 1 | 1 |
| 0503 | 20,001–50,000 | Covilhã | 3 | 2 | 2 | 1 |
| 1206 | > 7500 | Crato | 2 | 2 | 2 | 1 |
| 0207 | > 7500 | Cuba | 2 | 2 | 2 | 2 |
| 1207 | 7501–20,000 | Elvas | 2 | 2 | 2 | 2 |
| 1410 | 7501–20,000 | Entroncamento | 2 | 2 | 2 | 1 |
| 0107 | 20,001–50,000 | Espinho | 2 | 1 | 2 | 2 |
| 0306 | 20,001–50,000 | Esposende | 2 | 2 | 2 | 1 |
| 0108 | 20,001–50,000 | Estarreja | 2 | 1 | 2 | 2 |
| 0704 | 7501–20,000 | Estremoz | 2 | 3 | 2 | 2 |
| 0705 | 20,001–50,000 | Évora | 1 | 1 | 2 | 1 |
| 0307 | 20,001–50,000 | Fafe | 1 | 2 | 3 | 1 |

(continued)

(continued)

| DICOFRE | Population | Municipalities | EAP | MSACC/ MPACC/ AAP | CNR/ CNS | MPAC |
|---|---|---|---|---|---|---|
| 0805 | 50,001–100,000 | Faro | 2 | 1 | 2 | 2 |
| 1303 | 50,001–100,000 | Felgueiras | 2 | 2 | 2 | 1 |
| 0208 | > 7500 | Ferreira do Alentejo | 2 | 1 | 2 | 2 |
| 1411 | > 7500 | Ferreira do Zêzere | 2 | 2 | 2 | 1 |
| 0605 | 20,001–50,000 | Figueira da Foz | 1 | 1 | 2 | 1 |
| 0904 | > 7500 | Figueira de Castelo Rodrigo | 1 | 2 | 2 | 2 |
| 1008 | > 7500 | Figueiró dos Vinhos | 2 | 2 | 2 | 2 |
| 0905 | > 7500 | Fornos de Algodres | 2 | 2 | 2 | 2 |
| 0404 | > 7500 | Freixo de Espada à Cinta | 1 | 2 | 2 | 2 |
| 1208 | > 7500 | Fronteira | 2 | 2 | 2 | 1 |
| 0504 | 20,001–50,000 | Fundão | 2 | 2 | 2 | 1 |
| 1209 | > 7500 | Gavião | 2 | 2 | 2 | 1 |
| 0606 | > 7500 | Góis | 2 | 2 | 2 | 1 |
| 1412 | > 7500 | Golegã | 2 | 1 | 2 | 2 |
| 1304 | > 100,001 | Gondomar | 1 | 1 | 2 | 1 |
| 0906 | 7501–20,000 | Gouveia | 2 | 2 | 2 | 2 |
| 1505 | 7501–20,000 | Grândola | 2 | 2 | 2 | 2 |
| 0907 | 20,001–50,000 | Guarda | 2 | 1 | 2 | 2 |
| 0308 | > 100,001 | Guimarães | 1 | 1 | 1 | 1 |
| 0505 | > 7500 | Idanha-a-Nova | 3 | 2 | 2 | 2 |
| 0110 | 20,001–50,000 | Ílhavo | 3 | 1 | 2 | 2 |
| 0806 | 7501–20,000 | Lagoa | 3 | 2 | 2 | 2 |
| 0807 | 20,001–50,000 | Lagos | 2 | 1 | 2 | 2 |
| 1805 | 20,001–50,000 | Lamego | 1 | 1 | 2 | 2 |
| 1009 | > 100,001 | Leiria | 2 | 1 | 2 | 1 |
| 1106 | > 100,001 | Lisboa | 1 | 1 | 1 | 2 |
| 0808 | 50,001–100,000 | Loulé | 1 | 1 | 2 | 1 |
| 1107 | > 100,001 | Loures | 1 | 1 | 2 | 1 |
| 1108 | 20,001–50,000 | Lourinhã | 1 | 1 | 2 | 2 |
| 0607 | 7501–20,000 | Lousã | 2 | 2 | 2 | 1 |
| 1305 | 20,001–50,000 | Lousada | 2 | 2 | 2 | 1 |
| 1413 | > 7500 | Mação | 2 | 2 | 2 | 2 |
| 0405 | 7501–20,000 | Macedo de Cavaleiros | 2 | 2 | 2 | 2 |

(continued)

(continued)

| DICOFRE | Population | Municipalities | EAP | MSACC/ MPACC/ AAP | CNR/ CNS | MPAC |
|---|---|---|---|---|---|---|
| 1109 | 50,001–100,000 | Mafra | 2 | 1 | 2 | 1 |
| 1306 | > 100,001 | Maia | 1 | 1 | 1 | 3 |
| 1806 | 7501–20,000 | Mangualde | 2 | 2 | 2 | 2 |
| 0908 | > 7500 | Manteigas | 2 | 2 | 2 | 2 |
| 1307 | 20,001–50,000 | Marco de Canaveses | 2 | 2 | 2 | 1 |
| 1010 | 20,001–50,000 | Marinha Grande | 1 | 2 | 2 | 1 |
| 1210 | > 7500 | Marvão | 2 | 2 | 2 | 1 |
| 1308 | > 100,001 | Matosinhos | 1 | 1 | 1 | 2 |
| 0111 | 7501–20,000 | Mealhada | 1 | 2 | 2 | 2 |
| 0909 | > 7500 | Mêda | 2 | 2 | 2 | 2 |
| 1603 | > 7500 | Melgaço | 2 | 2 | 2 | 2 |
| 0209 | > 7500 | Mértola | 1 | 2 | 2 | 2 |
| 1704 | > 7500 | Mesão Frio | 2 | 2 | 2 | 2 |
| 0608 | 7501–20,000 | Mira | 2 | 2 | 2 | 1 |
| 0609 | 7501–20,000 | Miranda do Corvo | 2 | 2 | 2 | 1 |
| 0406 | > 7500 | Miranda do Douro | 2 | 2 | 2 | 2 |
| 0407 | 7501–20,000 | Mirandela | 2 | 2 | 2 | 3 |
| 0408 | > 7500 | Mogadouro | 2 | 2 | 2 | 2 |
| 1807 | 7501–20,000 | Moimenta da Beira | 3 | 2 | 2 | 2 |
| 1506 | 50,001–100,000 | Moita | 1 | 2 | 2 | 3 |
| 1604 | 7501–20,000 | Monção | 2 | 2 | 2 | 2 |
| 0809 | > 7500 | Monchique | 2 | 2 | 2 | 3 |
| 1705 | > 7500 | Mondim de Basto | 2 | 2 | 2 | 2 |
| 1211 | > 7500 | Monforte | 2 | 2 | 2 | 1 |
| 1706 | > 7500 | Montalegre | 2 | 1 | 2 | 1 |
| 0706 | 7501–20,000 | Montemor-o-Novo | 2 | 1 | 2 | 2 |
| 0610 | 20,001–50,000 | Montemor-o-Velho | 2 | 2 | 2 | 1 |
| 1507 | 20,001–50,000 | Montijo | 1 | 2 | 2 | 2 |
| 0707 | > 7500 | Mora | 2 | 3 | 2 | 2 |
| 1808 | > 7500 | Mortágua | 2 | 2 | 2 | 1 |
| 0210 | 7501–20,000 | Moura | 1 | 2 | 2 | 2 |
| 0708 | > 7500 | Mourão | 2 | 3 | 2 | 2 |
| 1707 | > 7500 | Murça | 1 | 2 | 2 | 2 |
| 0112 | 7501–20,000 | Murtosa | 2 | 2 | 2 | 2 |
| 1011 | 7501–20,000 | Nazaré | 1 | 1 | 2 | 2 |
| 1809 | 7501–20,000 | Nelas | 2 | 2 | 2 | 2 |

(continued)

(continued)

| DICOFRE | Population | Municipalities | EAP | MSACC/ MPACC/ AAP | CNR/ CNS | MPAC |
|---|---|---|---|---|---|---|
| 1212 | > 7500 | Nisa | 2 | 2 | 2 | 1 |
| 1012 | 7501–20,000 | Óbidos | 2 | 1 | 2 | 2 |
| 0211 | 7501–20,000 | Odemira | 2 | 1 | 2 | 2 |
| 1116 | > 100,001 | Odivelas | 2 | 3 | 2 | 2 |
| 1110 | > 100,001 | Oeiras | 1 | 1 | 2 | 2 |
| 0506 | > 7500 | Oleiros | 2 | 2 | 2 | 1 |
| 0810 | 20,001–50,000 | Olhão | 2 | 2 | 2 | 3 |
| 0113 | 50,001–100,000 | Oliveira de Azeméis | 2 | 1 | 2 | 2 |
| 1810 | 7501–20,000 | Oliveira de Frades | 2 | 2 | 2 | 1 |
| 0114 | 7501–20,000 | Oliveira do Bairro | 1 | 1 | 2 | 2 |
| 0611 | 7501–20,000 | Oliveira do Hospital | 2 | 2 | 2 | 1 |
| 1421 | 20,001–50,000 | Ourém | 2 | 2 | 2 | 1 |
| 0212 | > 7500 | Ourique | 2 | 2 | 2 | 2 |
| 0115 | 20,001–50,000 | Ovar | 1 | 2 | 2 | 1 |
| 1309 | 20,001–50,000 | Paços de Ferreira | 2 | 2 | 2 | 2 |
| 1508 | 50,001–100,000 | Palmela | 2 | 1 | 1 | 2 |
| 0612 | > 7500 | Pampilhosa da Serra | 2 | 2 | 2 | 1 |
| 1310 | 50,001–100,000 | Paredes | 2 | 1 | 2 | 1 |
| 1605 | > 7500 | Paredes de Coura | 1 | 2 | 2 | 2 |
| 1013 | > 7500 | Pedrógão Grande | 2 | 2 | 2 | 2 |
| 0613 | 7501–20,000 | Penacova | 2 | 2 | 2 | 1 |
| 1311 | 50,001–100,000 | Penafiel | 2 | 2 | 2 | 1 |
| 1811 | > 7500 | Penalva do Castelo | 2 | 2 | 2 | 2 |
| 0507 | > 7500 | Penamacor | 2 | 2 | 2 | 2 |
| 1812 | > 7500 | Penedono | 2 | 2 | 2 | 1 |
| 0614 | > 7500 | Penela | 1 | 2 | 2 | 2 |
| 1014 | 20,001–50,000 | Peniche | 2 | 1 | 2 | 2 |
| 1708 | 7501–20,000 | Peso da Régua | 2 | 2 | 2 | 2 |
| 0910 | > 7500 | Pinhel | 2 | 2 | 2 | 2 |
| 1015 | 20,001–50,000 | Pombal | 2 | 2 | 2 | 2 |
| 1606 | 7501–20,000 | Ponte da Barca | 1 | 2 | 2 | 2 |
| 1607 | 20,001–50,000 | Ponte de Lima | 1 | 2 | 2 | 2 |
| 1213 | 7501–20,000 | Ponte de Sor | 2 | 2 | 2 | 1 |
| 1214 | 7501–20,000 | Portalegre | 2 | 2 | 2 | 1 |
| 0709 | > 7500 | Portel | 2 | 3 | 2 | 2 |
| 0811 | 20,001–50,000 | Portimão | 2 | 2 | 2 | 2 |

(continued)

(continued)

| DICOFRE | Population | Municipalities | EAP | MSACC/ MPACC/ AAP | CNR/ CNS | MPAC |
|---|---|---|---|---|---|---|
| 1312 | > 100,001 | Porto | 1 | 1 | 1 | 1 |
| 1016 | 7501–20,000 | Porto de Mós | 2 | 2 | 2 | 2 |
| 0309 | 7501–20,000 | Póvoa de Lanhoso | 2 | 2 | 2 | 1 |
| 1313 | 50,001–100,000 | Póvoa de Varzim | 1 | 1 | 2 | 2 |
| 0508 | > 7500 | Proença-a-Nova | 2 | 2 | 2 | 1 |
| 0710 | > 7500 | Redondo | 2 | 3 | 2 | 2 |
| 0711 | 7501–20,000 | Reguengos de Monsaraz | 2 | 3 | 2 | 2 |
| 1813 | 7501–20,000 | Resende | 2 | 2 | 2 | 1 |
| 1709 | > 7500 | Ribeira de Pena | 1 | 2 | 2 | 1 |
| 1414 | 7501–20,000 | Rio Maior | 2 | 1 | 2 | 2 |
| 1710 | > 7500 | Sabrosa | 2 | 2 | 2 | 2 |
| 0911 | 7501–20,000 | Sabugal | 2 | 2 | 2 | 2 |
| 1415 | 7501–20,000 | Salvaterra de Magos | 2 | 1 | 2 | 2 |
| 1814 | 7501–20,000 | Santa Comba Dão | 2 | 2 | 2 | 2 |
| 0109 | > 100,001 | Santa Maria da Feira | 2 | 1 | 2 | 2 |
| 1711 | > 7500 | Santa Marta de Penaguião | 2 | 2 | 2 | 2 |
| 1416 | 20,001–50,000 | Santarém | 1 | 1 | 2 | 2 |
| 1509 | 20,001–50,000 | Santiago do Cacém | 1 | 2 | 2 | 2 |
| 1314 | 50,001–100,000 | Santo Tirso | 1 | 1 | 2 | 2 |
| 0812 | 7501–20,000 | São Brás de Alportel | 2 | 2 | 2 | 2 |
| 0116 | 7501–20,000 | São João da Madeira | 1 | 1 | 2 | 2 |
| 1815 | > 7500 | São João da Pesqueira | 1 | 1 | 2 | 2 |
| 1816 | 7501–20,000 | São Pedro do Sul | 2 | 2 | 2 | 2 |
| 1417 | > 7500 | Sardoal | 1 | 2 | 2 | 2 |
| 1817 | 7501–20,000 | Sátão | 2 | 2 | 2 | 2 |
| 0912 | 7501–20,000 | Seia | 1 | 1 | 2 | 2 |
| 1510 | > 100,001 | Seixal | 1 | 2 | 2 | 2 |
| 1818 | > 7500 | Sernancelhe | 2 | 2 | 2 | 2 |
| 0213 | 7501–20,000 | Serpa | 2 | 2 | 2 | 2 |
| 0509 | 7501–20,000 | Sertã | 2 | 2 | 2 | 1 |
| 1511 | 20,001–50,000 | Sesimbra | 3 | 1 | 1 | 2 |
| 1512 | > 100,001 | Setúbal | 1 | 1 | 1 | 1 |
| 0117 | 7501–20,000 | Sever do Vouga | 1 | 2 | 2 | 2 |

(continued)

(continued)

| DICOFRE | Population | Municipalities | EAP | MSACC/ MPACC/ AAP | CNR/ CNS | MPAC |
|---|---|---|---|---|---|---|
| 0813 | 20,001–50,000 | Silves | 2 | 2 | 2 | 2 |
| 1513 | 7501–20,000 | Sines | 2 | 2 | 2 | 2 |
| 1111 | > 100,001 | Sintra | 1 | 2 | 2 | 2 |
| 1112 | 7501–20,000 | Sobral de Monte Agraço | 2 | 1 | 2 | 2 |
| 0615 | 7501–20,000 | Soure | 2 | 2 | 2 | 1 |
| 1215 | > 7500 | Sousel | 1 | 2 | 2 | 1 |
| 0616 | 7501–20,000 | Tábua | 1 | 2 | 2 | 1 |
| 1819 | > 7500 | Tabuaço | 2 | 2 | 2 | 2 |
| 1820 | > 7500 | Tarouca | 1 | 2 | 2 | 2 |
| 0814 | 20,001–50,000 | Tavira | 2 | 2 | 2 | 3 |
| 0310 | > 7500 | Terras de Bouro | 2 | 2 | 2 | 1 |
| 1418 | 20,001–50,000 | Tomar | 2 | 1 | 2 | 1 |
| 1821 | 20,001–50,000 | Tondela | 2 | 1 | 2 | 2 |
| 0409 | > 7500 | Torre de Moncorvo | 2 | 2 | 2 | 2 |
| 1419 | 20,001–50,000 | Torres Novas | 2 | 2 | 2 | 1 |
| 1113 | 50,001–100,000 | Torres Vedras | 1 | 1 | 2 | 1 |
| 0913 | > 7500 | Trancoso | 2 | 2 | 2 | 2 |
| 1318 | 20,001–50,000 | Trofa | 1 | 1 | 2 | 2 |
| 0118 | 7501–20,000 | Vagos | 1 | 2 | 2 | 2 |
| 0119 | 7501–20,000 | Vale de Cambra | 2 | 1 | 2 | 2 |
| 1608 | 7501–20,000 | Valença | 2 | 2 | 2 | 2 |
| 1315 | 50,001–100,000 | Valongo | 1 | 1 | 2 | 1 |
| 1712 | 7501–20,000 | Valpaços | 2 | 2 | 2 | 1 |
| 0712 | 7501–20,000 | Vendas Novas | 2 | 3 | 2 | 2 |
| 0713 | > 7500 | Viana do Alentejo | 2 | 2 | 2 | 2 |
| 1609 | 50,001–100,000 | Viana do Castelo | 2 | 1 | 2 | 1 |
| 0214 | > 7500 | Vidigueira | 2 | 1 | 2 | 2 |
| 0311 | 7501–20,000 | Vieira do Minho | 2 | 2 | 2 | 2 |
| 0510 | > 7500 | Vila de Rei | 1 | 1 | 2 | 1 |
| 0815 | > 7500 | Vila do Bispo | 2 | 2 | 2 | 2 |
| 1316 | 50,001–100,000 | Vila do Conde | 2 | 1 | 2 | 1 |
| 0410 | > 7500 | Vila Flor | 2 | 2 | 2 | 2 |
| 1114 | > 100,001 | Vila Franca de Xira | 1 | 1 | 2 | 2 |
| 1420 | > 7500 | Vila Nova da Barquinha | 2 | 2 | 2 | 1 |

(continued)

(continued)

| DICOFRE | Population | Municipalities | EAP | MSACC/ MPACC/ AAP | CNR/ CNS | MPAC |
|---|---|---|---|---|---|---|
| 1610 | > 7500 | Vila Nova de Cerveira | 1 | 2 | 2 | 2 |
| 0312 | > 100,001 | Vila Nova de Famalicão | 2 | 3 | 1 | 1 |
| 0914 | > 7500 | Vila Nova de Foz Côa | 1 | 2 | 2 | 2 |
| 1317 | > 100,001 | Vila Nova de Gaia | 1 | 1 | 2 | 2 |
| 1822 | > 7500 | Vila Nova de Paiva | 2 | 2 | 2 | 2 |
| 0617 | > 7500 | Vila Nova de Poiares | 1 | 1 | 2 | 2 |
| 1713 | 7501–20,000 | Vila Pouca de Aguiar | 1 | 2 | 2 | 1 |
| 1714 | 20,001—50,000 | Vila Real | 1 | 1 | 2 | 2 |
| 0816 | 7501–20,000 | Vila Real de Santo António | 2 | 2 | 2 | 2 |
| 0511 | > 7500 | Vila Velha de Ródão | 2 | 2 | 2 | 1 |
| 0313 | 20,001–50,000 | Vila Verde | 2 | 2 | 2 | 2 |
| 0714 | > 7500 | Vila Viçosa | 2 | 1 | 2 | 2 |
| 0411 | > 7500 | Vimioso | 2 | 2 | 2 | 2 |
| 0412 | > 7500 | Vinhais | 2 | 2 | 2 | 2 |
| 1823 | 50,001–100,000 | Viseu | 2 | 2 | 2 | 2 |
| 0314 | 20,001–50,000 | Vizela | 1 | 2 | 2 | 2 |
| 1824 | 7501–20,000 | Vouzela | 2 | 2 | 2 | 2 |

1. Present TMI.
2. Does not present TMI.
3. In the process of making.

# Appendix 2—MCPC and CC Relationship, Questionnaire Survey

**Policies adopted in the context of the effects of Climate Change**

**Introduction:** As part of the master's thesis "The Adaptation of Civil Protection Agents to Climate Change in Mainland Portugal", of the master's Degree in Geography: Risks and Civil Protection, of the Department of Geography of the University of Minho, I hereby ask you to participate in this Questionnaire, which we will make to all municipal government units of the Portuguese mainland.

**Objectives:** This Questionnaire Survey aims to answer pertinent questions of policies adopted in the context of the effects of Climate Change in Portuguese municipalities.

**Information:** We ask you to answer only 1 (one) questionnaire, identifying the name of the municipality in question and the respondent's data. Any question related to the questionnaire (questions or completion), I would appreciate it if you could contact us at the following e-mail: franciscosilva258@gmail.com

**Estes dados serão utilizados apenas para fins académicos.**

## Consentient

## I Have Read the Description of Use of the Form and Authorize the Processing and Dissemination of the Result

Yes ☐ No ☐

**Part I - Characterization and Municipal Policies**

1. From 1 to 5, how do you consider the current performance of the Municipal Civil Protection Service (SMPC) of the municipality where you work?

   ○ Very week ○ Week ○ Reasonable ○ Good ○ Very good

2. Does the municipality have a Municipal Emergency Plan (PME)?

   Yes ☐ No ☐ I don't know ☐

3. In what year was the PME prepared?
   *(Only Answer This Question if You Answered "Yes" in the Prior Question)*

   ☐

4. Has the PME review already been made?

   Yes ☐ No ☐ It's under review or approval ☐ I don't know ☐

5. In what year was it reviewed?
   (Only answer this question if you answered "yes" in the prior question).

   ☐

6. Has the Municipal Strategy for Adaption to Climate Change been developed?

   Yes ☐ No ☐ I don't know ☐

7. In the absence of the plan, is there any document that can answer the question posed in the previous question?

   Yes ☐ No ☐ It's under review or approval ☐ I don't know ☐

8. If yes, is it possible to indicate which?

   **Type your answer:**

   ______________________________

9. In a scale of 1 to 5, how would you classify the priority that the municipality gives to climate change issues? *(e.g. awareness campaigns, plan development…)*

   ○ Not a priority ○ Low priority ○ Reasonable ○ Priority ○ Highly priority

10. Is there a Senior Tech (ST), specialized in climate change, currently working at the city hall?

Yes ☐ No ☐ I don't know ☐

11. According to Article 14, paragraph 2, of the Portuguese Climate Framework Law (Law n.º 98/2021, December 31st). By January 2023, all municipalities must prepare a Municipal Climate Action Plan. At the time of the response to this survey, has your municipality already prepared the plan?

Yes ☐ No ☐ I don't know ☐

**Part II – Risk assessment**

12. Based on studies or plans, a scale of 1 to 5, how vulnerable do you consider the municipality to Climate Change?

| | Very little vulnerable (1) | Slightly vulnerable (2) | Vulnerable (3) | Something vulnerable (4) | Extremely vulnerable (5) |
|---|---|---|---|---|---|
| Vulnerability | ○ | ○ | ○ | ○ | ○ |

13. What is the level of relevance regarding the following climate risks in the municipality and considering the impacts of climate change in the future?

| | Nothing relevant (1) | Not very relevant (2) | Relevant (3) | Very relevant (4) | Extremely relevant (5) |
|---|---|---|---|---|---|
| Relevance | ○ | ○ | ○ | ○ | ○ |

14. What are the natural risks associated with climate change identified in Climate TMI?

| Risks | Yes | No |
|---|---|---|
| Extreme heat events | ○ | ○ |
| Floods | ○ | ○ |
| Extreme precipitation | ○ | ○ |
| Extreme cold | ○ | ○ |

(continued)

(continued)

| Risks | Yes | No |
|---|---|---|
| Landslide | O | O |
| Sea levels rise | O | O |
| Wildfires | O | O |
| Strong winds | O | O |
| Drought | O | O |
| Spread of disease vectors | O | O |
| Other, which? ____________________ | | |

**Part III – Prevention, Adaption and Mitigation measures**

15. Is there any measure in force or in Approval carried out by the City Council that considers the risks of climate change?

Yes ☐ No ☐ In approval ☐

16. If yes, please indicate which one(s) and what is the forecast for their completion?

**Write your answer:**

______________________________

______________________________

______________________________

______________________________

______________________________

______________________________

17. Among the measures listed, was any type of dissemination campaign carried out by the municipality on climate change?

Yes ☐ No ☐

18. Please indicate how they were implemented (Target Audience, Location, Duration, Format, Recurrence).

**Write your answer:**

19. What is the role of Civil Protection entities (agents, municipal service, fire brigade…), in the application and definition of these measures?

**Write your answer:**

20. What is the relationship between measures and risk assessment identified in the IPACC/ MPACC, and or the MPAC and the practical application by the municipality?

**Write your answer:**

**Part IV – Information**

**Municipality identification:**

**Position of the respondent:**

**N. ° of years in the service of the Municipality:**

**Respondent Age:**

**If you want to ask any kind of question/comment/suggestion you will be grateful**

*Thank you again for your collaboration and availability,*

# Appendix 3—Questionnaire Survey Answers List

Municipalities

| | | | | | | | | | | |
|---|---|---|---|---|---|---|---|---|---|---|
| A | Albergaria-a-Velha | Albufeira | Alcoutim | Almada | Alvaiázere | Amadora | Amarante | Armamar | Arraiolos | Avis |
| B | Barcelos | Barreiro | Batalha | | | | | | | |
| C | Cadaval | Caldas da Rainha | Carezeda de Anciães | Celorico de Basto | Covilhã | | | | | |
| E | Espinho | | | | | | | | | |
| F | Faro | Felgueiras | Figueira de Castelo Rodrigo | Figueiró dos Vinhos | Fronteira | | | | | |
| G | Góis | Guarda | Guimarães | | | | | | | |
| I | Idanha-a-Nova | | | | | | | | | |
| L | Lamego | Loures | | | | | | | | |
| M | Maia | Mangualde | Matosinhos | Mirandela | Mondim de Basto | Moura | | | | |
| O | Oleiros | Olhão | Oliveira de Frades | Ovar | | | | | | |
| P | Palmela | Penacova | Penela | Ponte da Barca | Porto | Póvoa de Lanhoso | | | | |
| R | Redondo | Reguengos de Monsaraz | | | | | | | | |
| S | Santa Comba Dão | Santa Maria da Feira | Santarém | Sardoal | Seixal | Sernancelhe | Sertã | Soure | | |
| T | Terras de Bouro | Tomar | Tondela | Torres Vedras | Trofa | | | | | |
| V | Viana do Alentejo | Vidigueira | Vila do Conde | Vila Nova de Famalicão | Vila Nova de Poiares | Vila Pouca de Aguiar | Vila Real | Vila Velha de Ródão | Vila Viçosa | Viseu Vouzela |

## Appendix 4—Regional and Intermunicipal Scope TMI Withdraw

| Population | Intermunicipal Community (IC)* | IPACC/ MPACC | RCAP | EAP |
|---|---|---|---|---|
| 100,001–150,000 | Alentejo Central | Yes | No | No |
| > 100,000 | Alentejo Litoral | Yes | No | No |
| 150,001–300,000 | Algarve | Yes | No | No |
| > 100,000 | Alto Alentejo | Yes | No | No |
| 150,001–300,000 | Alto Minho | Yes | In elaboration | Yes |
| > 100,000 | Alto Tâmega e Barroso | Yes | In elaboration | No |
| 150,001–300,000 | Área Metropolitana de Lisboa (AML) | Yes | No | No |
| 100,001–150,000 | Área Metropolitana do Porto (AMP) | Yes | In elaboration | No |
| < 300,001 | Ave | Yes | In elaboration | No |
| > 100,000 | Baixo Alentejo | Yes | No | No |
| > 100,000 | Beira Baixa | Yes | No | No |
| 150,001–300,000 | Beiras e Serra da Estrela | Yes | No | No |
| < 300,001 | Cávado | Yes | In elaboration | No |
| 150,001–300,000 | Douro | Yes | In elaboration | No |
| 150,001–300,000 | Lezíria do Tejo | Yes | No | No |
| 150,001–300,000 | Médio Tejo | Yes | No | No |
| < 300,001 | Oeste | Yes | No | No |
| < 300,001 | Região da Aveiro | No | No | No |
| < 300,001 | Região de Coimbra | Yes | No | Yes |
| 150,001–300,000 | Região de Leiria | Yes | No | No |
| < 300,001 | Tâmega e Sousa | Yes | In elaboration | No |
| > 100,000 | Terras de Trás-os-Montes | Yes | In elaboration | No |
| 150,001–300,000 | Viseu Dão Lafões | Yes | No | No |

* IC data from CAOP 2022.

## References

Aguiar FC, Bentz J, Silva JMN, Fonseca AL, Swart R, Santos FD, Penha-Lopes G (2018) Adaptation to climate change at local level in Europe: an overview. Environ Sci Policy 86:38–63

Almeida M (2012) Fundamentação teórica para a criação de um sistema de alerta e resposta online durante episódios térmicos de calor extremo para uma unidade de saúde da GAMP [Master

Thesis] Faculdade de Letras da Universidade do Porto, Portugal https://sigarra.up.pt/flup/pt//pub_geral.pub_view?pi_pub_base_id=27922
Alves M, Leal C, Vidal DG (2023) Doenças transmitidas por vetores num contexto de alterações climáticas: antecipando riscos para uma melhor preparação dos territórios. Estudos de caso da Região de Coimbra, Portugal. Recima21—Revista Científica Multidisciplinar 4(10):2675–6218
Amaro A (2023) A Proteção Civil em Portugal-Visão. Os Desafios (Geográficos) da Governação Territorial 85
Anastasopoulou M (2023) Stakeholder engagement for nature-based solutions–a collaborative approach for climate adaptation at the urban environment
Arnell NW (2022) The implications of climate change for emergency planning. Int J Disaster Risk Reduct 83:103425
Bento-Gonçalves A (2021) Os incêndios florestais em Portugal. Fundação Francisco Manuel dos Santos
Birchall SJ, Bonnett N, Kehler S (2023) The influence of governance structure on local resilience: enabling and constraining factors for climate change adaptation in practice. Urban Climate 47:101348
Bittencourt JCN, Costa DG, Portugal P, Vasques F (2024) A data-driven clustering approach for assessing spatiotemporal vulnerability to urban emergencies. Sustain Cities Soc 108:105477
Campbell-Lendrum D, Manga L, Bagayoko M, Sommerfeld J (2015) Climate change and vector-borne diseases: what are the implications for public health research and policy? Philos Trans R Soc B: Biol Sci 370(1665):20130552
Campos I, Guerra J, Gomes JF, Schmidt L, Alves F, Vizinho A, Lopes GP (2017) Understanding climate change policy and action in Portuguese municipalities: a survey. Land Use Policy 62:68–78
Carvalho A, Pereira E, Rodrigues A, Silveira A, Cabecinhas R, Lázaro A (2011) As Alterações Climáticas, os Media e os Cidadãos. Coimbra: Grácio
Cook AD, Dorussen H (2021) Civil protection and disaster management. In: The European Union's security relations with Asian partners, 185–207. Springer
Cossu M (2020) Spatial planning and climate change adaptation in Mainland Portugal Politecnico di Torino
Costa FS, Lopes HS (2024) The Portuguese dams of the international Douro, climate change and adaptation strategies: perspectives within the framework of the Albufeira convention and the water framework directive. Water Int 49(3–4):417–428
Figueiredo R, Romao X, Paupério E (2020) Flood risk assessment of cultural heritage at large spatial scales: framework and application to mainland Portugal. J Cult Herit 43:163–174
Githeko A, Lindsay S, Confalonieri U, Patz J (2000) Climate change and vector-borne diseases: a regional analysis. Bull World Health Organ 78(9):00429686, 1136–1147
Groven K, Aall C, Van de Berg M, Carlsson-Kanyama A, Coenen F (2012) Integrating climate change adaptation into civil protection: comparative lessons from Norway, Sweden and the Netherlands. Local Environ Int J Justice Sustain 17:6–7, Nordic Climate Change. Reino Unido. https://doi.org/10.1080/13549839.2012.665859
Haque CE, Doberstein B (2021) Adaptive governance and community resilience to cyclones in coastal Bangladesh: addressing the problem of fit, social learning, and institutional collaboration. Environ Sci Policy 124:580–592
Hua J, Zhang X, Ren C, Shi Y, Lee TC (2021) Spatiotemporal assessment of extreme heat risk for high-density cities: a case study of Hong Kong from 2006 to 2016. Sustain Cities Soc 64:102507
Huang X, Khan YA, Arshed N, Salem S, Shabeer MG, Hanif U (2023) Increasing social resilience against climate change risks: a case of extreme climate affected countries. Int J Clim Chang Strat Manag 15(3):412–431
Instituto Português do Mar e da Atmosfera (2011) Normais Climatológicas. https://www.ipma.pt/pt/oclima/normais.clima/

Lopes HS, Remoaldo PC, Ribeiro V, Martin-Vide J (2022) Pathways for adapting tourism to climate change in an urban destination–evidences based on thermal conditions for the Porto Metropolitan Area (Portugal). J Environ Manage 315:115161

Lopes HS, Remoaldo PC, Ribeiro V, Martín-Vide J (2023) The impacts of climate change on human wellbeing in the municipality of Porto—an analysis based on remote sensing. In: Climate change and health hazards: addressing hazards to human and environmental health from a changing climate 135–172 Springer

Miranda P, Valente M, Tomé A, Trigo R, Coelho M, Aguiar A, Azevedo EB (2006) O clima de Portugal nos séculos XX e XXI the Portuguese climate in the 20th and 21st centuries. In: Alterações Climáticas em Portugal. Cenários impactos e medidas de adaptação. Projeto SIAM II, 47–11, Lisbon, Portugal

Naess L, Bang G, Eriksen S, Vevatne J (2005) Institutional adaptation to climate change: flood responses at the municipal level in Norway. Glob Environ Chang 15:125–138. https://doi.org/10.1016/j.gloenvcha.2004.10.003

Sá P, Costa AP, Moreira A, Alves ATA, Nascimento A, Ulhôa A, Batista B, (...) Silva PCB (2021) Reflexões em torno de Metodologias de Investigação: recolha de dados. *In ria.ua.pt volume 2, UA Editora, Aveiro, Portugal.* https://ria.ua.pt/handle/10773/30772

Salimi M, Al-Ghamdi SG (2020) Climate change impacts on critical urban infrastructure and urban resiliency strategies for the Middle East. Sustain Cities Soc 54:101948

Santos JR, Henriques S (2021) Inquérito por questionário: contributos de conceção e utilização em contextos educativos. *In repositorioaberto.uab.pt. Universidade Aberta.* http://hdl.handle.net/10400.2/10696

Santos S, Bento-Gonçalves A, Vieira A, Teixeira G (2023) Grandes Incêndios Florestais no noroeste de Portugal: explorando padrões espaciais entre 2001 e 2020, com base em dados Landsat. Sociedade e natureza 35(1):1982–4513. https://doi.org/10.14393/SN-v35-2023-68265

Schmidt L, Gomes C, Guerreiro S, O'Riordan T (2014) Are we all on the same boat? the challenge of adaptation facing Portuguese coastal communities: risk perception, trust-building and genuine participation. Land Use Policy 38:355–365

Van Ginkel KC, Haasnoot M, Botzen WW (2022) A stepwise approach for identifying climate change induced socio-economic tipping points. Clim Risk Manag 37:100445

Velosa N (2016) As alterações climáticas e a protecção civil. [Master Thesis] ISEC Lisbon, Portugal. http://hdl.handle.net/10400.26/21730

Zandvoort M, Campos IS, Vizinho A, Penha-Lopes G, Lorencová EK, van der Brugge R, van der Vlist MJ, van den Brink A, Jeuken AB (2017) Adaptation pathways in planning for uncertain climate change: applications in Portugal, the Czech Republic and the Netherlands. Environ Sci Policy 78:18–26

**P. F. Silva** da Silva is a 22-year-old student with a bachelor's degree in Civil Protection and Territorial Management from the Geography Department at the University of Minho. He is currently in the second year of a master's degree in Geography, specializing in Risks and Civil Protection, at the same institution. At present, he is developing his master's thesis on municipal tourism planning tools in relation to climatic events. This is part of the research grant project "WEAQ-TourPor: The impact of extreme climatic-meteorological conditions and air quality on the space-time behavior of tourists in an urban environment," funded by FCT through Lab2PT/IN2PAST.

**R. Dantas** has a Bachelor's Degree in Civil Protection and Territory Management. She is currently pursuing a Master's Degree in Geography, specializing in Geographic Information Systems (GIS). Since 2022, she has been a member of the Association of Students of Geography and Planning at the University of Minho, where she has served as President since 2023.

**H. S. Lopes** holds a Ph.D. in Geography and Regional Planning from the University of Minho and a Ph.D. in Geography, Territorial Planning and Environmental Management from the University of Barcelona. Researcher at the Landscape, Heritage, and Territory Laboratory (Lab2PT). He has carried out doctoral research funded by the Portuguese Foundation for Science and Technology and is currently Invited Assistant Professor at the Department of Geography, Institute of Social Sciences, University of Minho. Recent research has focused on the design of adaptation and mitigation strategies for urban areas to face climate change through citizen participation techniques (namely Delphi techniques and workshops). His main research interests are geographic information systems; citizen science; geostatistics; health environment; remote sensing; urban planning; environmental management; and climate change.

**P. Remoaldo** is a full professor at the University do Minho, in the Department of Geography, has extensive experience in researching tourism themes (Scopus Index h12). She has a PhD in Human Geography with co-direction of the University of Louvain-la-Neuve (Belgium). Her research interests are Regional and Local Development, Cultural Tourism, Creative Tourism and Health Geography. Between 2017 and 2020 she was the Head of the Department of Geography and in the period 2017–2020 the Director of Lab2PT (Laboratory of Landscape, Heritage and Territory—Excellent Evaluation). From 2021 to 2023 is Co-Coordinator of SpaceR (Space and Representation of Lab2PT). She is the author of 155 papers dealing with Cultural Tourism, Creative Tourism, Regional and Local Development, and Climate Change. She has been involved in several territorial and tourism planning projects.

# Urban Green Spaces, New Forms of Participation, and New Protagonists: From Crisis to (Re)Centralization

**Diogo Guedes Vidal, Hélder Silva Lopes, and Rui Leandro Maia**

**Abstract** Green spaces are social projects that reflect power relations, culture, class, and gender. As they are socially constructed, they also serve as arenas for struggles over environmental, social, and urban spatial justice. Considering the diversity of green spaces and the multitude of actors involved, this work seeks to identify and analyse how these spaces serve as platforms for new forms of participation that call upon new actors in the fight for socio-environmental justice in an urban context through two case studies. The current complexity of the climate narrative and contemporary socio-ecological issues, in which social inequalities are both a structuring and structured element, requires an analysis of these phenomena as a response to such challenges that may influence decisions on public policies at the local level.

**Keywords** Urban green spaces · Social justice · New social movements · Public policies

D. G. Vidal (✉)
Department of Life Sciences (DCV), TERRA Associate Laboratory, University of Coimbra (UC), Coimbra, Portugal
e-mail: diogo.vidal@uc.pt

Department of Social Sciences and Management, Universidade Aberta, Lisbon, Portugal

H. S. Lopes
Heritage and Territory Laboratory/IN2PAST, Department of Geography, Institute of Social Sciences, University of Minho, Braga, Portugal

Department of Geography, University of Barcelona, Barcelona, Spain

Department of Teacher Training, Higher School of Education Paula Frassinetti (ESEPF), Porto, Portugal

R. L. Maia
RISE-Health, FP-I3ID, Faculty of Human and Social Sciences, University Fernando Pessoa (UFP), Porto, Portugal
e-mail: rlmaia@ufp.edu.pt

Transdisciplinary Culture, Space and Memory Research Centre (CITCEM), University of Porto, Porto, Portugal

Faculty of Medicine, University of Porto, Porto, Portugal



W. Leal Filho et al. (eds.), *Composing Worlds: Humanities, Health and Wellbeing in the XXI Century Towards a More Sustainable World*, World Sustainability Series,
https://doi.org/10.1007/978-3-031-87108-5_20

# 1 Introduction

Since the dawn of civilisation, cities have been regarded as centres of social, economic, cultural, religious, and political dimensions. During the medieval period, cities combined commerce, religious activity, public life, and private life. Max Weber (1966) argued that cities created the idea of urban citizens, a new population no longer under the control of the "lord," which inspired the expression "the air of the city makes one free," suggesting cities as guarantors of freedom, though this ideal was not universally or democratically realised. This expectation led to a migration to cities in the twentieth century, with people drawn by hopes of better living conditions, resulting in the growth of megacities (Kleer and Nawrot 2018; Oliveira et al. 2020).

The timeline of cities and their understanding is shaped by modifications and transformations in landscapes that reflect their dynamism and volatility, mirroring how society organises, consumes, moves, and responds to complex social phenomena. The plurality of urban space, initially characterised by principles contrary to those of rural life, featured population density, diversity, and, consequently, heterogeneity (Wirth 1938). In his work "Urbanism as a Way of Life" (1938), Louis Wirth observed that contemporary cities are sites of detachment, lacking a sense of community and place identity—a concept that, derived from Durkheim's (1893) work, recognises a social anomie specific to urban environments. At the same time, contemporary cities are shaped by globalisation and the constant flow of people, goods, and services that configure their social space (Featherstone 1997).

However, the social anomie that Durkheim associates with cities can and should be contested, especially when viewed against a contemporary backdrop. While "old" challenges remain active, they have taken on new forms that demand fresh perspectives, and "new" challenges are emerging, highlighting the need to rethink cities and urban spaces. These challenges are no longer simply "old" or "new" as they intertwine within social structures, complicating understanding and calling for innovative approaches that embrace diversity and plurality. These socio-ecological challenges involve complex issues arising from the interaction between social and ecological systems, reflecting the interconnectedness and interdependence of human societies and the natural environment. This interaction includes social, economic, and environmental factors, producing varied impacts on individuals, communities, and ecosystems. Since the mid-twentieth century, following two world wars that devastated landscapes, natural heritage, and numerous biodiversity habitats, threatening species' reproduction, there has been increasing social awareness regarding humanity's relationship with Nature and non-human elements (Catton and Dunlap 1978; Carson 2000; Aldeia and Alves 2019; Alves and Vidal 2024). The relationship between humans and Nature is not, however, innocent; humanity's desire to dominate and control Nature has driven efforts that, in pursuit of this goal, have had and will continue to have harmful effects on ecosystems and the remaining resources on Earth. Much of this desire is rooted in the belief that Nature's course can be altered to satisfy human ambitions. In line with this thinking, the scientific community has proposed the concept of the "Anthropocene" as a new geological era characterised by unprecedented anthropogenic influence (Crutzen and Stoermer 2000; Steffen et al. 2011). This context calls for a transformative approach to the relationships between Nature, society, and culture—one that is pluralistic and transdisciplinary (Alves et al. 2012, 2013).

A large part of the debate that has evolved since then has focused on urban areas, as they are the primary form of organisation in contemporary societies. It is expected that the population living in these areas will increase from the current 53.9% to approximately 68.4% by 2050 (United Nations, Department of Economic and Social Affairs 2019), presenting new challenges in land management, exacerbated by the rising pressure of population growth (Soja 2000). Some authors already position themselves within a framework of societal collapse concerning a potential overpopulation crisis, with the exponential increase in population being linked to emerging risks to the maintenance of both human and non-human life. These risks are indisputably connected to the threat to wildlife and the depletion of natural resources (Abegão 2022). Contemporary urban spaces, particularly cities, face multiple, complex, and intertwined challenges. Growing urbanisation, the expansion of buildings, daily consumption practices, and the habits characteristic of urban lifestyles, such as intense traffic, represent challenges that urban planners and decision-makers, grounded in scientific knowledge, are called upon to address. The way contemporary society is organised does not account for the finitude of resources. The current fluidity of consumption, the "liquidity" of human relationships, and the disposability and substitutability of material goods contribute to the current landscape of unsustainability. "Liquid modernity" (Bauman 2005) is the era of privatisation and individualisation, severing individuals from their foundational structures—family, groups of friends, communities—discouraging a sense of community, and emptying public spaces, which were once places for socialisation and identity construction.

When we place the issue of climate change under the umbrella of exacerbating contemporary socio-ecological challenges, we find that urban populations are among the most affected by its undesirable effects, with impacts on both human health and ecosystems. At this level, it is important to bring into the debate the perspective of Ulrich Beck (2016), particularly when the author states that cities are now at the forefront of a "metamorphosis of the world," with the goal of "...finding responses to a world at risk" (2016, p. 205). As he argues, cities are part of a logic of "cosmopolitan risk communities," for instead of sharing common norms and values (traditionally associated with the concept of community), they now share crises, risks, and threats—in other words, the "evils" of society. Their purpose is driven by the risk of anticipating catastrophe. In response to such challenges, the presence of urban nature, in a structural and integrated form, has come to be seen as vital for the sustainability of cities and their human and non-human populations. Some of the solutions proposed to mitigate the negative environmental impacts in cities, as well as to provide a balanced combination of human growth and nature, are known as "nature-based solutions" (NbS) (Pauleit et al. 2017), offering environmental, social, and economic benefits for the entire ecosystem, both human and non-human (Salata et al. 2023).

Urban Green Spaces (UGSs) are among the most recognised nature-based solutions (NbS) due to their contribution to environmental quality, enhancing local resilience, promoting sustainability in cities, and improving the health and well-being of their direct or indirect users (de Vries 2010; Braubach et al. 2017; MacKinnon et al. 2019; Vidal et al. 2022). The European Commission's green infrastructure strategy (European Comission 2013) already highlighted the benefits of UGSs in combating social isolation and strengthening communities. The contribution of UGSs to global sustainability is widely recognised in an international framework and within European policies, framed by Goal 11.7 of the United Nations 2030 Agenda for Sustainable Development—to provide universal access to safe, inclusive, and accessible green and public spaces (United Nations 2015). Alongside the 2030 Agenda, the European Union's biodiversity strategy for 2030 encourages bringing nature back to cities (Rewilding), creating biodiverse green areas and accessible infrastructures, while emphasising the importance of developing urban green plans in large cities and towns (European Commission 2020). Equally important is the role of UGSs in climate resilience, recognised by the European Union in its climate change adaptation strategy (European Commission 2021). Recently, signatories of the Green City Accord, an initiative by the European Commission for cities striving for sustainability, have committed to safeguarding and enhancing biodiversity by increasing the extent and quality of green areas in urban contexts (European Comission 2020). Due to the specific dynamics and stimuli of cities, such as stress and disorientation (chaos), alongside environmental risks that are felt more acutely, UGSs can serve as escape points from urban lifestyles. They can also stimulate a desire to connect with nature, associated with a form of biophilic behaviour (Wilson 1984; Lencastre et al. 2022, 2023), that is, an affinity or love for living elements. In light of the above, this work seeks to identify and analyse how these spaces serve as a stage for new forms of participation, calling upon new protagonists in the fight for socio-environmental justice in urban contexts.

## 2 Public Space and Urban Green Space

Public space is where the right to the city is exercised, serving as the paradigmatic example of urbanity. As noted by Arantes (1995), and reinforced by Menezes (2009), public space is a key element in the process of consolidating the city as a "public place." In addition to the right to the city, characteristics of public space include accessibility and visibility, which are naturally linked to enjoyment (Graff 1996; Castro 2002), and it is also a space for the exercise of democracy (Costa et al. 2020). Francesco Indovina (2002) further argues that public spaces represent the condition for effectively living in an urban context and for one's own existence, whether they are spaces or infrastructures of mobility, such as roads, roundabouts, squares, and sidewalks, or spaces with specific functions and uses, such as gardens and parks, among others.

The notion of "public" has been defined, most significantly, in opposition to that of the "private," implying the total accessibility that the former provides, allowing anyone, at any time, to enter a space. According to Rémy and Voyé (1981), public space offers complete freedom of movement and circulation, where the dynamics of interaction are not controlled, assuming a degree of autonomy for individuals. Such spaces carry a physical dimension, as social interactions occur within them, but they also have a symbolic dimension, reflected in the forms and contents of these interactions, or in their meaning. According to Borja (2000), public space must be structured to meet the uses and functions of its potential users, without neglecting its role as a collective expression of life in society, serving as a stage for encounters among different social actors who converge in this shared space, thus becoming an indicator of urban values. Public space can, therefore, be understood as a space intended for public use, characterised by total accessibility and autonomy, with functions related to the promotion of sociability and human and non-human interaction. Its value, recognised by society, is crucial for safeguarding environmental, social, and cultural heritage for current and future generations.

According to Le Bras (1997), urban public space has led to a certain distancing of city inhabitants, resulting from an "imposition" of behaviours or practices that were previously codified, which results in a lack of freedom of use. In fact, experiencing and using public space are closely linked to forms of accessibility, a relationship extensively explored by Joseph (1998). The same author, along with Remy and Voyé (1997), argues that one of the fundamental values of urbanity is the possibility of social mobility, but equally important is spatial mobility. Regarding this mobility, linked to accessibility, it is important to highlight both its physical dimension, which is directly visible in the architecture and functionality of the space, and its symbolic dimension, which enables its use, exploration, and the offering of freedom in routes, paths, and movements.

Much of the public space, visible in its crisis, has been vulnerable to being classified, according to Marc Augé (2007), as "non-places." Such spaces are predominantly used for transient and anonymous relationships, where the logic of public space—associated with contact and convergence of various social actors—is replaced by individuality and anonymity. However, and without disregarding the actual existence of these spaces, the truth is that they are not reduced to such relationships but instead reflect transformations in the way people relate to each other and to the space itself. It is, however, identifiable that the social cohesion function of public space has weakened (Castro 2002), due to processes of institutionalisation of social practices, the spread of a sense of fear and insecurity that prevails in cities (Indovina 2002; Gill 2007) and, above all, the transfer of the "physical" social arena to the "virtual," mostly visible among younger social groups (Tomás 2007). The weakening of public space may also be linked to what Levy (1997) identifies as "uncivil public space," permeated by deficiencies that condition and restrict its full exploitation. These deficiencies are linked to the weakness of services and facilities, when present, the lack of aesthetic care, restricted or non-existent accessibility, and, equally important, a lack of alignment between the space, the opportunities it offers, and local specificities. Alongside its physical and material conditions, which ensure use and enjoyment, the symbolic and sociability conditions are also crucial for ensuring that such spaces reflect vitality and dynamism.

Public spaces are multifunctional, continuously interactive, and serve as a stage for the expression of diverse people (UN-Habitat, 2018). In this regard, urban planning must be capable of designing and organising these spaces to enable encounters and full enjoyment. However, in light of various contemporary phenomena that trigger transformations in the city's landscapes, it becomes crucial, as Menezes (2009) argues, to understand the multiple dynamics that shape public space, its experiences, uses, and appropriation phenomena in different social contexts. It is, above all, about "deconstructing" public space, as the same author suggests. This involves looking beyond the physical, built dimension, the immediately visible, so that the dynamics taking place in the city's gardens and parks can be effectively understood. This perspective is an effort of "deconstruction," easily linked to the sociological view defended by Peter Berger (1963) of seeing reality through a different lens than one believes to know. In fact, this is a true exercise in "sociological imagination"(Mills 1959), as it requires the ability to view phenomena and dynamics not as isolated facts, but as interactions that trigger new phenomena and extend beyond what is immediately visible. In this regard, Menezes (2009) notes that the "deconstruction" of a public space is essential in the process of understanding it, since the social practices that occur within it are responsible for shaping and reshaping its meanings. This attentive and reflective gaze on space is important in the process of identifying the regularities and patterns of social practices. The author, drawing on the work of key scholars like Barker and Shoggen (1973), Low (Low 2000a, b), Noschis (1984) and Wicker (1979), argues that this process allows such regularities to emerge as dynamics of space appropriation, leading to the creation of "...behavioural scenarios, pathways, points, landscapes, microgeographies, and/or forms that are ritualised." This reflects Certeau's (1990) view that daily space is continually reappropriated by individuals through creative practices. Therefore, the appropriation and reappropriation of both the physical and symbolic space are an unfinished process.

The devaluation and neglect of public space, whether by local authorities or through the transformation of daily social practices, fueled by the processes of globalization and rapid technological and scientific advances, was concomitant with a distancing from Nature by humanity. Throughout much of the twentieth century, there was a continuous disconnection from the natural world (Kesebir and Kesebir 2017), driven by the progress of science and economic development, supported by the belief in Nature's eternal resilience to cope with human actions (Farinha-Marques et al. 2011). The intense urbanization process in Portugal, alongside this, led to the supremacy of "concrete" over natural elements and territorial planning with serious deficiencies (Campos and Ferrão 2015; Schmidt 2016). This process was not properly planned and resulted in diffuse and fragmented urbanization (Fernandes and Seixas 2018).

Alongside public space, UGS have been undervalued (Magalhães 1992; Campos and Ferrão 2015) or maintained solely in their "romantic" and untouchable vision. The process of forgetting and the "death" of public space was widely discussed by

Sennett (1993) and Davis (1993), stemming from the decline in public participation, isolation, and exclusion from the public sphere, leading to the dominance of the private sphere and a persistent distance between these spaces and socialization networks. On the other hand, Arantes (2000) points out that public space is predominantly used by the "poor" and marginalized members of society, as more dominant social classes moved into private spheres of activity. This assumption is historically framed by the emergence of urban parks at the end of the nineteenth century.[1] Concerned with aesthetic and sanitary issues due to urban growth at the time, urban parks were primarily recreational (Jordan 1994; Lencastre and Farinha Marques 2021), specifically designed for leisure for the working class and older groups in society (Gardeners' Chronicle 1906). Concurrently, parks and gardens shifted from being seen purely from an aesthetic perspective to a more utilitarian and functional one, taking on roles in health, recreation, and performing ecological and environmental functions (Quintas 2014). In the Portuguese context, thanks to the pioneering work of Gonçalo Ribeiro Telles, not only as a staunch critic of land use planning during the Estado Novo era (such as in the case of the 1967 Lisbon floods), but also as the author of the "ecodevelopment" idea[2] (Telles 1985; Soromenho-Marques 2005) and the Green Plan for the city of Lisbon[3] (Telles 1997), the 1980s saw the introduction of new territorial instruments such as the National Ecological Reserve and the

---

[1] Although the first urban parks emerged in the late 18th century in England, as a response to the negative impacts of the Industrial Revolution on forests and the disorganized urban growth (Bovo and Conrado 2012), these spaces were seen as "escapes" from the unhealthy and poor-quality urban environment. This led to the creation of the ideal of English gardens, characterized by symmetry, romantic aesthetics, and organic forms. These parks were designed not only to provide a refuge from the industrialized city but also to promote public health, offering a space where citizens could reconnect with nature. The design principles of these early parks emphasized natural beauty, incorporating features like winding paths, lakes, and strategically placed trees, all meant to evoke a sense of tranquility and escape from the stresses of urban life.

[2] According to Gonçalo Ribeiro Telles, ecodesenvolvimento (ecodevelopment) is a line of thought based on the possibility of improving the living standards of populations while simultaneously ensuring the preservation of the environment and natural resources. In this context, economic and industrial activities are subjected to the principles of ecological balance, meaning that development must not come at the expense of the environment. Ribeiro Telles emphasizes the need for a harmonious relationship between human activities and nature, advocating for urban planning and policies that promote sustainable growth, protect ecosystems, and reduce environmental degradation. This concept underpins his work on urban green spaces, advocating for a planning approach that integrates nature and human settlements in a mutually beneficial way.

[3] The Green Plan for the city of Lisbon, developed by Gonçalo Ribeiro Telles, was based on four key guiding principles for the planning and development of urban green spaces. These principles aimed to transform the urban structure into an integrated and functional green network. The principles are as follows: 1. Empty spaces without use or meaning are not sufficient for a useful green structure: The creation of green spaces must be intentional and not based merely on the occupation of land without a clear purpose. These spaces should play an active role in the city and its organisation; 2. Green spaces should not be residual spaces: They cannot be simply neglected or sidelined in urban planning. They must be central to the organisation of urban space, fulfilling a role that contributes to the health and quality of life of citizens; 3. City planning should go beyond conventional practices: The simplistic notion that green spaces are confined to plants in roundabouts and separating strips must be avoided. Urban planning needs to be more integrated, considering sustainability, biodiversity, and social well-being; 4. Ecological and social balance must be considered in city planning: Sustainability is not

National Agricultural Reserve, as well as Regional Land Use Plans. However, the need for more sustainable urban planning that considered the multidimensionality of the urban green structure only saw its implementation more than a decade later (Campos and Ferrão 2015).

## 3 New Forms of Participation and New Protagonists

In Lefebvre Right to the City (2001), here are, albeit not deeply explored, elements that hint at a Right to Nature in urban contexts. This concept touches on inequalities of access to Nature and contrasts between humans and non-human beings or natural elements. Although Lefebvre himself does not extensively develop this idea, it resonates with broader debates on how urban environments and public spaces should be more inclusive of natural elements, not just as aesthetic or leisure spaces, but as integral parts of urban life. More recently, this idea has been further explored and expanded, particularly in the context of children and youth (Seixas et al. 2023), leading to the creation of a conceptual model of child-friendly green infrastructure, which acknowledges the rights of children (Vidal and Castro Seixas 2022). Lefebvre touches upon the idea of the right to rest and leisure in his work, acknowledging that urban spaces should cater to these needs. However, he did not fully develop the idea of a more expansive Right to Nature that goes beyond recreation. The Right to Nature entails a broader understanding that Nature should not only be seen as a place for rest or leisure. Beyond these important functions, which should not be reduced to mere instrumentalisation of nature, the Right to Nature advocates for social and personal rights that extend to a wider range of beings, not just humans. In this perspective, the Right to Nature is seen as a call for harmony between human and non-human elements in urban spaces. It stresses the importance of having the right to engage with and benefit from nature in ways that are holistic, recognising its intrinsic value and not merely as a tool for human well-being. This right is still not universally recognised, but the growing eco-social movements and the increasing awareness of environmental justice are pushing for a more inclusive and sustainable approach to urban planning and policy, where the rights of nature can be integrated alongside human rights.

Although the deterioration of public space, including urban gardens and parks, was significant at the turn of the twentieth and twenty-first centuries, this decline has, in contemporary times, given rise to movements advocating a return to the public sphere. Public squares, gardens, and urban parks are now spaces for encounter and socialisation. As a result of the exhaustion of a consumerist and postmodern societal model, these spaces have gained prominence on both the national and international public agendas, driven by ecological values and the right to the city. These new social

only environmental but also social. Planning should ensure that green spaces contribute to quality of life, urban resilience, and food supply, while respecting ecological balance and promoting social inclusion.

movements, often linked to urban activism, are characterised by a decentralised, horizontal, and participatory approach. They frequently focus on specific and emergent issues, such as environmental, climate, and social justice (Domaradzka 2018). These movements stand in opposition to established power structures and seek to promote social change through collective actions and popular mobilisation. They emphasise the intersection of environmental concerns with social, economic, and political inequalities, which result in disproportionate impacts on marginalised and vulnerable communities (Armiero et al. 2023). Alongside their approach and configuration, these movements also aim to give visibility and "protagonism" to actors traditionally marginalised or underrepresented in activism and civic participation contexts. This includes young people, ethnic minorities, women, people with disabilities, and other groups who have historically had limited access to spaces of power and decision-making. Therefore, as will be seen in the case studies that follow, their inclusion also serves as a call for representativeness and diversity within urban social movements, ensuring a variety of perspectives that should be considered in strategies for change. Figure 1 highlights the importance of participatory discourses, protest, and active citizenship in a period characterised by emerging trends in governance and bottom-up strategies. It also reflects how academia itself seeks to integrate the perspectives of communities and their stakeholders, particularly in the context of citizen science.

This approach promotes the inclusion of a variety of perspectives and experiences, enriching governmental policies and strategies with a broader and more representative view of society.

## 4 Case Studies: Territory and Methods

Two green spaces have been selected as case studies that have been subject to public discussion and community participation: the Jardim Público in Boavista (public garden), Porto, and Rua Luís Soares Barbosa in Braga (public green space) (Fig. 2).

### *4.1 Jardim Público in Boavista (Porto)—Porto-Boavista Railway Station*

Located in the Union of Parishes of Cedofeita, Santo Ildefonso, Sé, Miragaia, São Nicolau, and Vitória, in the municipality of Porto, NUTS III of the Metropolitan Area of Porto (AMP), the former Porto-Boavista Railway Station is a historically and culturally significant landmark for the city of Porto. The station, opened in 1875, played a major role in Portugal's railway history and culture. Being the first railway station in Porto and one of the first narrow-gauge stations in the country, its cultural heritage value is immeasurable. Despite this legacy, the station is currently under threat due to the lack of preservation efforts. The station ceased passenger services

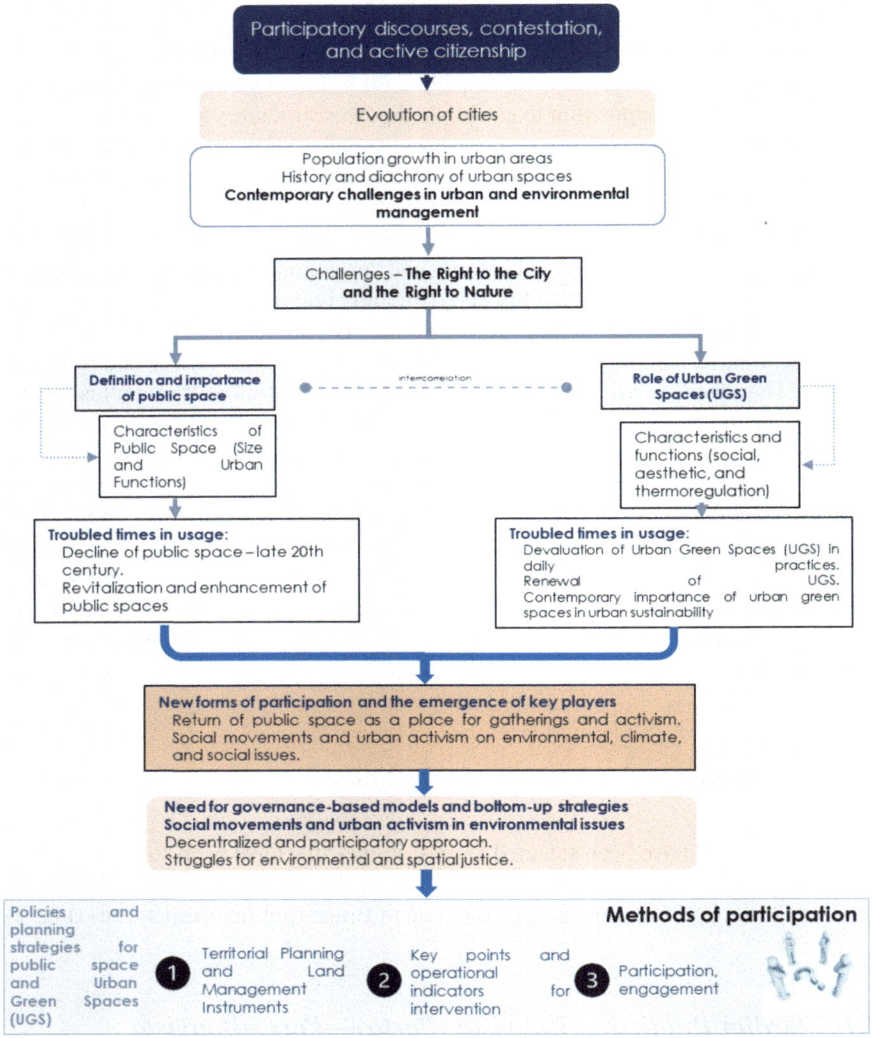


**Fig. 1** Models of intervention based on participatory discourses, protest, and active citizenship. *Source* Authors' elaboration

in 1938, although it continued to operate until the closure of the Porto to Póvoa de Varzim line in 2001. The imminent risk of the station's destruction has increased due to a real estate project from a large international chain on the surrounding land, which is currently managed by IP—Infraestruturas de Portugal. The land transfer to this international chain, promised since 2000 through a contract for the sale, continues to be renewed without an adequate assessment of the impact on the city of Porto and its historical-cultural heritage.

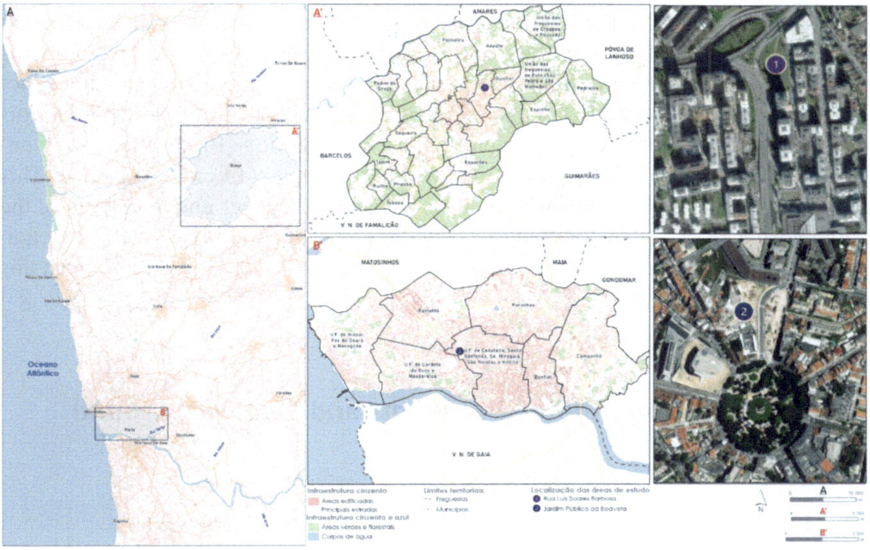


**Fig. 2** Location of case studies: The public garden at Boavista, Porto, and Luís Soares Barbosa Street, Braga. *Source* Authors' own elaboration

## *4.2 Rua Luís Soares Barbosa (Braga)*

In the parish of São Victor, located in the municipality of Braga and NUTS III of the Cávado region, Rua Luís Soares Barbosa is where the construction of a sports complex near the Hospital de Braga's access roads is being contested. The urban area of Rua Luís Soares Barbosa faces challenges similar to those of Jardim Público in Boavista due to a planned construction of a sports complex near the hospital. This project has raised concerns among a group of citizens who oppose the construction, advocating for the preservation of the green space and questioning the potential impacts on traffic, safety, and the quality of life for residents. The urban enclave formed by the João Paulo II, Padre Júlio Fragata, and Clermont Ferrant avenues, situated on the slopes of Sete Fontes, is under scrutiny due to the proposed development and its potential implications for the environment and local community.

### 4.2.1 Methodology

This study uses secondary sources and non-participant observation, through social media (Facebook), of the two movements under analysis. The research does not aim to analyze the specific processes of urban land occupation or debate strategic options. Instead, it seeks to investigate behaviors, participatory options, and the identification of new actors in urban areas.

The methodology adopted in this article is based on the synthesis of information from online field observations, news, and the use of images to illustrate developments associated with citizen participation in public construction processes. The aim is not to take a specific position but to contrast the problems and advantages related to this type of action. The intention is to explore potential directions for political intervention and interaction within communities, within the context of citizen science. This methodological approach aims to provide a broader and deeper understanding of the challenges faced by urban communities, encouraging reflection on citizen participation in public decision-making processes and on the possible future directions in this field.

## 5 A Reflection on Public Movements and Contests Around Green Spaces

### *5.1 Citizens' Movement for a Public Garden at Boavista*

The citizens' movement for the preservation and transformation of the former Boavista railway garden, located in the city of Porto, emerges as a collective voice advocating for a green space in the heart of the city. With a special emphasis on the reuse of the region's historical railway heritage, the movement firmly supports the creation of a public garden accessible to all — a place to relax, breathe, and enjoy the essence of the community, as opposed to the construction of an El Corte Inglés shopping center. The proposal presented by this group of citizens is supported by several fundamental reasons that not only echo present needs but also address future concerns about the central Boavista area. The city lacks high-quality, tranquil green spaces, which are essential to balance risks associated with heat and noise, as highlighted in previous studies (Vidal et al. 2021a; Lopes et al. 2022, 2023). With the imminent threat of transforming the area into another shopping center, the movement points out the potential negative impacts on already congested traffic, as well as the unfair competition posed to small local businesses.

Beyond these pragmatic concerns, the group also values the historical and symbolic importance of the former Boavista railway station as a living testament to the city's past. Defending the preservation of this building is not just an issue of memory, but also a commitment to the community's identity and culture. The revitalization of this abandoned space into a public garden, complete with local services, represents a unique opportunity to rejuvenate a central area of the city and ignite civic pride among its residents. Indeed, according to Bandarin and Van Oers (2012), the importance of conserving historical buildings and sites as fundamental elements of cultural identity and collective memory is emphasized. Preservation not only serves as a source of historical education but also strengthens community ties and contributes to the sustainable development of urban areas, as highlighted by Avrami et al. (2000).

The citizens' movement for the creation of this public garden is more than just a fight for a green space. It is a testament to the collective will to shape the future of the city, to redefine its center, and to preserve its historical identity. With a clear vision for the future, this movement is inspired by the current needs of the community while committing to creating a lasting legacy for future generations. Its struggle reflects the desire to promote a greener, more inclusive city centered on the needs of its citizens. Amidst the growing complexity of the climate narrative and contemporary socio-ecological challenges, the attention given to social inequalities as a structural element reflects a comprehensive understanding of the intersection between environmental and social issues.

With the inclusion of a variety of voices and participants previously not traditionally involved in decision-making spheres, the movement highlights the need for a holistic approach to addressing socio-environmental challenges. It calls for new protagonists in the struggle for socio-environmental justice, recognizing the importance of considering the interconnection between social justice and environmental preservation as a whole, thus emphasizing the need for public policies sensitive to these interrelations. Note, for example, the phrase "Porto needs everyone! And everyone needs to be heard," which reflects a strong call for inclusive and collaborative participation in urban decision-making processes (Fig. 3). This message emphasizes the importance of giving a voice to all stakeholders and communities involved, highlighting the need for democratic and participatory approaches in shaping the future of the city. By underscoring the significance of representation and inclusion, the image emphasizes the need for urban governance that is sensitive to the diverse interests and desires of the population.

**Fig. 3** A mural painted near the contested site reads, "Porto needs everyone! And everyone needs to be heard." *Source* https://www.facebook.com/umjardimnaboavista/photos/pb.100066463424307.-2207520000/323484346285276/?type=3. Published in November 21th 2021

Thus, the movement becomes not only a voice for the preservation of a green and historical space but also a key player in the construction of a broader narrative of socio-environmental justice, encompassing the complexity of contemporary urban dynamics and advocating for the importance of inclusive and comprehensive approaches to address current challenges. In Fig. 4, with the statement "More green to live," it conveys a clear aspiration for a healthier and more sustainable urban environment, emphasizing the fundamental role of green spaces in the community's quality of life. The image reinforces the need to integrate green elements into the urban fabric, highlighting the tangible benefits of green spaces for both physical and mental health, as well as promoting the environmental resilience of urban spaces. Additionally, the reference to grass at the base symbolizes the foundation of life, play, and exploration in (and of) the city, with small images above the letters depicting children playing, physical activities, and dog walking—activities that evoke a sense of freedom and vitality.

Finally, Fig. 5, with the phrase "No to El Corte Inglés, yes to a public garden," reveals a stance of resistance towards the allocation of spaces for purely economic (and commercial) purposes at the expense of green and public spaces. This image reflects the movement's concern with preserving historical heritage and promoting a balanced urban environment, one that values not only economic development but also the quality of life and well-being of the local community— a strong, long-term sustainability vision, considering future generations. By highlighting the importance of a public garden instead of more commercial developments, the image underscores

**Fig. 4** Poster placed near the contested location that says "More Green to Live." *Source* https://www.facebook.com/umjardimnaboavista/photos/pb.100066463424307.-2207520000/274812297819148/?type=3. Published on 4th September 2021

**Fig. 5** Poster placed near the contested location that says "No to El Corte Inglés, yes to a public garden."

the need to balance urban growth with the preservation of open, accessible spaces for the benefit of all city inhabitants.

Source: https://www.facebook.com/umjardimnaboavista/photos/pb.100066463424307.-2207520000/377659754201068/?type=3. Published on February 19, 2022.

This movement emphasizes the importance of considering the complexity of contemporary challenges as an integrated whole, recognizing the interdependent social and ecological structures that shape the urban environment, while also addressing the structural socio-environmental inequalities embedded in territories and institutions. By approaching socio-ecological issues as part of a connected system, the movement highlights the need for local public policies that consider both socioeconomic factors and environmental impacts at their core.

## *5.2 Opposition to the Construction of the Sports Complex—Rua Luís Soares Barbosa*

The "Moradores Rua Luís Soares Barbosa" collective is opposing the construction of a sports complex in the São Victor district of Braga, highlighting key concerns regarding sensitive urban planning and the preservation of green spaces, particularly in already densely urbanized areas. The dispute over the proposed use of a 6,476 m2 plot, which includes a green area, reflects concerns about the environmental, social, and mobility impacts resulting from the proposed development. The arguments put forth by the collective reveal the needs arising from the socio-environmental challenges faced by the local community. The emphasis on preserving the green space on Rua Luís Soares Barbosa underscores the importance of recreational areas for residents' well-being and urban quality of life, as discussed in academic studies such as those by Fuller and Gaston (2009), Jennings and Bamkole (2019) and Vidal et al. (2020, 2021b) which highlight the importance of green spaces for social cohesion and community health. Furthermore, concerns about environmental impacts, such as disrupting natural drainage and soil impermeabilization, emphasize the need to consider sustainability and ecological resilience in urban planning decisions, following the recommendations of Smit and Wandel (2006) and Pickett et al. (2020).

The collective's message also highlights the importance of civic participation and defending community interests in urban decision-making processes, as discussed in literature on participatory governance by Baiocchi (2003) and Gaventa and Barrett (2010). The emphasis on a respectful and civilized approach to expressing community concerns demonstrates a commitment to dialogue and cooperation in seeking solutions that benefit all stakeholders, as discussed by Innes and Booher (2010). This movement represents a concrete example of how social mobilization and civic participation can directly influence local public policy decisions, emphasizing the need for inclusive approaches that are sensitive to socio-environmental issues in decision-making.

The movement has been actively defending the green space, as evidenced by Fig. 6. The messages contained in the image reveal concerns about political decision-making that, according to the residents, may compromise natural elements and community well-being in favor of private interests. The statement "Environmental crime" emphasizes the movement's perception that the destruction of green spaces and the degradation of the natural environment violate the principles of environmental justice and sustainability. On the other hand, the message "We want green spaces" highlights the movement's aspiration for a more balanced and healthy urban environment, with access to natural and recreational spaces for community well-being (Braubach et al. 2017). Additionally, the statement "Anything goes?" reflects the movement's stance in defense of ethical and responsible practices in urban planning and development, underscoring the importance of considering ethical values and long-term impacts on the environment and community (Taylor et al. 2020).

In Fig. 7, there is a visible effort to mobilise, calling for a collective demonstration titled "In defence of our green space," where the central objective of the demonstration is clearly highlighted: the protection and preservation of urban green spaces for the benefit of the entire community. This aligns with the understanding of green spaces as essential commons, discussed by authors such as Tidball and Krasny (2014), who emphasise the cultural, environmental, and social values of urban green spaces and their importance for community health and resilience. By adopting public engagement strategies, such as organising demonstrations and using social media, the movement strengthens its message and broadens its support base, demonstrating the significance of active citizen participation in shaping the future of their urban environments.

**Fig. 6** Posters placed near the contested area reading: "Council grants green space to private developer for construction"; "Environmental crime"; "We want green spaces"; "Not everything is acceptable." *Source* https://www.facebook.com/photo.php?fbid=118628584637600&set=pb.100094713770412.-2207520000&type=3. Published on 28 July 2023

Another example is illustrated in Fig. 8. In this context, a photo of a public vigil stands out, aiming to highlight the solidarity and involvement of the community in defending socio-environmental issues. The image conveys a powerful message of unity and determination in the face of the challenges being faced, demonstrating the collective commitment to protect the threatened green space and to promote positive changes in the urban environment. The holding of the public vigil reflects the importance of raising awareness and mobilizing support around common causes, showcasing the movement's ability to draw public attention to crucial issues of environmental preservation and social equity.

The image of the public vigil also highlights the importance of collective expression and community solidarity in defending shared interests, relying on the significance of attributes such as social capital and community cohesion in promoting social and political change. Thus, it is not just a documentation of a vigil action, but primarily the dissemination of the power of community mobilization in promoting social change. In this way, the importance of public events and visual initiatives in constructing powerful narratives and raising public awareness around critical issues

**Fig. 7** Poster posted on the movement's Facebook page calling for a collective demonstration titled "In Defence of Our Green Space". *Source* https://www.facebook.com/photo.php?fbid=154650711035387&set=pb.100094713770412.-2207520000&type=3. Published on 22nd September 2023

becomes clear, aiming at the valorization of socio-environmental justice and environmental preservation, which often elude the decisions made for the very territories themselves.

**Fig. 8** Public vigil organized by the movement in defense of the green space. *Source* https://www.facebook.com/photo.php?fbid=162670676900057&set=pb.100094713770412.-2207520000&type=3. Published on 5 October 2023

## 6 Conclusions

Based on the analysis of two case studies, it is possible to argue that urban green spaces represent social projects that reflect complex dynamics of power, struggle, and resistance, where local communities come together to defend their natural environments and promote socio-environmental justice. In these movements, there is a clear opposition to the processes of "space appropriation" driven by the "capital" of certain groups, grounded in values that, in some way, are based on the defense of the common good, in response to the growing awareness of the need for a return to Nature. Understanding the ownership and value of space involves a distinction between private access to space and public access to space, and in this sense, the selectivity exercised by groups in the use and social occupation of spaces, as described by the Chicago School and encapsulated in the expression "human ecology," seems to be surpassed by the need to defend a greater good—the public urban spaces, which are of transversal social use, and, given current challenges, fostering a closer connection to Nature, in contrast to all the harmful aspects of the city, such as pollution (Bourdieu 1993).

The analysis of the movements of the Residents of Rua Luís Soares Barbosa and the Citizens' Movement for the Ferroviário Garden of Boavista underscores the crucial role of new actors in defending community interests and promoting meaningful social change, especially through inclusive and participatory approaches in urban planning and public policy formulation. These findings indicate that current public participation exercises, based on existing governance models, are often unable to capture the interest of communities. It is only when unilateral decisions affecting specific projects are made that the real territorial consequences are recognized.

The mobilization around green spaces and socio-environmental justice can influence local public policy decisions, shaping more inclusive and sustainable urban development practices, and emphasizing the importance of civic participation in local governance. By uniting these key points and highlighting the relevance of collective actions and civic participation in defending green spaces, it seems plausible to emphasize the central role of green spaces as places of resistance and hope in the creation of more sustainable and equitable urban communities. These efforts, though local, are situated within a broader context of current socio-ecological challenges, including the complexity of the climate narrative and structural social inequalities.

**Acknowledgements** Diogo Guedes Vidal expresses gratitude for the funding provided by the European Social Fund, the National Funds of MCTES, and the Human Capital Operational Programme through the Scholarship reference SFRH/BD/143238/2019 awarded by the Foundation for Science and Technology, I.P. He also thanks the support from the R&D Unit Functional Ecology Center—Science for People and the Planet (CFE), with the reference UIDB/04004/2020, funded by FCT/MCTES through national funds (PIDDAC), and from the Associated Laboratory TERRA with the reference LA/P/0092/2020.

## References

Abegão JLR (2022) The limits of sustainability: lessons from past societal collapse and transformation, for a civilization currently defying humanity's safe operating space. In: Leal Filho W, Vidal DG, Dinis MAP, Dias RC (eds) Sustainable Policies and Practices in Energy, Environment and Health Research. Springer, Cham, pp 439–454

Aldeia J, Alves F (2019) Against the environment. Problems in society/nature relations. Front Sociol 4. https://doi.org/10.3389/fsoc.2019.00029

Alves F, Vidal DG (2024) Plural nature (s): an overview of their sociocultural construction. Encyclopedia 4:1–12. https://doi.org/10.3390/encyclopedia4010001

Alves F, Araújo MJ, Azeiteiro U (2012) Cidadania ambiental e participação: o diálogo e articulação entre distintos saberes-poderes. Saúde Em Debate 36:46–54

Alves F, Filho WL, Araújo MJ, Azeiteiro UM (2013) Crossing borders and linking plural knowledge: biodiversity conservation, ecosystem services and human well-being. Int J Innov Sustain Dev 7:111–125. https://doi.org/10.1504/IJISD.2013.053323

Arantes O (1995) O lugar da arquitetura depois dos modernos. Edusp, São Paulo

Arantes A (2000) Paisagens paulistanas—transformações do espaço público. Editora UNICAMP, Campinas

Armiero M, de Rosa SP, Turhan E, Price (2023) Urban movements and climate change: loss. Amsterdam University Press, Amsterdam, Damage and Radical Adaptation

Augé M (2007) Não-Lugares, introdução a uma antropologia da sobremodernidade. Editora 90°, Lisboa

Avrami EC, Mason R, Torre MD la (2000) Values and heritage conservation: research report. Getty Conservation Institute, Los Angeles, CA

Baiocchi G (2003) Emergent public spheres: talking politics in participatory governance. Am Sociol Rev 68:52–74. https://doi.org/10.2307/3088902

Bandarin F, van Oers R (2012) The historic urban landscape: managing heritage in an urban century. Wiley, Hoboken

Barker R, Shoggen P (1973) Qualitaties of community life. Jossey- Bass Publishers, London
Bauman Z (2005) Liquid life. Polity Press, Cambridge
Beck U (2016) The metamorphosis of the world how climate change is transforming our concept of the world. Polity Press, Cambridge
Berger PL (1963) Invitation to sociology: a humanistic perspective. Anchor Books, New York
Borja J (2000) Fazer cidade na cidade actual. Centros e espaços públicos como oportunidades. In: Brandão P RA (eds) Espaço Público e Interdisciplinaridade. Centro Português do Design, Lisboa, pp 79–89
Bourdieu P (1993) La misère du monde. Éditions du Seuil, Paris
Bovo MC, Conrado D (2012) O Parque Urbano no Contexto da Organização do Espaço da Cidade de Campo Mourão. Cad Prudentino Geogr 1:50–71
Le Bras H (1997) Ville et hospitalité. Fondation Maison dês Sciences de l'Homme, Plan Construction et Architecture, Paris
Braubach M, Egorov A, Mudu P et al (2017) Effects of urban green space on environmental health, equity and resilience. In: Kabisch N, Korn H, Stadler J, Bonn A (eds) Nature-Based Solutions to Climate Change Adaptation in Urban Areas. Springer International Publishing, Cham, pp 187–205
Campos V, Ferrão J (2015) O Ordenamento do território: uma perspetiva genealógica. ICS Work Pap 1:1–45
Carson R (2000) Silent Spring. Penguin Books LTD, Londres
Castro A (2002) Espaços Públicos, Coexistência Social e Civilidade. Contributos para uma Reflexão sobre os Espaços Públicos Urbanos. Cid Comunidades e Territ 5:53–67
Catton WR, Dunlap RE (1978) Environmental sociology: a new paradigm. Am Sociol 13:41–49
Certeau M (1990) L'invention du quotidien. Gallimard, Paris
European Comission (2013) Communication from the commission to the european parliament, the council, the european economic and social committee and the committee of the regions 'Green infrastructure (GI)—enhancing Europe's natural capital' (COM(2013) 249 final of 6 May 2013). European Commission, Bruxelas
European Comission (2020) Green city accord. European Union, Bruxelas
European Commission (2020) Communication from the commission to the European parliament, the european council, the council and the european economic and social committee and the committee of the regions 'EU biodiversity strategy for 2030—bringing nature back into our lives' (COM). European Commission, Bruxelas
European Commission (2021) Communication from the commission to the European parliament, the council, the european economic and social committee and the committee of the regions 'Forging a climate-resilient Europe—the new EU strategy on adaptation to climate change' (COM(2021) 82. European Commission, Bruxelas
Costa CS, Menezes M, Mačiulienè M, Marušić BG (2020) Foreword. In: Costa CS, Mačiulienė M, Menezes M, Marušić BG (eds) Co-Creation of Public Open Spaces: Practice, Reflection and Learning. Edições Universitárias Lusófonas, Lisboa, pp 7–15
Crutzen P, Stoermer E (2000) The 'Anthropocene.' Glob Chang Newsl 41:17–18
Davis M (1993) A cidade de quartzo—Escavando um futuro em Los Angeles. Editora Página Aberta, São Paulo
Domaradzka A (2018) Urban social movements and the right to the city: an introduction to the special issue on urban mobilization. Voluntas 29:607–620. https://doi.org/10.1007/s11266-018-0030-y
Durkheim É (1893) The division of labour in society. Presses Universitaires de France, Paris
Farinha-Marques P, Lameiras JM, Fernandes C et al (2011) Urban biodiversity: a review of current concepts and contributions to multidisciplinary approaches. Innov Eur J Soc Sci Res 24:247–271. https://doi.org/10.1080/13511610.2011.592062
Featherstone M (1997) Culturas Globais e Culturas Locais. In: Fortuna C (ed) Cidade, Cultura e Globalização: Ensaios de Sociologia. Celta Editora, Oeiras, pp 82–102

Fernandes JAR, Seixas J (2018) Cities and urbanisation in democratic Portugal. Méditerranée 130. https://doi.org/10.4000/mediterranee.10698
Fuller RA, Gaston KJ (2009) The scaling of green space coverage in European cities. Biol Lett 5:352–355. https://doi.org/10.1098/rsbl.2009.0010
Gardeners' Chronicle, (1906) Parks and squares of the United States. Gard Chron 40:279
Gaventa J, Barrett G (2010) So what difference does it make? mapping the outcomes of citizen engagement. IDS Work Pap 2010:01–72. https://doi.org/10.1111/j.2040-0209.2010.00347_2.x
Gill T (2007) No Fear: Growing up in a risk society. Fundação Calouste Gulbenkian, Lisboa
Graff P (1996) Approche analytique de l´espace public urbain contemporain. In: L'espace public dans la ville méditerranéenne : actes du colloque de Montpellier, 14–15–16 mars 1996. Volume 2. Éd. de l'Espérou, Montpellier
Indovina F (2002) O Espaço Público: Tópicos sobre a sua Mudança. Cid Comunidades e Territ 5:119–123
Innes JE, Booher DE (2010) Planning with complexity: an introduction to collaborative rationality for public policy. Routledge, London
Jennings V, Bamkole O (2019) The Relationship between social cohesion and urban green space: an avenue for health promotion. Int J Environ Res Public Health 16. https://doi.org/10.3390/ijerph16030452
Jordan H (1994) Public parks, 1885–1914. Gard Hist 22:85. https://doi.org/10.2307/1587004
Joseph I (1998) La ville sans qualités. Editions de l'Aube, Paris
Kesebir S, Kesebir P (2017) A growing disconnection from nature is evident in cultural products. Perspect Psychol Sci 12:258–269. https://doi.org/10.1177/1745691616662473
Kleer J, Nawrot KA (2018) The rise of megacities: Challenges, opportunities and unique characteristics. World Scientific Europe, Londres
Lefebvre H (2001) O direito à cidade. Centauro, São Paulo
Lencastre MPA, Farinha Marques P (2021) Da Biofilia à Ecoterapia. A Importância dos Parques Urbanos para a Saúde Mental. Trab Antropol e Etnol 61:131–155
Lencastre MPA, Guedes Vidal D, Estrada R, et al (2022) The biophilia hypothesis explored: regenerative urban green spaces and well-being in a Portuguese sample. Int J Environ Stud 1–15. https://doi.org/10.1080/00207233.2022.2067411
Lencastre MPA, Vidal DG, Lopes HS, Curado MJ (2023) Biophilia in pieces: Critical approach of a general concept. Environ Soc Psychol 8:1869. https://doi.org/10.54517/esp.v8i3.1869
Levy A (1997) Explorer l'espace public. Laboratoire Theorie des Mutations Urbaines en Pays Developpes, Institut Français d'Urbanisme, Marne-La-Vallée
Lopes HS, Remoaldo PC, Ribeiro V, Martín-Vide J (2022) A comprehensive methodology for assessing outdoor thermal comfort in touristic city of Porto (Portugal). Urban Clim 45:101264. https://doi.org/10.1016/j.uclim.2022.101264
Lopes H, Remoaldo P, Ribeiro V, Martín-Vide J (2023) The impacts of climate change on human wellbeing in the municipality of Porto—an analysis based on remote sensing. In: Leal Filho W, Vidal DG, Dinis MAP (eds) Climate Change Management. Springer, Cham, pp 135–172
Low SM (2000a) Culture in the modern city: the microgeographies of gender, class, and generation in the Costa Rican plaza. Horizontes Antropológicos 6:31–64
Low SM (2000b) On the plaza: the politics of public space and culture. University of Texas Press, Austin
MacKinnon K, van Ham C, Reilly K, Hopkins J (2019) Nature-based solutions and protected areas to improve urban biodiversity and health. In: Marselle MR, Stadler J, Korn H et al (eds) Biodiversity and Health in the Face of Climate Change. Springer International Publishing, Cham, pp 363–380
Magalhães MM (1992) Espaços Verdes Urbanos. Direcção Geral do Ordenamento do Território, Lisboa
Menezes M (2009) A praça do Martim Moniz: etnografando lógicas socioculturais de inscrição da praça no mapa social de Lisboa. Horizontes Antropológicos 15:301–328. https://doi.org/10.1590/s0104-71832009000200013
Mills CW (1959) The sociological imagination. Oxford University Press, Oxford

Noschis K (1984) Signification affective du quartier. Libraire des Méridiens, Paris
Oliveira GM, Vidal DG, Ferraz MP (2020) Urban Lifestyles and Consumption Patterns. In: Filho WL, Azul AM, Brandli L, et al (eds) Sustainable cities and communities. Encyclopedia of the UN Sustainable Development Goals. Springer Nature Switzerland AG, Cham, pp 851–860
Pauleit S, Zölch T, Hansen R, et al (2017) Nature-based solutions and climate change—Four Shades of Green
Pickett STA, Cadenasso ML, Baker ME et al (2020) Theoretical perspectives of the Baltimore ecosystem study: conceptual evolution in a social-ecological research project. Bioscience 70:297–314. https://doi.org/10.1093/biosci/biz166
Quintas AV (2014) Génese e evolução dos modelos de estrutura verde urbana na estratégia de desenvolvimento das cidades. A Obra Nasce Rev Arquit Da Univ Fernando Pessoa 8:153–167
Remy J, Voyé L (1997) Cidade: rumo a uma nova definição? Edições Afrontamento
Rémy J, Voyé L (1981) Ville, ordre et violence. PUE, Paris
Salata S, Vidal DG, Alves F, et al (2023) Editorial: shaping healthier cities—Ecosystem services and health for responsive human–nature relations. Front Environ Sci 10
Schmidt L (2016) Portugal: Ambientes de Mudança. Erros, Mentiras e Conquistas. Temas e Debates—Círculo dos Leitores, Lisboa
Seixas PC, Dias RC, Vidal DG (2023) Políticas públicas e direito à cidade: uma realidade em mudança? In: Castro Seixas E, Seixas PC, Lopes JT (eds) O direito das crianças à cidade. Estudos sobre as cidades de Lisboa e do Porto. Mundos Sociais, Lisboa, pp 55–67
Sennett R (1993) A decadência do Homem público—As tiranias da intimidade. Companhia das Letras, São Paulo
Smit B, Wandel J (2006) Adaptation, adaptive capacity and vulnerability. Glob Environ Chang 16:282–292. https://doi.org/10.1016/j.gloenvcha.2006.03.008
Soja E (2000) Postmetropolis. Blackwell Publishers Ltd., USA, Critical Studies of Cities and Regions
Soromenho-Marques V (2005) Raízes do Ambientalismo em Portugal. Metamorfoses. Entre o Colapso e o Desenvolvimento Sustentável. Publicações Europa-América, Mem-Martins, pp 127–144
Steffen W, Persson A, Deutsch L et al (2011) The anthropocene: from global change to planetary stewardship. Ambio 40:739–761. https://doi.org/10.1007/s13280-011-0185-x
Taylor B, Chapron G, Kopnina H et al (2020) The need for ecocentrism in biodiversity conservation. Conserv Biol 34:1089–1096. https://doi.org/10.1111/cobi.13541
Telles GR (1985) Para além da Revolução. Edições Salamandra, Lisboa
Telles GR (1997) Plano Verde de Lisboa: Componente do plano director municipal de Lisboa. Edições Colibri, Lisboa, Portugal
Tidball KG, Krasny ME (2014) Introduction: greening in the red zone. In: Tidball KG, Krasny ME (eds) Greening in the Red Zone. Springer, Netherlands, Dordrecht, pp 3–24
Tomás C (2007) Paradigmas, imagens e concepções da infância em sociedades mediatizadas. Media J 11:119–134
United Nations (2015) Transforming our world: the 2030 Agenda for Sustainable Development. Resolution adopted by the General Assembly on 25 September 2015, A/RES/70/1. Geneva
United Nations, Department of Economic and Social Affairs PD (2019) World Urbanization Prospects: The 2018 Revision (ST/ESA/SER.A/420). United Nations, New York
Vidal DG, Castro Seixas E (2022) Children's green infrastructure: children and their rights to nature and the city. Front Sociol 7:804535. https://doi.org/10.3389/fsoc.2022.804535
Vidal DG, Fernandes CO, Viterbo LMF et al (2021a) Combining an evaluation grid application to assess ecosystem services of urban green spaces and a socioeconomic spatial analysis. Int J Sustain Dev World Ecol 28:291–302. https://doi.org/10.1080/13504509.2020.1808108
Vidal DG, Salata S, Vilaça H et al (2021b) Envisioning the future in public urban green spaces planning and design. Lessons from Porto. Urban Inf 297:52–57
Vidal DG, Dias RC, Oliveira GM et al (2022) A review on the cultural ecosystem services provision of urban green spaces: perception, use and health benefits. In: Leal Filho W, Vidal DG, Dinis

MAP, Dias RC (eds) Sustainable Policies and Practices in Energy, Environment and Health Research. Springer, Cham, pp 287–331

Vidal DG, Barros N, Maia RL (2020) Public and green spaces in the context of sustainable development. In: Leal Filho W, Azul AM, Brandli L, et al (eds) Sustainable cities and communities, encyclopedia of the UN sustainable development goals. Springer Nature Switzerland AG, Cham, pp 479–487

de Vries S (2010) Nearby nature and human health: looking at mechanisms and their implications. Innov Approaches to Res Landsc Heal 77–96

Weber M (1966) The city. Free Press, Michigan

Wicker A (1979) An introduction to ecological psychology. Brooks: Cole Pub., San Francisco

Wilson EO (1984) Biophilia. Harvard University Press, Cambridge, The Human Bond with Other Species

Wirth L (1938) Urbanism as a way of life. Am J Sociol 44:1–24

# From Crisis to Collaboration: Tackling Climate Change Health Impacts with One Health and Ecohealth

**Maria Pia Ferraz and Diogo Guedes Vidal**

**Abstract** Climate change represents a profound and escalating threat to global public health, undermining decades of progress over the past 50 years and exacerbating vulnerabilities worldwide. Its impacts span direct health effects, such as those from extreme heat and weather events, to indirect challenges, including ecosystem disruptions, food insecurity, and socio-economic crises like displacement, poverty, and conflict. The disproportionate burden on marginalized populations, particularly in low-income regions, underscores the urgent need for equitable and inclusive strategies. However, existing research on the health implications of climate change is often siloed, limiting the integration of insights and hindering the identification of critical gaps. Compounding this challenge is the rapid proliferation of literature, which outpaces traditional evidence review methods. Recent extreme weather events—record-breaking heatwaves, catastrophic floods, widespread wildfires, and prolonged droughts—have further highlighted the interconnectedness of environmental, animal, and human health. To confront these complex, multi-dimensional issues effectively, interdisciplinary frameworks such as One Health and Ecohealth offer promising pathways for collaborative action. This paper explores the intersections of climate change and health, advocating for the application of these integrative approaches to foster timely, comprehensive, and actionable public health responses.

M. P. Ferraz (✉)
Departamento de Engenharia Mecânica, Faculdade de Engenharia, Universidade do Porto, s/n, R. Dr. Roberto Frias, 4200-465, Porto, Portugal
e-mail: mpferraz@ineb.up.pt; mpferraz@fe.up.pt

i3S—Instituto de Investigação e Inovação em Saúde, Universidade do Porto, R. Alfredo Allen 208, 4200-135 Porto, Portugal

INEB—Instituto de Engenharia Biomédica, Universidade do Porto, R. Alfredo Allen 208, 4200-135 Porto, Portugal

D. G. Vidal
Centre for Functional Ecology—Science for People and the Planet (CFE), TERRA Associate Laboratory, Department of Life Sciences (DCV), University of Coimbra (UC), Coimbra, Portugal
e-mail: diogo.vidal@uc.pt

Department of Social Sciences and Management, Universidade Aberta, Lisbon, Portugal



W. Leal Filho et al. (eds.), *Composing Worlds: Humanities, Health and Wellbeing in the XXI Century Towards a More Sustainable World*, World Sustainability Series,
https://doi.org/10.1007/978-3-031-87108-5_21

**Keywords** Climate change · One Health · Ecohealth · Mental health · Food security · Human health · Infectious disease

## 1 Introduction

Climate change is causing negative impacts on global public health, evident in many populations worldwide, and is likely to undo relevant progress achieved in the past 50 years (Intergovernmental Panel on Climate Change. Working 2022). The changing climate affects health in various ways, including direct effects from extreme weather events and heat, effects on ecosystems through food systems changes and disease vectors, and socioeconomic effects such as poverty, population displacement, and conflict (Jolly et al. 2015; Gautier et al. 2016; Sheridan and Allen 2018; Haines and Ebi 2019; Intergovernmental Panel on Climate Change. Working 2022). Moreover, coping mechanisms, especially for susceptible communities residing in impoverished nations, are anticipated to be compromised by climate change (Bhatta and Aggarwal 2016; Gautier et al. 2016; Mycoo 2018).

To identify the most important connections between climate change and health, and respond with actions that promote health, it's necessary to regularly review the scientific evidence (Lesnikowski et al. 2011, Araos et al. 2016, Ebi et al. 2018). However, there are two main obstacles to having regular and timely reviews of the evidence on the connections between climate change and health. Firstly, research on this topic is spread out across different fields, leading to a fragmented landscape that makes it difficult to bring together important insights and identify gaps in evidence. Secondly, there is so much literature being produced on this topic that traditional methods of synthesizing evidence, which involve human review and screening, are no longer adequate or practical (Lamb et al. 2019, Callaghan et al. 2020, Haddaway et al. 2020).

The impact of climate change is characterized by scorching temperatures, frequent extreme weather conditions, drought, rising sea levels, soil and water salinization, and scarcity of food and fresh water (Intergovernmental Panel on Climate Change. Working 2022). The effects of these changes are far-reaching and interrelated, affecting human health and increasing the risk of disease (Rakib et al. 2019; Ebi et al. 2021a, b; Jay et al. 2021). A comprehensive understanding of the issue can be found in the 2022 Intergovernmental Panel on Climate Change 6th Assessment Report (Intergovernmental Panel on Climate Change. Working 2022). Recent events, such as the unprecedented heatwaves of the summer and fall of 2022 in various regions across the globe, serve as a testament to the severity of the climate crisis (Intergovernmental Panel on Climate Change. Working 2022; Williams et al. 2022).

The year 2022 saw several record-breaking heatwaves across different continents. Several examples can be pointed out: Pakistan reached temperatures exceeding 50 °C (COPERNICUS 2022); the western United States experienced a heat dome and the most severe drought in the last 1200 years, leading to dangerously low water levels in underground aquifers and lakes (COPERNICUS 2022). The drought had severe consequences for water availability and quality, putting pressure on socio-ecological systems and human health (Williams et al. 2022). In the Greater Horn of Africa, a period of drought lasting four years resulted in a broad famine and insecurity regarding access to food and water, leading to the demise of hundreds of thousands of livestock (WHO). In August 2022 Western Europe suffers a period of extreme heat and prolonged drought, leading to a heatwave and wildfires in Portugal (van der Wiel et al. 2022). Pakistan experienced a monsoon rainfall event that affected two-thirds of the country and resulted in the deaths of over 1500 people, the destruction of more than one million houses and the displacement of fifty million individuals (Sarkar 2022). Eastern Kentucky and St. Louis were struck by severe flooding, resulting in fatalities and damage on infrastructures (Nwanaji-Enwerem and Casey 2022). The city of Jackson, Mississippi, which was already grappling with water-related issues, experienced flooding that rendered the water supply unusable. As a result, the lack of access to potable water led to a national emergency (Nwanaji-Enwerem and Casey 2022). This situation highlights the impact of climate and social disasters, especially in communities facing lack of investment in water infrastructure (Nwanaji-Enwerem and Casey 2022). During the autumn of 2022, two hurricanes of category 4 magnitude wreaked havoc in Puerto Rico, the Caribbean, and the southeastern region of the United States, serving as a poignant reminder of the heightened vulnerability of coastal communities to the devastating impact of storms and floods (Roque et al. 2021).

The recent string of intense storms serves as a wake-up call to the disproportionate impact that coastal communities face in terms of extreme weather and its associated consequences. Hurricanes Fiona and Ian caused significant damage, particularly to areas still struggling to recover from previous hurricanes. The increasing frequency of "once in a thousand year" events, such as droughts, floods, and storms, suggests that current models and projections may need to be revised (Kent et al. 2022).

The contribution of this review is to list the human health impacts of climate highlighting the importance of interdisciplinary methods for evaluating and mitigating the impacts of climate change on human health.

One Health is an integrated approach that recognizes the interconnectedness of human, animal, and environmental health. It emphasizes collaboration across sectors to prevent and manage health risks, particularly those that arise from climate change, such as zoonotic diseases, ecosystem degradation, and food insecurity. EcoHealth focuses on the dynamic relationships between ecosystems and human health, acknowledging that changes in the environment—such as those caused by climate change—can directly affect health outcomes, particularly through biodiversity loss, altered disease patterns, and environmental contamination (Ferraz 2024). Both approaches are critical in addressing the health impacts of climate change by promoting sustainable solutions that protect ecosystems, prevent disease outbreaks, and improve health outcomes across species and environments. Climate change is not only an environmental issue but also a profound public health crisis, with interconnected effects on human, animal, and environmental health. The increasing frequency of extreme weather events, such as heatwaves, floods, and droughts, exacerbates vulnerabilities and challenges traditional public health models. This review highlights the urgent need for interdisciplinary approaches, particularly One Health and Ecohealth, to effectively address the complex and evolving health impacts of climate change. These frameworks enable integrated solutions by recognizing the deep interconnections between human well-being, biodiversity, and ecosystem integrity.

## 2 Materials and Methods

The present study is a narrative literature review based on the analysis of scientific publications related to the health impacts of climate change, aiming for an integrated up-date of scientific knowledge, through a holistic approach based on the updated knowledge. This review includes studies published in the following databases: Web of Science, Scopus, and PubMed, which are relevant scientific databases concerning the focus of the study. The search terms used were "Climate change" AND "Health" in "Title" (3561); "heat waves" OR "drought" OR "extreme weather" AND "Health" in "Keywords" (4380); "Climate change" AND "One Health" in "Title" (50); "Climate change" in "Title" and "Ecohealth" in "Title" OR "Abstract" OR "Keywords" (12); "Climate change" in "Title" and "Mental health" in "Title" OR "Abstract" OR "Keywords" (952) and "Climate change" in "Title" and "Infectious diseases" in "Title" OR "Abstract" OR "Keywords" (850); "Climate change" in "Title" and "Temperature-related AND death AND illness" in "Title" OR "Abstract" OR "Keywords" (2); "Climate change" in "Title" and "respiratory AND disease" in "Title" OR "Abstract" OR "Keywords" (478); "Climate change" in "Title" and "cardiovascular AND diseases" in "Title" OR "Abstract" OR "Keywords" (277) and "Climate change" in "Title" and "diabetes" in "Title" OR "Abstract" OR "Keywords" (117).

Neither language nor date restrictions were imposed during the search. All the studies published and in press until 18 September 2024 were considered and selected based on the most relevant information from the different aspects. Only peer-reviewed articles were considered. From all the analyzed documents, the selection was made based on the relevance of information included in the Title, Keywords, and Abstract by the author.

This review focuses on providing a broad overview rather than an in-depth analysis, covering a range of climate change issues and their human health impacts. Articles were included in the manuscript based on the following organization: (i) The study was based on climate change's influence on human health (170); (ii) The study was based on climate change's influence on mental health (16); (iii) The study was based on climate change's influence on infectious diseases (32); (iv) The study presents One Health and Ecohealth approaches concerning climate changes (59). Studies may belong to more than one group (i; ii; ii, or iv) once the information is interconnected, therefore, a total of 176 studies were selected.

# 3 Human Health Impacts of Climate Change

Climate change and its impact on health is a vast and rapidly changing field, with many questions to be explored. It is crucial to address the effects of wildfires on air and water quality, as well as their impact on respiratory and cardiovascular health, and the resulting food and water insecurity (Raoelison, Valenca et al. 2023). As the number of extremely hot days increases, so does the risk of wildfires, which can have a major impact on health. The recent fires in Portugal (September 2024) will certainly have severe health consequences that are not yet possible to quantify. Another important area of research is the salinization of water sources and its effects on drinking water and health, particularly in terms of cardiovascular and kidney health (Mueller et al. 2024). The study of how rising sea levels affect island populations, including their access to clean water and food and overall health is also important (Roy et al. 2023). Also, the effects of climate change on mental health, including exploring how uncertainty about the future is impacting anxiety, and depression (Kenstler and Shelley 2024; Ruiz-Páez et al. 2024). It is crucial to focus on marginalized populations, as these communities are often most vulnerable to the extreme weather events effects, due to environmental injustice and past policies (Parsons et al. 2024).

Increased frequency and intensity of heat waves lead to heat exhaustion, heatstroke, and exacerbated pre-existing conditions, particularly affecting the elderly, children, and those with chronic illnesses. Higher temperatures and air pollution (such as ground-level ozone and particulate matter) can worsen respiratory diseases like asthma and chronic obstructive pulmonary disease (COPD), and increase the risk of cardiovascular problems. However, an immense amount of health problems have arisen from climate change as therefore described.

Climate change poses one of the greatest public health challenges of our time, affecting human health through a wide range of mechanisms. While the consequences of climate change are diverse, its health effects can be classified into several key categories: temperature-related deaths and illnesses, respiratory diseases, cardiovascular diseases, diabetes, infectious diseases, mental health and wellbeing, and injuries and mortality from extreme weather events. Each of these health impacts is critical in its own right, and none can be considered more important than another once they are interconnected and represent various ways climate change exacerbates vulnerabilities in human health.

Listing the effects of climate change on health is important because it allows us to understand the wide range of consequences and how they impact different aspects of human well-being. However, it is impossible to order these effects by importance for several reasons: (i) interconnectedness of Health Impacts. Many of these health effects are interrelated and often exacerbate one another. For example, extreme heat can worsen respiratory and cardiovascular conditions, while also increasing mental health stress and the risk of infectious diseases. Trying to isolate and rank one issue over another ignores the complex interactions that make each effect part of a larger web of health risks; (ii) variability in vulnerability: Different populations are affected by climate change in different ways. Vulnerable groups such as the elderly, children, low-income communities, and people with pre-existing health conditions might experience these effects more severely. For example, while one community might suffer predominantly from infectious diseases due to flooding, another might be more affected by mental health challenges after extreme weather events. These differences make it impossible to generalize the importance of one health issue over another; (iii) context-specific impacts. Climate change impacts vary depending on geography, infrastructure, and social conditions. Coastal regions might prioritize the health impacts of hurricanes and rising sea levels, while urban areas might focus on air pollution and heat-related illnesses. Therefore, the relative importance of each health effect changes depending on the context in which it occurs; (iv) equal threats to well-being. All the listed effects—whether physical, like cardiovascular diseases, or mental, like anxiety—are essential components of human health and well-being. One cannot say that physical health is more important than mental health, or that immediate injuries from extreme weather events outweigh the long-term consequences of infectious diseases. Each has its own critical role in maintaining overall health, and neglecting any aspect weakens the holistic approach to addressing health risks from climate change. In summary, listing the health effects of climate change gives a clear understanding of the diverse challenges we face, but ordering them by importance is neither feasible nor productive. Every effect is part of a complex system, and the priority shifts depending on context, vulnerability, and interconnections between different health outcomes. Therefore, all impacts must be addressed simultaneously to ensure a comprehensive and effective response to the health risks of climate change.

## *3.1 Temperature-Related Death and Illness*

Temperature changes directly impact human health (Program 2024). Extreme temperatures directly impact health by disrupting the body's ability to maintain its internal temperature. When this regulation fails, it can lead to various illnesses such as heat cramps, heat exhaustion, heatstroke, and hyperthermia in extreme heat, and hypothermia and frostbite in extreme cold. Additionally, extreme temperatures can exacerbate chronic conditions like cardiovascular, respiratory, cerebrovascular diseases, and diabetes. Prolonged high temperatures are linked to more hospital admissions for cardiovascular, kidney, and respiratory issues. High minimum temperatures can also hinder the body's recovery from high daily maximum temperatures (Martínez-Solanas and Basagaña 2019; Ricciardi et al. 2019; Aydin-Ghormoz et al. 2022) being temperature extremes associated with various illnesses reported in emergency rooms and hospitals (Dring et al. 2022).

High ambient heat has been linked to various illnesses, including cardiovascular, respiratory, and kidney disorders; diabetes; hyperthermia; mental health issues; and preterm births (Faurie et al. 2022). Children, who spend more time outdoors and have limited physiological adaptation, are particularly vulnerable during heat waves (Malmquist, Lundgren et al. 2021). Respiratory illnesses among the elderly are commonly reported during extreme heat (Dring et al. 2022). Although there is uncertainty about how heat-related illness levels will change with rising summer temperatures due to climate change, progress has been made in monitoring heat-related illnesses. For instance, tracking emergency ambulance calls during heat waves can help establish real-time surveillance systems to identify extreme heat events. The increase in emergency visits for various illnesses during the 2006 California heat wave demonstrates the potential for using such data in real-time health surveillance systems (Knowlton et al. 2009). People living in cities are particularly vulnerable due to the Urban Heat Island (UHI) effect, where urban areas are warmer than rural ones. This has resulted in extensive research on the UHI during summer and its consequences, while much less attention has been given to the UHI during winter. Cold exposure also causes numerous health issues, and in the UK, cold-related deaths annually surpass those linked to heat. The extent to which the wintertime UHI might help reduce cold-related deaths is still not well understood (Macintyre et al. 2021).

The rise in temperature and heatwaves is expected to have a direct impact on human mortality (Perkins-Kirkpatrick and Lewis 2020; Watts et al. 2020). Many research studies predict that as greenhouse gas emissions cause the Earth to warm, there will be more deaths caused by excessive heat and fewer deaths caused by extreme cold in different places around the world. However, whether this overall effect will be positive or negative in the long term depends on the specific location and how much the temperature rises (Gasparrini et al. 2017). The degree to which heat-related deaths will increase due to global warming is uncertain and can be affected by many factors, including how well societies can adjust to hotter temperatures. This means that there are significant uncertainties in estimating the impact of rising temperatures on mortality rates (Ballester et al. 2011; Huang et al. 2011). In fact, in many parts of the world, particularly in developed countries, there have been documented decreases in the risk of death and overall burden associated with high temperatures over the past few decades, although temperatures have already risen and the population has aged (Achebak et al. 2018, 2019, 2020; Vicedo-Cabrera et al. 2018).

Europe has become one of the regions most affected by climate change, experiencing both increases in average temperatures and more extreme weather conditions. Over the past few decades, Europe has experienced a more rapid rise in temperature than any other continent, culminating in 2020 being the hottest year on record, exceeding 1.6 °C above the 1981–2010 average (VER). During the summer of 2003, temperatures reached record-breaking levels, causing thousands of premature deaths in Europe alone (Robine et al. 2008). However, the impact of these extreme temperatures was not evenly distributed across the continent, with some countries being more vulnerable to the effects of heat than others. For example, although temperature anomalies were twice higher in France than in Spain, the relative increase in mortality was higher in Spain, highlighting the differing levels of vulnerability of European societies to extreme heat (Ballester et al. 2011).

The main impacts of extreme temperature in health are depicted in Table 1.

**Table 1** Extreme temperature-related death and illness

| Hazard | Health impact | | | References |
|---|---|---|---|---|
| Extreme temperatures | Disrupt body's ability to maintain its internal temperature | Extreme heat | Heat cramps, heat exhaustion, heatstroke, and hyperthermia | (Martínez-Solanas and Basagaña 2019; Ricciardi et al. 2019; Aydin-Ghormoz et al. 2022; Dring et al. 2022; Faurie et al. 2022) |
| | | Extreme cold | Hypothermia | (Macintyre et al. 2021; Aydin-Ghormoz et al. 2022) |
| | Exacerbation of chronic conditions like cardiovascular, respiratory, cerebrovascular diseases, and diabetes | | | (Malmquist et al. 2021; Faurie et al. 2022) |
| | Inability to maintain homeostasis | | | (Ballester et al. 2011; Huang et al. 2011; Gasparrini et al. 2017; Achebak et al. 2018; Vicedo-Cabrera et al. 2018; Achebak et al. 2019; Achebak et al. 2020; Perkins-Kirkpatrick and Lewis 2020; Watts et al. 2020; Macintyre et al. 2021) |

## *3.2 Respiratory Diseases*

Climate change is linked to air pollution via several complex pathways (Afifa et al. 2024) and significantly worsen respiratory infections, asthma, allergic rhinitis, lung cancer, and chronic obstructive pulmonary disease (COPD) (Hoffmann, Maglakelidze et al. 2022; Song et al. 2023). Understanding the relationship between these factors is critical for developing effective public health strategies and mitigating their impacts on vulnerable populations (Maung et al. 2022). Warmer temperatures, rising levels of carbon dioxide ($CO_2$), and shifting precipitation patterns are influencing air quality and triggering processes that worsen respiratory diseases such as COPD and asthma (Eguiluz-Gracia et al. 2020).

The simultaneous exposure to rising temperatures, increased ozone levels, and higher concentrations of particulate matter exacerbate these chronic respiratory diseases in several ways (Jin et al. 2021; Yao et al. 2022). Both COPD and asthma are characterized by episodes of exacerbation, where symptoms worsen rapidly and often require medical intervention. Higher air temperatures and more pollution lead to increased hospital admissions for acute exacerbations of both diseases (Chung et al. 2022). Wildfires, heatwaves, and high-ozone days are strongly correlated with spikes in emergency room visits and hospitalizations for respiratory issues (Yuan et al. 2023). Asthma patients may experience more frequent and severe asthma attacks, as well as long-term structural changes in the airways, making the disease more difficult to manage (Tran et al. 2022). This disproportionately affect vulnerable populations, including children, the elderly, and those with preexisting health conditions. Children with asthma are particularly at risk due to their developing lungs and higher

**Table 2** Main impacts of climate change on respiratory diseases

| Hazard | Health impact | References |
|---|---|---|
| Warmer temperatures, increased $CO_2$ levels, shifting precipitation patterns affect air quality | Exacerbates the severity, frequency of exacerbations, and long-term health outcomes of respiratory diseases like COPD and asthma | (Eguiluz-Gracia et al. 2020; Hoffmann et al. 2022; Maung et al. 2022; Song et al. 2023) |
| Wildfires, heatwaves, and high-ozone days | Spikes in emergency room visits and hospitalizations for respiratory issues | (Yuan et al. 2023) |
| Combined effects of climate change and air pollution | Exposure to pollutants and extreme weather conditions accelerates lung function decline in COPD patients and worsens asthma severity, resulting in faster disease progression and lower quality of life asthma, allergic rhinitis, lung cancer, and chronic obstructive pulmonary disease | (Jin et al. 2021; Chung et al. 2022; Li et al. 2022; Mata et al. 2022; Tran et al. 2022; Wu et al. 2022; Yao et al. 2022; Han et al. 2023) |

exposure to outdoor air pollution (Wu et al. 2022; Han et al. 2023). Older adults with COPD are also more susceptible to the effects of heat stress, respiratory infections, and pollution, leading to poorer health outcomes (Mata et al. 2022; Han et al. 2023).

Allergic rhinitis is a widespread condition worldwide, and its occurrence is strongly associated with exposure to air pollution. A thorough meta-analysis found a positive relationship between the incidence of rhinitis and the concentration of air pollutants such as PM10, PM2.5, $NO_2$, $SO_2$, and $O_3$. The analysis revealed that children and adolescents face a higher risk than adults (Li et al. 2022). The main impacts of climate change on diseases diseases and the details involved are depicted in Table 2.

## 3.3 *Cardiovascular Diseases*

Studying the number of deaths has revealed that both very hot and very cold temperatures are linked to higher occurrences of deaths caused by cardiovascular disease (Watts et al. 2019). While this trend is consistently seen worldwide, the local climate and how people in a given region adapt to hot and cold temperatures affect the temperature range in which the fewest deaths occur. The Global Burden of Disease estimation now includes the impact of non-optimal temperatures, which accounts for 1.96 million deaths globally. So far, the majority of these deaths are caused by low temperatures rather than high temperatures (Murray et al. 2020). In the context of climate change, the significance of severe weather events is undergoing a shift. In fact, we have been able to witness alterations in the danger of heart attacks related to high temperatures, even in Augsburg, Germany (Chen et al. 2019). In a recent study, researchers examined data from the KORA Myocardial Infarction Registry to assess how cold and heat affected the occurrence of heart attacks from 1987–2000 and 2001–2014. In the first period, only cold exposure was linked to heart attacks. However, in the second period, temperatures above 18 °C significantly increased the risk of heart attacks, particularly in individuals with type 2 diabetes or hypertension (Peters and Schneider 2021). This finding suggests that the impacts of changes in temperature on health are already noticeable and have an additional negative impact on vulnerable groups within the population.

It is scientifically reasonable to believe that higher temperatures can increase the risk of sudden onset heart disease due to the physiological effects of heat exposure. When exposed to high temperatures, the body responds by sweating, increasing heart and breathing rates, widening blood vessels, and potentially affecting blood clotting. These alterations may disrupt the autonomic regulation of the heart, raise arterial pressure levels, trigger systemic inflammation, and impede clotting mechanisms, which can result in the rupture of atherosclerotic plaques and subsequent heart attacks. Moreover, people with pre-existing heart failure and reduced heart function may be especially vulnerable to heat exposure. The link between increased respiratory disease mortality and heat exposure may also be related to the underlying effects of heat exposure on right ventricular heart failure (Schneider et al. 2017).

Regrettably, climate change not only impacts temperatures but also has negative consequences for other environmental factors such as air pollution, which can increase due to factors like wildfires. Fine particulate matter, which is linked to cardiovascular disease, is a significant contributor to air pollution (Brook et al. 2010). Research has revealed that episodes of exposure to both high temperatures and particulate matter have a severe and combined effect on mortality due to cardiovascular disease. For instance, in Russia in 2010, a joint and significant impact of heat and particulate matter was observed (Shaposhnikov et al. 2014). Additionally, a nationwide study in Italy showed that the effect of heat on mortality is three times higher when fine particulate matter concentrations are high compared to when they are low (Scortichini et al. 2018). Despite the most optimistic projections, avoidable acute cardiovascular disease incidents are already resulting from environmental alterations. Consequently, addressing these changes must be an integral component of our broader initiatives aimed at preventing and managing cardiovascular diseases (Peters and Schneider 2021).

### *3.4 Diabetes*

The increase in temperature caused a variety of negative consequences such as reduced agricultural output, damage to soil quality, the transformation of land into desert, harm to ecosystems, the acidification of oceans, a decrease in the availability of fresh water, the loss of biodiversity, and the thinning of the ozone layer (Rockstrom et al. 2009). Climate change exacerbates health risks, leading to greater morbidity and mortality rates among non-communicable diseases like type 2 diabetes (Zilbermint 2020).

Brown adipose tissue (BAT) is a natural component of the human body that becomes active when exposed to cold temperatures. Its primary function is to burn a significant amount of lipids and stimulate the sympathetic nervous system to produce heat (Hoeke et al. 2016; Symonds et al. 2019).

When brown fat (BAT) burns lipids, it causes an increase in the amount of glucose that enters the muscles, leading to improved insulin sensitivity (Schilperoort et al. 2016). Research shows that spending time in a moderately cold environment can improve insulin sensitivity, even after just 10 days (Hanssen et al. 2015). There is also a connection between lower blood glucose and HbA1C levels after exposure to cold, regardless of body fat (Matsushita et al. 2014). The increasing global temperatures due to global warming can negatively impact the role of brown fat in regulating glucose levels, which may increase the risk of diabetes among the population. Studies show that for every additional degree Celsius increase, there is a corresponding increase in obesity-adjusted diabetes incidence rate and diabetes prevalence (Ratter-Rieck et al. 2023). Conversely, in cold environments, individuals with diabetes who have autonomic neuropathy may be at a higher risk of hypothermia, especially in the elderly (Blauw et al. 2017).

The changing climate puts food production at risk, which can cause food shortages and insecurity. This insecurity can then lead to negative impacts on the health of people with diabetes and the environment (Nguyen et al. 2023; Meo and Meo 2024). As fresh products become scarcer and more expensive due to climate change, people may turn to processed and imported foods with less nutritional value because they are more easily accessible and affordable (Tchonkouang et al. 2024). However, the increase in demand for processed foods can further harm the environment. Overall, the relationship between climate change, food insecurity, and health creates a challenging cycle. The production of processed foods leads to higher greenhouse gas emissions due to the carbon-intensive processes involved in transportation, retail processing, and storage. In addition, consuming processed foods can increase the risk of developing obesity and type 2 diabetes, which then places more demand on the healthcare system (Dain and Hadley 2012). Overall, the production and consumption of processed foods have negative impacts on both the environment and public health. Furthermore, the increasing prevalence of obesity is likely to result in more sedentary lifestyles, greater reliance on cars, and higher demand for food production, all of which contribute to the depletion of the earth's resources and exacerbate the problem of climate change. Ultimately, all of these factors are interconnected and lead to a negative impact on the environment (Edwards and Roberts 2009).

As urban areas continue to grow, the issue of food insecurity is amplified, as these areas are often disconnected from traditional sources of fresh food and must rely heavily on imported food. Thus, urbanization plays a role in both food insecurity and environmental degradation (Oliveira et al. 2019). It's important to recognize that livestock production has a significant and negative impact on the planet, as it produces the highest amount of greenhouse gas emissions. The "carbon cost" of producing livestock is seven times greater than that of producing vegetables. This means that reducing or changing our consumption of meat and animal products could have a significant positive impact on the environment (Berners-Lee et al. 2012).

### 3.5 *Vector-Borne Diseases*

Changes in temperature, humidity, and rainfall patterns can expand the habitat and lifecycle of vectors such as mosquitoes and ticks, spreading diseases like malaria, dengue fever, Zika virus, Lyme disease, and West Nile virus to new regions (Semenza and Paz 2021; Tavares and Oliveira 2023). Climate change significantly impacts vector-borne diseases by altering the habitats, life cycles, and population dynamics of vectors such as mosquitoes, ticks, and fleas. The main impacts of climate change on vector-borne diseases and the details involved are depicted in Table 3.

Relevant specific disease impacts include: (i) Malaria, rising temperatures and changing rainfall patterns expand the range of Anopheles mosquitoes. Malaria could become endemic in areas where it was previously absent; (ii) Dengue Fever, Aedes mosquitoes, vectors of dengue, thrive in warmer temperatures and urban environments. Increased rainfall and higher temperatures boost their populations and extend transmission seasons; (iii) Zika Virus, similar to dengue, Zika virus transmission by Aedes mosquitoes is likely to expand with warmer and wetter conditions; (iv) Lyme Disease, warmer winters and longer growing seasons can increase tick populations and extend their geographic range, spreading Lyme disease to new areas; (v) Chikungunya, changes in temperature and rainfall patterns can support the spread of Aedes mosquitoes, increasing the risk of chikungunya outbreaks (Semenza and Paz 2021; Adepoju et al. 2023; Tavares and Oliveira 2023; Paz 2024).

### 3.6 *Water-Borne Diseases*

Climate change significantly impacts waterborne diseases through various mechanisms involving changes in temperature, precipitation, sea level, and extreme weather events. The main impacts of climate change on water-borne diseases and the details involved in health consequences are depicted in Table 4.

Relevant specific disease impacts include: (i) Cholera, increased temperatures and changing precipitation patterns can enhance the survival and proliferation of *V. cholerae* in water sources. Flooding can spread cholera by contaminating drinking water with fecal matter; (ii) Leptospirosis, more frequent and intense rainfall can lead to increased runoff and water contamination with urine from infected animals, leading to higher incidences of leptospirosis; (iii) Cryptosporidiosis and Giardiasis, heavy rainfall and flooding can increase the runoff of *Cryptosporidium* and *Giardia* from agricultural and urban areas into water supplies, causing outbreaks; (iv) Hepatitis A and E, poor sanitation conditions exacerbated by climate change-related disasters can lead to the spread of hepatitis A and E, which are transmitted through contaminated

**Table 3** Main impacts of climate change on vector-borne diseases

| Climate change in hazard on vector-borne disease | | Health impacts | References |
|---|---|---|---|
| Vector expanded geographic range | Temperature increases | Warmer temperatures allow vectors to survive in previously inhospitable regions. Diseases like malaria, dengue fever, and Zika virus are spreading to higher altitudes and latitudes | (Semenza and Paz 2021; Adepoju, Afinowi et al. 2023; Tavares and Oliveira 2023; Paz 2024) |
| | Seasonal changes | Longer warm seasons extend the breeding and transmission period of vectors, increasing the potential for disease transmission | (Ma et al. 2022; Adepoju et al. 2023) |
| Altered vector behavior and ecology | Reproduction rates | Higher temperatures can accelerate the reproduction rate of vectors, leading to larger populations | (Rupasinghe et al. 2022; Paz 2024) |
| | Feeding frequency | Increased temperatures can lead to more frequent feeding by vectors, heightening the risk of disease transmission | (Mojahed et al. 2022; Paz 2024) |
| Changes in precipitation patterns | Increased rainfall | Creates more breeding sites for mosquitoes, such as stagnant water pools, which can increase the spread of diseases like dengue fever and malaria | (Nosrat et al. 2021; Nabatanzi et al. 2022) |
| | Drought conditions | Can force vectors to seek out human-made water sources, bringing them into closer contact with human populations. However, educed water sources and arid landscapes can disrupt the life cycles of these vectors, leading to a decline in associated illnesses | (Brown et al. 2014) |
| Impact on pathogen development | Temperature-dependent pathogens | The development rate of many pathogens within vectors is temperature-dependent. Warmer temperatures can shorten the incubation period of diseases like malaria, making vectors infectious sooner | (Brown et al. 2014) |

(continued)

**Table 3** (continued)

| Climate change in hazard on vector-borne disease | | Health impacts | References |
|---|---|---|---|
| Altered ecosystems and biodiversity | Habitat changes | Climate change can alter habitats, affecting both vectors and their natural predators. Reduced biodiversity can lead to increased vector populations | (Rocklov and Dubrow 2020) |
| | Urbanization | Climate-induced migration and urbanization can create environments conducive to vectors, such as poorly managed water systems and increased human-vector contact | (Chaudhry 2024) |
| Human behavior and movement | Displacement and migration | Climate change-induced displacement can lead to the spread of vector-borne diseases as people move from affected areas to new regions, potentially carrying vectors and pathogens with them | (Lemon 2008; Ma et al. 2022; Paz 2024) |
| | Changes in agricultural practices | Altered farming practices and irrigation due to climate change can create new vector habitats | (Gale et al. 2009) |
| Public health and infrastructure strain | Increased disease burden | The spread of vector-borne diseases can strain public health systems, particularly in regions unprepared for new or increased disease burdens | (Ma et al. 2022) |
| | Economic impact | Increased disease prevalence can affect economic productivity and development, particularly in regions heavily dependent on agriculture and outdoor labor | |

water and food; (v) Norovirus, increased storm events and flooding can lead to the contamination of water supplies with norovirus, causing gastrointestinal illnesses (Stanke et al. 2013; Levy et al. 2018; Okaka and Odhiambo 2018; Griffith and Gobler 2020; Lassiter 2021; Dietrich et al. 2023; Jung et al. 2023; Mavrouli et al. 2023; Semenza and Ko 2023; Alied et al. 2024).

**Table 4** Main impacts of climate change on water-borne diseases

| Climate change in hazards on water-borne diseases | | Health impacts | References |
|---|---|---|---|
| Increased temperature | Pathogen proliferation | Higher water temperatures can accelerate the growth and reproduction of pathogens such as *Vibrio cholerae* (cholera), *Giardia*, and *Cryptosporidium* | (Dietrich et al. 2023) |
| | Seasonal changes | Warmer waters promote algal blooms, some of which can produce toxins harmful to humans and marine life, increasing the risk of diseases such as shellfish poisoning | (Griffith and Gobler 2020) |
| Changes in precipitation patterns | Heavy rainfall and flooding | Increased precipitation can lead to runoff that contaminates water sources with pathogens from human and animal waste. Floods can overwhelm sewage systems, leading to outbreaks of diseases like cholera, leptospirosis, and hepatitis A | (Jung et al. 2023; Semenza and Ko 2023) |
| | Drought conditions | Reduced water availability can concentrate contaminants and pathogens in water sources, making diseases more likely to occur. People may also resort to using unsafe water sources during droughts | (Stanke et al. 2013) |
| Sea level rise | Saltwater intrusion | Rising sea levels can lead to saltwater contamination of freshwater sources, impacting water quality and availability | (Lassiter 2021) |
| | Inundation of coastal areas | Coastal flooding can introduce pathogens into freshwater systems and disrupt sanitation infrastructure, leading to disease outbreaks | (Okaka and Odhiambo 2018) |
| Extreme weather events | Storm surges and hurricanes | These events can cause massive infrastructure damage, leading to contamination of drinking water supplies with pathogens and chemicals, increasing the risk of waterborne diseases | (Jung et al. 2023) |
| | Increased frequency of natural disasters | More frequent and severe natural disasters disrupt water and sanitation infrastructure, making it difficult to maintain clean water supplies and manage waste | (Mavrouli et al. 2023; Alied et al. 2024) |

(continued)

**Table 4** (continued)

| Climate change in hazards on water-borne diseases | | Health impacts | References |
|---|---|---|---|
| Impact on water infrastructure | Aging and inadequate infrastructure | Climate change can exacerbate the strain on aging or inadequate water and sewage systems, increasing the likelihood of contamination and disease outbreaks | (Levy et al. 2018) |
| | Adaptation challenges | Rapid changes in climate patterns make it difficult for existing water management systems to adapt, leading to increased vulnerability to waterborne diseases | (Levy et al. 2018) |

## *3.7 Food Security and Nutrition*

Climate change affects agricultural productivity, leading to food shortages and increased food prices. This can result in malnutrition and food insecurity, especially in vulnerable populations. Climate change significantly impacts food security and nutrition through a variety of mechanisms. The main impacts of climate change on food security and nutrition are depicted in Table 5.

The impacts of climate change on food security and nutrition are complex and multifaceted, affecting every aspect of the food system. Addressing these challenges requires comprehensive strategies, including improving agricultural practices, investing in resilient infrastructure, enhancing water management, promoting sustainable fisheries, and ensuring social safety nets for vulnerable populations. Global efforts to mitigate climate change and adapt to its effects are crucial to safeguard food security and nutrition for future generations (Tchonkouang et al. 2024).

## *3.8 Mental Health and Wellbeing*

Climate change poses a significant threat to mental health and well-being through a wide range of mechanisms, from the direct impacts of extreme weather events to the chronic stress associated with climate anxiety and economic instability. Addressing these challenges requires a multi-faceted approach, including mental health support services, community resilience building, economic and social support systems, and proactive climate mitigation and adaptation strategies. Recognizing and addressing the mental health impacts of climate change is crucial for fostering overall well-being and resilience in affected populations.

**Table 5** Main impacts of climate change on food security and nutrition

| Climate changes in hazard on food security and nutrition | | Impact on food security and nutrition | References |
|---|---|---|---|
| Crop yields and productivity | Temperature changes | Higher temperatures can reduce crop yields by accelerating plant development, reducing the time for grain filling, and increasing respiration rates. Heat stress affects staple crops like wheat, rice, and maize. However, a warmer climate might make cold and previously unfertile areas now accessible for farming | (Hatfield and Prueger 2015; Khan et al. 2020; Farhad et al. 2023) |
| | Altered rainfall patterns | Changes in precipitation can lead to droughts or excessive rainfall, both of which negatively affect crop productivity. Droughts can cause crop failures, while heavy rains can lead to soil erosion and flooding, damaging crops | (Furtak and Wolinska 2023) |
| | Extreme weather events | Increased frequency and severity of extreme weather events, such as hurricanes, floods, and droughts, can destroy crops and disrupt agricultural production | (Furtak and Wolinska 2023) |
| Soil health and fertility | Erosion and degradation | Increased rainfall intensity can cause soil erosion, while higher temperatures can lead to soil degradation and loss of organic matter, reducing soil fertility | (Masroor et al. 2022; Dash and Maity 2023) |
| | Salinization | Sea level rise and increased evaporation rates can lead to the salinization of soil and water resources, particularly in coastal areas, making the land less productive | (Mazhar et al. 2022; Tarolli et al. 2024) |
| Water availability | Reduced water supply | Changes in precipitation patterns and the melting of glaciers can reduce the availability of water for irrigation, crucial for agriculture in many regions | (Karaaslan 2024; Naderi et al. 2024; Rodrigues et al. 2024) |
| | Increased competition | Greater competition for water resources among agriculture, industry, and domestic use can exacerbate water scarcity issues | (Ingrao et al. 2023) |

(continued)

**Table 5** (continued)

| Climate changes in hazard on food security and nutrition | | Impact on food security and nutrition | References |
|---|---|---|---|
| Pests and diseases | Expanded ranges | Warmer temperatures can expand the geographical range of pests and pathogens, increasing the incidence of crop diseases and infestations | (Skendzic et al. 2021; Gullino et al. 2022) |
| | Increased resistance | Climate change can lead to the development of pest and pathogen resistance to traditional control methods, posing further challenges to crop protection | (Nnadi and Carter 2021; Grigorieva et al. 2023) |
| Livestock production | Heat stress | Higher temperatures can cause heat stress in livestock, reducing productivity, and fertility, and increasing mortality rates | (Thornton et al. 2022; Habeeb et al. 2023; Khan et al. 2023) |
| | Water and feed shortages | Droughts and changing precipitation patterns can reduce the availability of water and feed for livestock, impacting their health and productivity | (Godde et al. 2021) |
| | Disease spread | Climate changes can also alter the distribution and prevalence of livestock diseases | (Ali et al. 2020) |
| Fisheries and aquaculture | Ocean warming and acidification | Rising sea temperatures and ocean acidification can affect marine ecosystems, reducing fish stocks and impacting fisheries | (Issifu et al. 2022; Venegas et al. 2023) |
| | Habitat loss | Coral bleaching and the destruction of coastal habitats due to climate change can reduce biodiversity and fishery productivity | (Graham 2014) |
| Food prices and availability | Price volatility | Reduced crop yields and disruptions in supply chains due to extreme weather events can lead to increased food prices and greater price volatility | (Tchonkouang et al. 2024) |
| | Food shortages | Persistent adverse climate conditions can lead to food shortages, particularly in regions heavily dependent on local agriculture | (Mirzabaev et al. 2023; Tchonkouang et al. 2024) |

(continued)

**Table 5** (continued)

| Climate changes in hazard on food security and nutrition | | Impact on food security and nutrition | References |
|---|---|---|---|
| Nutritional quality | Reduced nutritional value | Elevated $CO_2$ levels can reduce the nutritional quality of crops, decreasing concentrations of essential nutrients like protein, zinc, and iron in staple crops | (Ebi and Loladze 2019; Ebi et al. 2021a, b; Ziska 2022) |
| | Dietary diversity | Climate change can reduce the availability of diverse food sources, leading to less varied diets and increased risk of malnutrition | (Mirzabaev et al. 2023) |

Climate change has significant impacts on mental health and well-being through a variety of direct and indirect mechanisms. The main impacts of climate change on mental health and well-being are depicted in Table 6.

### *3.9 Injuries and Mortality from Extreme Weather Events*

More frequent and severe storms, hurricanes, floods, and wildfires lead to direct injuries, deaths, and destruction of infrastructure, disrupting access to healthcare and essential services. Extreme weather events significantly impact injuries and mortality through various mechanisms. The main impacts of extreme weather events on injuries and mortality are depicted in Table 7.

## 4 The Role of One Health and Ecohealth in Adaptation and Mitigation

The rapidly changing climate poses significant challenges of this century, affecting not only the environment but also human health. As global temperatures rise, ecosystems shift, and extreme weather events become more frequent, the interconnectedness between human, animal, and environmental health becomes increasingly evident. In this context, the concepts of One Health and Ecohealth have emerged as vital frameworks for comprehension and mitigating the health consequences of climate change (Shafique et al. 2024).

Ecohealth expands on the principles of One Health by emphasizing the critical role ecosystems play in maintaining the health of all living beings. This approach highlights how the degradation of ecosystems—through deforestation, pollution, and climate change—leads to adverse health outcomes for humans and animals. By protecting and restoring ecosystems, it is possible to mitigate many of the health

**Table 6** Main impacts of climate change on mental health and well-being

| Climate change health effect | | Detailed cause/consequence effect | References |
|---|---|---|---|
| Direct psychological effects of extreme weather events | Trauma and stress | Natural disasters such as hurricanes, floods, wildfires, and heat waves can lead to immediate and long-term psychological trauma, stress, and anxiety. Survivors may experience post-traumatic stress disorder (PTSD), depression, and other anxiety disorders | (Makwana 2019; Meltzer et al. 2021) |
| | Displacement and loss | The destruction of homes and communities can result in displacement, loss of personal belongings, and a sense of identity, leading to increased mental health issues, including depression and anxiety | (Makwana 2019; Cianconi et al. 2020a, b) |
| Chronic stress and anxiety | Climate anxiety | Growing awareness of climate change and its potential impacts can lead to chronic stress and anxiety, often referred to as “climate anxiety” or “eco-anxiety.” This is particularly prevalent among young people who are concerned about the future of the planet | (Leger-Goodes, Malboeuf-Hurtubise et al. 2022; Ogunbode et al. 2022) |
| | Uncertainty and fear | The uncertainty and unpredictability of climate change-related events can contribute to ongoing stress and fear about the future | (Betro et al. 2020a, b) |
| Economic and social disruptions | Economic instability | Climate change can disrupt livelihoods, particularly in agriculture, fishing, and other climate-sensitive sectors, leading to financial stress and insecurity | (Tchonkouang et al. 2024) |
| | Social displacement | Communities affected by climate change may experience social displacement, leading to the loss of social networks and support systems, which are crucial for mental well-being | (Corvalan et al. 2022) |

(continued)

**Table 6** (continued)

| Climate change health effect | | Detailed cause/consequence effect | References |
|---|---|---|---|
| Health impacts and related stress | Physical health issues | Climate change can exacerbate physical health problems, which in turn can affect mental health. For instance, heatwaves can increase the incidence of heat-related illnesses, and the stress of managing these conditions can lead to mental health issues | (Cianconi et al. 2020a, b; White et al. 2023) |
| | Nutritional stress | Food insecurity and malnutrition, often linked to climate change, can lead to anxiety and depression, especially among vulnerable populations | (Myers 2020) |
| Community and cultural impacts | Loss of community cohesion | Climate change can disrupt communities, leading to a loss of social cohesion and support networks, which are essential for mental health and well-being | (Lawrance et al. 2022; Longman et al. 2023; White et al. 2023) |
| | Cultural loss | For Indigenous and other communities closely tied to their land, climate change can result in the loss of cultural heritage, traditions, and identity, leading to grief, sadness, and a sense of loss | (Lebel et al. 2022; Redvers et al. 2023) |

impacts of climate change. Climate change intensifies environmental stressors that harm ecosystems, such as rising sea levels, shifts in agricultural zones, and the loss of biodiversity. The depletion of ecosystems weakens natural defenses against climate-related health threats, such as floods, droughts, and the spread of disease vectors like mosquitoes. For instance, wetlands, forests, and coral reefs act as natural buffers against extreme weather events, and their degradation can leave human communities more vulnerable to disasters. In the context of Ecohealth, the protection and restoration of these ecosystems are crucial for enhancing resilience to the health impacts of climate change. Therefore, the health consequences of climate change are multifaceted and interconnected, highlighting the need for integrative approaches like One Health and Ecohealth (Durand et al. 2022; Grigorieva 2024).

The One Health and Ecohealth frameworks are crucial for developing adaptive strategies to mitigate the health impacts of climate change. By recognizing the interconnectedness of human, animal, and environmental health, these approaches promote the integration of health into climate policies and vice versa. For example, improving veterinary services and wildlife monitoring can enhance our ability to detect and control emerging diseases early. Similarly, restoring ecosystems and

**Table 7** Main impacts of extreme weather events on health

| Climate change health effect | | Causes | References |
|---|---|---|---|
| Immediate physical injuries | Trauma and injury | Extreme weather events such as hurricanes, tornadoes, floods, and wildfires can cause direct physical injuries, including fractures, lacerations, head trauma, and crush injuries due to falling debris, collapsing structures, and accidents during evacuations | (Bartholdson and von Schreeb 2018; Sever et al. 2023) |
| | Burns and respiratory injuries | Wildfires and extreme heat events can result in burns and respiratory injuries due to smoke inhalation and exposure to high temperatures | (Finlay, Moffat et al. 2012; Xu et al. 2020) |
| | Heat waves | Extreme heat is one of the deadliest weather phenomena, leading to heat-related illnesses such as heatstroke and hyperthermia. Vulnerable populations, including the elderly, children, and those with preexisting health conditions, are at higher risk of mortality during heat waves | (Meade et al. 2020; Arsad et al. 2022) |
| | Cold waves | Severe cold weather can lead to hypothermia and frostbite, increasing mortality rates, especially among the homeless and those without adequate heating | (Zhang et al. 2019) |
| | Floods | Drowning is a leading cause of death during floods. Additionally, waterborne diseases and infections can increase mortality following flooding events | (Yari et al. 2020) |
| | Storms and hurricanes | High winds, flying debris, storm surges, and flooding associated with storms and hurricanes can lead to numerous fatalities | (Marchigiani et al. 2013) |
| Secondary health impacts | Infectious diseases | Floods and storms can lead to outbreaks of waterborne diseases such as cholera, typhoid, and hepatitis A, increasing mortality rates. Stagnant water left after floods can also become breeding grounds for mosquitoes, leading to diseases such as malaria and dengue fever | (Liang and Messenger 2018) |

(continued)

**Table 7** (continued)

| Climate change health effect | | Causes | References |
|---|---|---|---|
| | Chronic disease exacerbation | Extreme weather can exacerbate existing chronic conditions such as cardiovascular disease, respiratory conditions, and diabetes, increasing the risk of severe complications and death | (Ratter-Rieck et al. 2023; De Vita et al. 2024) |
| Healthcare system strain | Overwhelmed emergency services | Extreme weather events can overwhelm emergency services and healthcare systems, delaying medical treatment for injuries and illnesses and increasing mortality | (Bell et al. 2018) |
| | Disrupted access to care | Damage to infrastructure can disrupt access to healthcare facilities, making it difficult for injured or ill individuals to receive timely treatment | (Khirekar et al. 2023; Salam et al. 2023) |

promoting biodiversity can help buffer the effects of climate change on vulnerable communities (Durand et al. 2022; Grigorieva 2024; Shafique et al. 2024).

Furthermore, One Health and Ecohealth emphasise collaboration between sectors, disciplines, and communities. This is essential in the fight against climate change, where solutions require coordinated action across agriculture, energy, urban planning, and health systems. Public health interventions informed by these frameworks are more likely to be effective and sustainable, as they address the root causes of health risks rather than just their symptoms (Zinsstag et al. 2018).

## *4.1 Implications of Ecohealth Thinking for Research Priorities and Public Health Interventions Concerning Health Effects of Climate Change*

When applied to climate change and its impact on public health, Ecohealth offers a comprehensive framework to rethink research priorities and public health interventions. Ecohealth thinking demands that researchers prioritize studies that transcend traditional disciplinary boundaries. The effects of climate change on health are not merely biomedical; they encompass environmental, social, and economic dimensions. For example, research on vector-borne diseases like malaria or dengue must consider not only changing weather patterns but also deforestation, urbanization, and migration. Therefore, Ecohealth thinking for research priorities and public health interventions concerning the health effects of climate change are described thereafter (Grigorieva 2024).

Key research areas include: (i) Emerging infectious diseases. Understanding how climate change alters habitats for disease vectors (e.g., mosquitoes and ticks) is crucial. Research should also focus on how ecosystem disruptions increase zoonotic disease risks; (i) Food security and nutrition. Investigating the effects of drought, changing crop yields, and ocean acidification on food availability and nutrition, especially in vulnerable populations; (iii) Mental health. The psychological toll of climate-related events (e.g., floods, wildfires) is rising. There is a need for interdisciplinary studies that explore the links between mental health, displacement, and socio-ecological loss; (iv) Resilience of ecosystems: Research on how natural ecosystems can support human health in the face of climate stressors. This includes understanding how biodiversity and ecosystem services can mitigate negative health impacts (Harper et al. 2012).

Systems-oriented Public Health Interventions include (i) Climate adaptation strategies. Public health interventions should prioritize ecosystem-based adaptation, which involves restoring natural systems to reduce climate vulnerability and support human health (e.g., mangroves for coastal protection). These approaches not only safeguard the environment but also protect communities from extreme weather events and associated health risks; (ii) Urban planning and green infrastructure. Interventions should focus on integrating green spaces, improving air quality, and enhancing water management systems in cities. These efforts simultaneously improve public health (e.g., by reducing respiratory illnesses and heat-related deaths) and strengthen environmental sustainability; (iii) Community involvement and empowerment. Ecohealth emphasizes local knowledge and community-based approaches. Public health interventions should therefore prioritize community-led climate adaptation, where vulnerable populations actively shape strategies that address both environmental and health challenges; (iv) One Health approach: Health interventions should be framed in terms of the "One Health" concept, which links human, animal, and environmental health. For example, mitigating the impact of climate on agricultural health requires coordinated interventions across human, livestock, and crop systems (Harper et al. 2012; Weatherly et al. 2023).

Climate change disproportionately affects low-income populations, Indigenous communities, and countries with limited resources. Ecohealth thinking highlights the need for equity-centered interventions and research, particularly in areas where climate vulnerability is highest. Public health strategies should focus on vulnerable populations. Special attention should be paid to populations that are most at risk of climate-related health impacts, including children, the elderly, and those living in climate-sensitive regions and adapt health systems to climate risks: Health services must be resilient to climate-related shocks. This includes preparing for the increased burden of disease following extreme weather events and ensuring that healthcare infrastructure is capable of responding to new environmental challenges (Harper et al. 2012; Aparicio-Effen et al. 2022; Grigorieva et al. 2023; Weatherly et al. 2023; Grigorieva 2024).

The application of Ecohealth thinking in addressing the health effects of climate change encourages a comprehensive, integrative, and equity-driven approach. Research must focus on the complex interactions between human health, ecosystems, and social structures, while interventions need to prioritize sustainable, community-based, and system-wide strategies. This way, public health initiatives can be more effective in mitigating the profound health impacts of climate change and fostering long-term resilience (Harper et al. 2012; Aparicio-Effen et al. 2022; Grigorieva et al. 2023; Weatherly et al. 2023; Grigorieva 2024).

## *4.2 Key Challenges Under the Climate Crisis Through an Ecohealth Lens*

As the climate crisis intensifies, the impacts on human health and well-being are becoming increasingly evident. From worsening air quality to food insecurity and water scarcity, the health of populations is deeply intertwined with environmental changes. Viewing these issues through an Ecohealth lens—which integrates human, animal, and ecosystem health—helps bring a more holistic understanding to the complex and interconnected challenges posed by climate change. Four critical areas that demand closer scrutiny through this perspective include urban health, food security, sustainable water supplies, and population mobility (Harper et al. 2012; Aparicio-Effen et al. 2022; Grigorieva et al. 2023; Weatherly et al. 2023; Grigorieva 2024).

### 4.2.1 Urban Health

Cities are particularly vulnerable to the effects of climate change due to their high population densities, extensive infrastructure, and environmental degradation. Rising temperatures, air pollution, and increased frequency of extreme weather events exacerbate health risks, especially for marginalized communities. The Ecohealth framework encourages urban planners and public health experts to work together, ensuring that cities are designed to promote ecological sustainability while safeguarding public health (Harper et al. 2012; Oliveira et al. 2020; Aparicio-Effen et al. 2022).

### 4.2.2 Food Security

Climate change threatens food systems globally, affecting the availability, accessibility, and quality of food. Droughts, floods, and shifting growing seasons are already leading to crop failures and food shortages in many regions, with direct consequences for nutrition and public health. By viewing food security through an Ecohealth lens, interventions can be designed to address not just the ecological and economic dimensions of food systems, but their profound implications for human health and well-being (Harper et al. 2012; Aparicio-Effen et al. 2022).

### 4.2.3 Sustainable Water Supplies

Climate change is also drastically affecting the availability and quality of freshwater resources. Increasing droughts, glacial melt, and pollution of water bodies are creating severe water scarcity in many parts of the world. As access to clean and sustainable water supplies diminishes, public health risks such as waterborne diseases, malnutrition, and sanitation challenges rise. The Ecohealth approach encourages sustainable water management that considers the needs of both humans and ecosystems, recognizing that protecting water sources is essential for sustaining public health in a changing climate (Harper et al. 2012; Aparicio-Effen et al. 2022).

### 4.2.4 Population Mobility

Climate change is already a driving force behind population mobility, with millions of people displaced by rising sea levels, desertification, and extreme weather events. This phenomenon, often referred to as "climate migration," presents unique challenges to public health, as displaced populations face risks such as infectious diseases, mental health stress, and lack of access to healthcare. Through an Ecohealth lens, population mobility must be understood as a complex interplay between environmental change, social systems, and health, requiring coordinated efforts across sectors to mitigate its impacts (Harper et al. 2012; Aparicio-Effen et al. 2022).

The climate crisis brings with it a host of challenges that directly impact human health, from urban environments to global food systems and water supplies. By examining these issues through an Ecohealth lens, researchers and policymakers can better understand the intricate connections between ecological systems and human health, leading to more effective, equitable, and sustainable interventions. As we confront these pressing questions, the Ecohealth approach offers a path forward that prioritizes both the health of our planet and the well-being of all its inhabitants.

## 5 Conclusions

Climate change poses a profound threat to global public health, undoing decades of progress and disproportionately affecting vulnerable populations, especially in impoverished regions. The wide-ranging effects, from extreme weather events and ecosystem disruptions to socioeconomic challenges, highlight the urgent need for comprehensive and interdisciplinary approaches to understanding and mitigating these impacts. The increasing frequency and intensity of extreme weather events, such as the extreme heatwave with wildfires in Portugal or the extreme rainfall in Pakistan, underscore the reality that climate change is no longer a future threat but a present crisis. To effectively address the health challenges posed by climate change, it is essential to overcome the fragmentation in research and the overwhelming volume of literature by adopting innovative methods for evidence synthesis. This will enable timely and informed public health responses, crucial for protecting global health in the face of a rapidly changing climate.

As climate change continues to disrupt ecosystems, threaten biodiversity, and affect the health of both humans and animals, the importance of One Health and Ecohealth cannot be overstated. These frameworks offer a holistic approach to understanding and addressing the health impacts of climate change, recognizing that the health of people is inseparable from the health of animals and the environment. By integrating the principles of One Health and Ecohealth into climate adaptation and mitigation strategies, we can enhance resilience, protect ecosystems, and safeguard the health of future generations.

## References

Achebak H, Devolder D, Ballester J (2019) Trends in temperature-related age-specific and sex-specific mortality from cardiovascular diseases in Spain: a national time-series analysis. Lancet Planet Health 3(7):E297–E306

Achebak H, Devolder D. Ballester J (2018) Heat-related mortality trends under recent climate warming in Spain: A 36-year observational study. Plos Med 15(7)

Achebak H, Devolder D, Ingole V, Ballester J (2020) Reversal of the seasonality of temperature-attributable mortality from respiratory diseases in Spain. Nat Commun 11(1)

Adepoju OA, Afinowi OA, Tauheed AM, Danazumi AU, Dibba LBS, Balogun JB, Flore G, Saidu U, Ibrahim B, Balogun OO, Balogun EO (2023) Multisectoral perspectives on global warming and vector-borne diseases: a focus on Southern Europe. Curr Trop Med Rep 10(2):47–70

Afifa K, Arshad N, Hussain M, Ashraf H, Saleem MZ (2024) Air pollution and climate change as grand challenges to sustainability. Sci Total Environ 928

Ali MZ, Carlile G, Giasuddin M (2020) Impact of global climate change on livestock health: Bangladesh perspective. Open VetY J 10(2):178–188

Alied M, Salam A, Sediqi SM, Kwaah PA, Tran L, Huy NT (2024) Disaster after disaster: the outbreak of infectious diseases in Pakistan in the wake of 2022 floods. Ann Med Surg 86(2):891–898

Aparicio-Effen M, Paz-Rada O, Arana-Pardo I, Aparicio J, Ramallo C, Briançon E, Huanca X, Nagy G (2022) Climate change health impacts: the need for watershed and ecohealth approaches base for health adaptation strategies and policies. Handbook of Human and Planetary Health. W. Leal Filho, Springer, Cham

Araos M, Austin SE, Berrang-Ford L, Ford JD (2016) Public health adaptation to climate change in large cities: a global Baseline. Int J Health Serv 46(1):53–78

Arsad FS, Hod R, Ahmad N, Ismail R, Mohamed N, Baharom M, Osman Y, Radi MFM, Tangang F (2022) The impact of heatwaves on mortality and morbidity and the associated vulnerability factors: a systematic review. Int J Environ Res Public Health 19(23)

Aydin-Ghormoz H, Adeyeye T, Muscatiello N, Nayak S, Savadatti S, Insaf TZ (2022) Identifying risk factors for hospitalization with behavioral health disorders and concurrent temperature-related illness in New York State. Int J Environ Res Public Health 19(24)

Ballester J, Robine JM, Herrmann FR, Rodo X (2011) Long-term projections and acclimatization scenarios of temperature-related mortality in Europe. Nat Commun 2

Bartholdson S, von Schreeb J (2018) Natural disasters and injuries: what does a surgeon need to know? Curr Trauma Rep 4(2):103–108

Bell JE, Brown CL, Conlon K, Herring S, Kunkel KE, Lawrimore J, Luber G, Schreck C, Smith A, Uejio C (2018) Changes in extreme events and the potential impacts on human health. J Air Waste Manag Assoc 68(4):265–287

Berners-Lee M, Hoolohan C, Cammack H, Hewitt CN (2012) The relative greenhouse gas impacts of realistic dietary choices. Energy Policy 43:184–190

Bhatta GD, Aggarwal PK (2016) Coping with weather adversity and adaptation to climatic variability: a cross-country study of smallholder farmers in South Asia. Climate Dev 8(2):145–157

Blauw LL, Aziz NA, Tannemaat MR, Blauw CA, de Craen AJ, Pijl H, Rensen PC (2017) Diabetes incidence and glucose intolerance prevalence increase with higher outdoor temperature. BMJ Open Diabetes Res Care 5(1):e000317

Brook RD, Rajagopalan S, Pope CA, Brook JR, Bhatnagar A, Diez-Roux AV, Holguin F, Hong Y, Luepker RV, Mittleman MA, Peters A, Siscovick D, Smith SC Jr, Whitsel L, Kaufman JD, E. American Heart Association Council on, C. o. t. K. i. C. D. Prevention, P. A. Council on Nutrition and Metabolism (2010) Particulate matter air pollution and cardiovascular disease: An update to the scientific statement from the American Heart Association. Circulation 121(21):2331–2378

Brown L, Medlock J, Murray V (2014) Impact of drought on vector-borne diseases–how does one manage the risk? Public Health 128(1):29–37

Callaghan MW, Minx JC, Forster PM (2020) A topography of climate change research. Nat Clim Chang 10(2):118-+

Chaudhry D (2024) Climate change and health of the urban poor: the role of environmental justice. J Clim Chang Health 15

Chen K, Breitner S, Wolf K, Hampel R, Meisinger C, Heier M, von Scheidt W, Kuch B, Peters A, Schneider A, K.S. Group (2019) Temporal variations in the triggering of myocardial infarction by air temperature in Augsburg, Germany, 1987–2014. Eur Heart J 40(20):1600–1608

Chung CY, Yang J, Yang XG, He J (2022) Long-term effects of ambient air pollution on lung cancer and COPD mortalities in China: A systematic review and meta-analysis of cohort studies. Environ Impact Assess Rev 97

Cianconi P, Betro S, Janiri L (2020a) The impact of climate change on mental health: a systematic descriptive review. Front Psychiatry 11:74

Cianconi P, Betrò S, Janiri L (2020) The impact of climate change on mental health: a systematic descriptive review. Front Psychiatry 11

COPERNICUS (2022) Heatwaves grip parts of Europe, Asia and North America in the first half of 2022. Retrieved 12 september 2024, from https://climate.copernicus.eu/heatwaves-grip-parts-europe-asia-and-north-america-first-half-2022

Corvalan C, Gray B, Prats E, Sena A, Hanna F, Campbell-Lendrum D (2022) Mental health and the global climate crisis. Epidemiol Psychiatr Sci 31

Dain K, Hadley L (2012) Diabetes and climate change–two interconnected global challenges. Diabetes Res Clin Pract 97(2):337–339

Dash SS, Maity R (2023) Effect of climate change on soil erosion indicates a dominance of rainfall over LULC changes. J Hydrol-RegNal Stud 47

De Vita A, Belmusto A, Di Perna F, Tremamunno S, De Matteis G, Franceschi F, Covino M, C. Group (2024) The impact of climate change and extreme weather conditions on cardiovascular health and acute cardiovascular diseases. J Clin Med 13(3)

Dietrich J, Hammerl JA, Johne A, Kappenstein O, Loeffler C, Nockler K, Rosner B, Spielmeyer A, Szabo I, Richter MH (2023) Impact of climate change on foodborne infections and intoxications. J Health Monit 8(Suppl 3):78–92

Dring P, Armstrong M, Alexander R, Xiang H (2022) Emergency department visits for heat-related emergency conditions in the United States from 2008–2020. Int J Environ Res Public Health 19(22)

Durand T, Fleury O, Heuret M, Mougey T, Schwoehrer C, Vittecoq M (2022) One health and Ecohealth in protected areas in France. Environ Risques & Sante 21(3):227–234

Ebi KL, Capon A, Berry P, Broderick C, de Dear R, Havenith G, Honda Y, Kovats RS, Ma W, Malik A, Morris NB, Nybo L, Seneviratne SI, Vanos J, Jay O (2021b) Hot weather and heat extremes: health risks. Lancet 398(10301):698–708

Ebi KL, Loladze I (2019) Elevated atmospheric CO concentrations and climate change will affect our food's quality and quantity comment. Lancet Planet Health 3(7):E283–E284

Ebi KL, Boyer C, Bowen KJ, Frumkin H, Hess J (2018) Monitoring and evaluation indicators for climate change-related health impacts, risks, adaptation, and resilience. Int J Environ Res Public Health 15(9)

Ebi KL, Anderson CL, Hess JJ, Kim SH, Loladze I, Neumann RB, Singh D, Ziska L, Wood R (2021) Nutritional quality of crops in a high CO world: an agenda for research and technology development. Environ Res Lett 16(6)

Edwards P, Roberts I (2009) Population adiposity and climate change. Int J Epidemiol 38(4):1137–1140

Eguiluz-Gracia I, Mathioudakis AG, Bartel S, Vijverberg SJH, Fuertes E, Comberiati P, Cai YS, Tomazic PV, Diamant Z, Vestbo J, Galan C, Hoffmann B (2020) The need for clean air: the way air pollution and climate change affect allergic rhinitis and asthma. Allergy 75(9):2170–2184

Farhad M, Kumar U, Tomar V, Bhati PK, Krishnan NJ, Barek V, Brestic M, Hossain A (2023) Heat stress in wheat: a global challenge to feed billions in the current era of the changing climate. Front Sustain Food Syst 7

Faurie C, Varghese BM, Liu JW, Bi P (2022) Association between high temperature and heatwaves with heat-related illnesses: A systematic review and meta-analysis. Sci Total Environ 852

Ferraz MP (2024) Antimicrobial resistance: the impact from and on society according to one health approach. Societies 14(9)

Finlay SE, Moffat A, Gazzard R, Baker D, Murray V (2012) Health impacts of wildfires. PLoS Curr 4:e4f959951cce959952c

Furtak K, Wolinska A (2023) The impact of extreme weather events as a consequence of climate change on the soil moisture and on the quality of the soil environment and agriculture-A review. Catena 231

Gale P, Drew T, Phipps LP, David G, Wooldridge M (2009) The effect of climate change on the occurrence and prevalence of livestock diseases in great Britain: a review. J Appl Microbiol 106(5):1409–1423

Gasparrini A, Guo YM, Sera F, Vicedo-Cabrera AM, Huber V, Tong SL, Coelho MDZS, Saldiva PHN, Lavigne E, Correa PM, Ortega NV, Kan HD, Osorio S, Kysely J, Urban A, Jaakkola JJK, Ryti NRI, Pascal M, Goodman PG, Zeka A, Michelozzi P, Scortichini M, Hashizume M, Honda Y, Hurtado-Diaz M, Cruz JC, Seposo X, Kim H, Tobias A, Iniguez C, Forsberg B, Astrom DO, Ragettli MS, Guo YL, Wu CF, Zanobetti A, Schwartz J, Bell ML, Dang TN, Van DD, Heaviside C, Vardoulakis S, Hajat S, Haines A, Armstrong B (2017) Projections of temperature-related excess mortality under climate change scenarios. Lancet Planet Health 1(9):E360–E367

Gautier D, Denis D, Locatelli B (2016) Impacts of drought and responses of rural populations in West Africa: a systematic review. Wiley Interdiscip Rev-Clim Chang 7(5):666–681

Godde CM, Mason-D'Croz D, Mayberry DE, Thornton PK, Herrero M (2021) Impacts of climate change on the livestock food supply chain; a review of the evidence. Glob Food Secur-Agric Policy Econ Environ 28

Graham NAJ (2014) Habitat complexity: coral structural loss leads to fisheries declines. Curr Biol 24(9):R359–R361

Griffith AW, Gobler CJ (2020) Harmful algal blooms: A climate change co-stressor in marine and freshwater ecosystems. Harmful Algae 91

Grigorieva E, Livenets A, Stelmakh E (2023) Adaptation of agriculture to climate change: a scoping review. Climate 11(10)

Grigorieva EA (2024) Climate change and human health in the arctic: a review. Climate 12(7)

Gullino ML, Albajes R, Al-Jboory I, Angelotti F, Chakraborty S, Garrett KA, Hurley BP, Juroszek P, Lopian R, Makkouk K, Pan XB, Pugliese M, Stephenson T (2022) Climate change and pathways used by pests as challenges to plant health in agriculture and forestry. Sustainability 14(19)

Habeeb AA, Osman SF, Teama FEI, Gad AE (2023) The detrimental impact of high environmental temperature on physiological response, growth, milk production, and reproductive efficiency of ruminants. Trop Anim Health Prod 55(6)

Haddaway NR, Callaghan MW, Collins AM, Lamb WF, Minx JC, Thomas J, John D (2020) On the use of computer-assistance to facilitate systematic mapping. Campbell Syst Rev 16(4)

Haines A, Ebi K (2019) The imperative for climate action to protect health. N Engl J Med 380(3):263–273
Han AZ, Deng SZ, Yu JR, Zhang YL, Jalaludin B, Huang CR (2023) Asthma triggered by extreme temperatures: From epidemiological evidence to biological plausibility. Environ Res 216
Hanssen MJ, Hoeks J, Brans B, van der Lans AA, Schaart G, van den Driessche JJ, Jorgensen JA, Boekschoten MV, Hesselink MK, Havekes B, Kersten S, Mottaghy FM, van Marken Lichtenbelt WD, Schrauwen P (2015) Short-term cold acclimation improves insulin sensitivity in patients with type 2 diabetes mellitus. Nat Med 21(8):863–865
Harper SL, Edge VL, Cunsolo Willox A, G. Rigolet Inuit Community (2012) 'Changing climate, changing health, changing stories' profile: using an EcoHealth approach to explore impacts of climate change on inuit health. Ecohealth 9(1):89–101
Hatfield JL, Prueger JH (2015) Temperature extremes: effect on plant growth and development. Weather Clim Extrem 10:4–10
Hoeke G, Kooijman S, Boon MR, Rensen PC, Berbee JF (2016) Role of brown fat in lipoprotein metabolism and atherosclerosis. Circ Res 118(1):173–182
Hoffmann C,. Maglakelidze M, von Schneidemesser E, Witt C, Hoffmann P, Butler T (2022) Asthma and COPD exacerbation in relation to outdoor air pollution in the metropolitan area of Berlin, Germany. Respir Res 23(1)
Huang CR, Barnett AG, Wang XM, Vaneckova P, FitzGerald G, Tong SL (2011) Projecting future heat-related mortality under climate change scenarios: a systematic review. Environ Health Perspect 119(12):1681–1690
Ingrao C, Strippoli R, Lagioia G, Huisingh D (2023) Water scarcity in agriculture: an overview of causes, impacts and approaches for reducing the risks. Heliyon 9(8)
Intergovernmental Panel on Climate Change. Working, G., II (2022) Climate change 2022: impacts, adaptation and vulnerability: the Working Group II contribution to the Sixth Assessment Report of the Intergovernmental Panel on Climate Change. Switzerland, IPCC, Intergovernmental Panel on Climate Change, Geneva
Issifu I, Alava JJ, Lam VWY, Sumaila UR (2022) Impact of ocean warming, overfishing and mercury on european fisheries: a risk assessment and policy solution framework. Front Mar Sci 8
Jay O, Capon A, Berry P, Broderick C, de Dear R, Havenith G, Honda Y, Kovats RS, Ma W, Malik A, Morris NB, Nybo L, Seneviratne SI, Vanos J, Ebi KL (2021) Reducing the health effects of hot weather and heat extremes: from personal cooling strategies to green cities. Lancet 398(10301):709–724
Jin XT, Ren JJ, Li RH, Gao Y, Zhang HY, Li JM, Zhang JJ, Wang XC, Wang G (2021) Global burden of upper respiratory infections in 204 countries and territories, from 1990 to 2019. Eclinicalmedicine 37
Jolly WM, Cochrane MA, Freeborn PH, Holden ZA, Brown TJ, Williamson GJ, Bowman DMJS (2015) Climate-induced variations in global wildfire danger from 1979 to 2013. Nat Commun 6
Jung YJ, Khant NA, Kim H, Namkoong S (2023) Impact of climate change on waterborne diseases: directions towards sustainability. Water 15(7)
Karaaslan Y (2024) Examining crop yield losses in iğdir plain irrigation systems in Türkiye amidst water constraints. Sustainability 16(14)
Kenstler M, Shelley G (2024) Love of nature in the time of climate change: environmental science undergraduates report higher levels of nature-relatedness and eco-anxiety but not resilience. Ecopsychology
Kent C, Dunstone N, Tucker S, Scaife AA, Brown S, Kendon EJ, Smith D, McLean L, Greenwood S (2022) Estimating unprecedented extremes in UK summer daily rainfall. Environ Res Lett 17(1)
Khan A, Ahmad M, Ahmed M, Iftikhar Hussain M (2020) Rising atmospheric temperature impact on wheat and thermotolerance strategies. Plants (Basel) 10(1)
Khan I, Mesalam A, Heo YS, Lee SH, Nabi G, Kong IK (2023) Heat stress as a barrier to successful reproduction and potential alleviation strategies in cattle. Animals 13(14)

Khirekar J, Badge A, Bandre GR, Shahu S (2023) Disaster preparedness in hospitals. Cureus 15(12):e50073

Knowlton K, Rotkin-Ellman M, King G, Margolis HG, Smith D, Solomon G, Trent R, English P (2009) The 2006 California heat wave: impacts on hospitalizations and emergency department visits. Environ Health Perspect 117(1):61–67

Lamb WF, Creutzig F, Callaghan MW, Minx JC (2019) Learning about urban climate solutions from case studies. Nat Clim Chang 9(4):279

Lassiter A (2021) Rising seas, changing salt lines, and drinking water salinization. Curr Opin Environ Sustain 50:208–214

Lawrance EL, Thompson R, Le Vay JN, Page L, Jennings N (2022) The impact of climate change on mental health and emotional wellbeing: a narrative review of current evidence, and its implications. Int Rev Psychiatry 34(5):443–498

Lebel L, Paquin V, Kenny TA, Fletcher C, Nadeau L, Chachamovich E, Lemire M (2022) Climate change and Indigenous mental health in the Circumpolar North: a systematic review to inform clinical practice. Transcult Psychiatry 59(3):312–336

Leger-Goodes T, Malboeuf-Hurtubise C, Mastine T, Genereux M, Paradis PO, Camden C (2022) Eco-anxiety in children: A scoping review of the mental health impacts of the awareness of climate change. Front Psychol 13

Lemon SM (2008) Vector-borne diseases. Vector-borne diseases: understanding the environmental, human health, and ecological connections: workshop summary. Washington (DC)

Lesnikowski AC, Ford JD, Berrang-Ford L, Paterson JA, Barrera M, Heymann SJ (2011) Adapting to health impacts of climate change: a study of UNFCCC Annex I parties. Environ Res Lett 6(4)

Levy K, Smith SM, Carlton EJ (2018) Climate change impacts on waterborne diseases: moving toward designing interventions. Curr Environ Health Rep 5(2):272–282

Li S, Wu W, Wang G, Zhang XY, Guo Q, Wang BB, Cao SZ, Yan ML, Pan XC, Xue T, Gong JC, Duan XL (2022) Association between exposure to air pollution and risk of allergic rhinitis: a systematic review and meta-analysis. Environ Res 205

Liang SY, Messenger N (2018) Infectious diseases after hydrologic disasters. Emerg Med Clin North Am 36(4):835–851

Longman J, Braddon M, Verlie B, Schlosberg D, Hampshire L, Hawke C, Noonan A, Saurman E (2023) Building resilience to the mental health impacts of climate change in rural Australia. J Clim Chang Health 12

Ma J, Guo Y, Gao J, Tang H, Xu K, Liu Q, Xu L (2022) Climate change drives the transmission and spread of vector-borne diseases: an ecological perspective. Biology (Basel) 11(11)

Macintyre HL, Heaviside C, Cai XM, Phalkey R (2021) The winter urban heat island: Impacts on cold-related mortality in a highly urbanized European region for present and future climate. Environ Int 154

Makwana N (2019) Disaster and its impact on mental health: a narrative review. J Fam Med Prim Care 8(10):3090–3095

Malmquist A, Lundgren T, Hjerpe M, Glaas E, Turner E, Storbjörk S (2021) Vulnerability and adaptation to heat waves in preschools: experiences, impacts and responses by unit heads, educators and parents. Clim Risk Manag 31

Marchigiani R, Gordy S, Cipolla J, Adams RC, Evans DC, Stehly C, Galwankar S, Russell S, Marco AP, Kman N, Bhoi S, Stawicki SP, Papadimos TJ (2013) Wind disasters: a comprehensive review of current management strategies. Int J Crit Illn Inj Sci 3(2):130–142

Martínez-Solanas E, Basagaña X (2019) Temporal changes in temperature-related mortality in Spain and effect of the implementation of a heat health prevention plan. Environ Res 169:102–113

Masroor M, Sajjad H, Rehman S, Singh R, Rahaman MH, Sahana M, Ahmed R, Avtar R (2022) Analysing the relationship between drought and soil erosion using vegetation health index and RUSLE models in Godavari middle sub-basin, India. Geosci Front 13(2)

Mata TM, Felgueiras F, Martins AA, Monteiro H, Ferraz MP, Oliveira GM, Gabriel MF, Silva GV (2022) Indoor air quality in elderly centers: pollutants emission and health effects. Environments 9(7)

Matsushita M, Yoneshiro T, Aita S, Kameya T, Sugie H, Saito M (2014) Impact of brown adipose tissue on body fatness and glucose metabolism in healthy humans. Int J Obes (Lond) 38(6):812–817
Maung TZ, Bishop JE, Holt E, Turner AM, Pfrang C (2022) Indoor air pollution and the health of vulnerable groups: a systematic review focused on particulate matter (PM), volatile organic compounds (VOCs) and their effects on children and people with pre-existing lung disease. Int J Environ Res Public Health 19(14)
Mavrouli M, Mavroulis S, Lekkas E, Tsakris A (2023) The impact of earthquakes on public health: a narrative review of infectious diseases in the post-disaster period aiming to disaster risk reduction. Microorganisms 11(2)
Mazhar S, Pellegrini E, Contin M, Bravo C, De Nobili M (2022) Impacts of salinization caused by sea level rise on the biological processes of coastal soils—a review. Front Environ Sci 10
Meade RD, Akerman AP, Notley SR, McGinn R, Poirier P, Gosselin P, Kenny GP (2020) Physiological factors characterizing heat-vulnerable older adults: a narrative review. Environ Int 144
Meltzer GY, Zacher M, Merdjanoff AA, Do MP, Pham NK, Abramson DM (2021) The effects of cumulative natural disaster exposure on adolescent psychological distress. J Appl Res Child 12(1)
Meo SA, Meo AS (2024) Climate change and diabetes mellitus—emerging global public health crisis: observational analysis. Pak J Med Sci 40(4):559–562
Mirzabaev A, Kerr RB, Hasegawa T, Pradhan P, Wreford A, Pahlen MCTV, Gurney-Smith H (2023) Severe climate change risks to food security and nutrition. Clim Risk Manag 39
Mojahed N, Mohammadkhani MA, Mohamadkhani A (2022) Climate crises and developing vector-borne diseases: a narrative review. Iran J Public Health 51(12):2664–2673
Mueller W, Zamrsky D, Essink GO, Fleming LE, Deshpande A, Makris KC, Wheeler BW, Newton JN, Narayan KMV, Naser AM, Gribble MO (2024) Saltwater intrusion and human health risks for coastal populations under 2050 climate scenarios. Sci Rep 14(1):15881
Murray CJL, Aravkin AY, Zheng P, Abbafati C, Abbas KM, Abbasi-Kangevari M, Abd-Allah F, Abdelalim A, Abdollahi M, Abdollahpour I, Abegaz KH, Abolhassani H, Aboyans V, Abreu LG, Abrigo MRM, Abualhasan A, Abu-Raddad LJ, Abushouk AI, Adabi M, Adekanmbi V, Adeoye AM, Adetokunboh OO, Adham D, Advani SM, Agarwal G, Aghamir SMK, Agrawal A, Ahmad T, Ahmadi K, Ahmadi M, Ahmadieh H, Ahmed MB, Akalu TY, Akinyemi RO, Akinyemiju T, Akombi B, Akunna CJ, Alahdab F, Al-Aly Z, Alam K, Alam S, Alam T, Alanezi FM, Alanzi TM, Alemu BW, Alhabib KF, Ali M, Ali S, Alicandro G, Alinia C, Alipour V, Alizade H, Aljunid SM, Alla F, Allebeck P, Almasi-Hashiani A, Al-Mekhlafi HM, Alonso J, Altirkawi KA, Amini-Rarani M, Amiri F, Amugsi DA, Ancuceanu R, Anderlini D, Anderson JA, Andrei CL, Andrei T, Angus C, Anjomshoa M, Ansari F, Ansari-Moghaddam A, Antonazzo IC, Antonio CAT, Antony CM, Antriyandarti E, Anvari D, Anwer R, Appiah SCY, Arabloo J, Arab-Zozani M, Ariani F, Armoon B, Arnlov J, Arzani A, Asadi-Aliabadi M, Asadi-Pooya AA, Ashbaugh C, Assmus M, Atafar Z, Atnafu DD, Atout MMW, Ausloos F, Ausloos M, Quintanilla BPA, Ayano G, Ayanore MA, Azari S, Azarian G, Azene ZN, Badawi A, Badiye AD, Bahrami MA, Bakhshaei MH, Bakhtiari A, Bakkannavar SM, Baldasseroni A, Ball K, Ballew SH, Balzi D, Banach M, Banerjee SK, Bante AB, Baraki AG, Barker-Collo SL, Barnighausen TW, Barrero LH, Barthelemy CM, Barua L, Basu S, Baune BT, Bayati M, Becker JS, Bedi N, Beghi E, Bejot Y, Bell ML, Bennitt FB, Bensenor IM, Berhe K, Berman AE, Bhagavathula AS, Bhageerathy R, Bhala N, Bhandari D, Bhattacharyya K, Bhutta ZA, Bijani A, Bikbov B, Bin Sayeed MS, Biondi A, Birihane BM, Bisignano C, Biswas RK, Bitew H, Bohlouli S, Bohluli M, Boon-Dooley AS, Borges G, Borzi AM, Borzouei S, Bosetti C, Boufous S, Braithwaite D, Breitborde NJK, Breitner S, Brenner H, Briant PS, Briko AN, Briko NI, Britton GB, Bryazka D, Bumgarner BR, Burkart K, Burnett RT, Nagaraja SB, Butt ZA, dos Santos FLC, Cahill LE, Camera LA, Campos-Nonato IR, Cardenas R, Carreras G, Carrero JJ, Carvalho F, Castaldelli-Maia JM, Castaneda-Orjuela CA, Castelpietra G, Castro F, Causey K, Cederroth CR, Cercy KM, Cerin E, Chandan JS, Chang KL, Charlson FJ, Chattu VK, Chaturvedi S, Cherbuin N, Chimed-Ochir O, Cho DY, Choi JYJ, Christensen H,

Chu DT, Chung MT, Chung SC, Cicuttini FM, Ciobanu LG, Cirillo M, Classen TKD, Cohen AJ, Compton K, Cooper OR, Costa VM, Cousin E, Cowden RG, Cross DH, Cruz JA, Dahlawi SMA, Damasceno AAM, Damiani G, Dandona L, Dandona R, Dangel WJ, Danielsson AK, Dargan PI, Darwesh AM, Daryani A, Das JK, Das Gupta R, Das Neves J, Davila-Cervantes CA, Davitoiu DV, De Leo D, Degenhardt L, DeLang M, Dellavalle RP, Demeke FM, Demoz GT, Demsie DG, Denova-Gutierrez E, Dervenis N, Dhungana GP, Dianatinasab M, da Silva DD, Diaz D, Forooshani ZSD, Djalalinia S, Do HT, Dokova K, Dorostkar F, Doshmangir L, Driscoll TR, Duncan BB, Duraes AR, Eagan AW, Edvardsson D, El Nahas N, El Sayed I, El Tantawi M, Elbarazi I, Elgendy IY, El-Jaafary SI, Elyazar IRF, Emmons-Bell S, Erskine HE, Eskandarieh S, Esmaeilnejad S, Esteghamati A, Estep K, Etemadi A, Etisso AE, Fanzo J, Farahmand M, Fareed M, Faridnia R, Farioli A, Faro A, Faruque M, Farzadfar F, Fattahi N, Fazlzadeh M, Feigin VL, Feldman R, Fereshtehnejad SM, Fernandes E, Ferrara G, Ferrari AJ, Ferreira ML, Filip I, Fischer F, Fisher JL, Flor LS, Foigt NA, Folayan MO, Fomenkov AA, Force LM, Foroutan M, Franklin RC, Freitas M, Fu WJ, Fukumoto T, Furtado JM, Gad MM, Gakidou E, Gallus S, Garcia-Basteiro AL, Gardner WM, Geberemariyam BS, Gebreslassie AAAA, Geremew A, Hayoon AG, Gething PW, Ghadimi M, Ghadiri K, Ghaffarifar F, Ghafourifard M, Ghamari F, Ghashghaee A, Ghiasvand H, Ghith N, Gholamian A, Ghosh R, Gill PS, Ginindza TG, Giussani G, Gnedovskaya EV, Goharinezhad S, Gopalani SV, Gorini G, Goudarzi H, Goulart AC, Greaves F, Grivna M, Grosso G, Gubari MIM, Gugnani HC, Guimaraes RA, Guled RA, Guo GR, Guo YM, Gupta R, Gupta T, Haddock B, Hafezi-Nejad N, Hafiz A, Haj-Mirzaian A, Haj-Mirzaian A, Hall BJ, Halvaei I, Hamadeh RR, Hamidi S, Hammer MS, Hankey GJ, Haririan H, Haro JM, Hasaballah AI, Hasan MM, Hasanpoor E, Hashi, Hassanipour S, Hassankhani H, Havmoeller RJ, Hay SI, Hayat K, Heidari G, Heidari-Soureshjani R, Henrikson HJ, Herbert ME, Herteliu C, Heydarpour F, Hird TR, Hoek HW, Holla R, Hoogar P, Hosgood HD, Hossain N, Hosseini M, Hosseinzadeh M, Hostiuc M, Hostiuc S, Househ M, Hsairi M, Hsieh VCR, Hu GQ, Hu KJ, Huda TM, Humayun A, Huynh CK, Hwang BF, Iannucci VC, Ibitoye SE, Ikeda N, Ikuta KS, Ilesanmi OS, Ilic IM, Ilic MD, Inbaraj LR, Ippolito H, Iqbal U, Irvani SSN, Irvine CMS, Islam MM, Islam SMS, Iso H, Ivers RQ, Iwu CCD, Iwu CJ, Iyamu IO, Jaafari J, Jacobsen KH, Jafari H, Jafarinia M, Jahani MA, Jakovljevic M, Jalilian F, James SL, Janjani H, Javaheri T, Javidnia J, Jeemon P, Jenabi E, Jha RP, Jha V, Ji JS, Johansson L, John O, John-Akinola YO, Johnson CO, Jonas JB, Joukar F, Jozwiak JJ, Jurisson M, Kabir A, Kabir Z, Kalani H, Kalani R, Kalankesh LR, Kalhor R, Kanchan T, Kapoor N, Matin BK, Karch A, Karim MA, Kassa GM, Katikireddi SV, Kayode GA, Karyani AK, Keiyoro PN, Keller C, Kemmer L, Kendrick PJ, Khalid N, Khammarnia M, Khan EA, Khan M, Khatab K, Khater MM, Khatib MN, Khayamzadeh M, Khazaei S, Kieling C, Kim YJ, Kimokoti RW, Kisa A, Kisa S, Kivimaki M, Knibbs LD, Knudsen AKS, Kocarnik JM, Kochhar S, Kopec JA, Korshunov VA, Koul PA, Koyanagi A, Kraemer MUG, Krishan K, Krohn KJ, Kromhout H, Defo BK, Kumar GA, Kumar V, Kurmi OP, Kusuma D, La Vecchia C, Ben Lacey, Lal DK, Lalloo R, Lallukka T, Lami FH, Landires I, Lang JJ, Langan SM, Larsson AO, Lasrado S, Lauriola P, Lazarus JV, Lee PH, Lee SWH, LeGrand KE, Leigh J, Leonardi M, Lescinsky H, Leung J, Levi M, Li SS, Lim LL, Linn S, Liu SW, Liu SM, Liu Y, Lo J, Lopez AD, Lopez JCF, Lopukhov PD, Lorkowski S, Lotufo PA, Lu A, Lugo A, Maddison ER, Mahasha PW, Mahdavi MM, Mahmoudi M, Majeed A, Maleki A, Maleki S, Malekzadeh R, Malta DC, Mamun AA, Manda AL, Manguerra H, Mansour-Ghanaei F, Mansouri B, Mansournia MA, Herrera AMM, Maravilla JC, Marks A, Martin RV, Martini S, Martins-Melo FR, Masaka A, Masoumi SZ, Mathur MR, Matsushita K, Maulik PK, McAlinden C, McGrath JJ, Mckee M, Mehndiratta MM, Mehri F, Mehta KM, Memish ZA, Mendoza W, Menezes RG, Mengesha EW, Mereke A, Mereta ST, Meretoja A, Meretoja TJ, Mestrovic T, Miazgowski B, Miazgowski T, Michalek IM, Miller TR, Mills EJ, Mini GK, Miri M, Mirica A, Mirrakhimov EM, Mirzaei H, Mirzaei M, Mirzaei R, Mirzaei-Alavijeh M, Misganaw AT, Mithra P, Moazen B, Mohammad DK, Mohammad Y, Mezerji NMG, Mohammadian-Hafshejani A, Mohammadifard N, Mohammadpourhodki R, Mohammed AS, Mohammed H, Mohammed JA, Mohammed S, Mokdad AH, Molokhia M, Monasta L, Mooney MD, Moradi G, Moradi M, Moradi-Lakeh M, Moradzadeh R, Moraga P, Morawska L, Morgado-da-Costa J, Morrison SD, Mosapour A,

Mosser JF, Mouodi S, Mousavi SM, Khaneghah AM, Mueller UO, Mukhopadhyay S, Mullany EC, Musa KI, Muthupandian S, Nabhan AF, Naderi M, Nagarajan AJ, Nagel G, Naghavi M, Naghshtabrizi B, Naimzada MD, Najafi F, V. Nangia, Nansseu JR, Naserbakht M, Nayak VC, Negoi I, Ngunjiri JW, Nguyen CT, Nguyen HLT, Nguyen M, Nigatu YT, Nikbakhsh R, Nixon MR, Nnaji CA, Nomura S, Norrving B, Noubiap JJ, Nowak C, Nunez-Samudio V, Oiu AO, Oancea B, Odell CM, Ogbo FA, Oh IH, Okunga EW, Oladnabi M, Olagunju AT, Olusanya BO, Olusanya JO, Omer MO, Ong KL, Onwujekwe OE, Orpana HM, Ortiz A, Osarenotor O, Osei FB, Ostroff SM, Otstavnov N, Otstavnov SS, Overland S, Owolabi MO, Section MPA, Padubidri JR, Palladino R, Panda-Jonas S, Pandey A, Parry CDH, Pasovic M, Pasupula DK, Patel SK, Pathak M, Patten SB, Patton GC, Toroudi HP, Peden AE, Pennini A, Pepito VCF, Peprah EK, Pereira DM, Pesudovs K, Pham HQ, Phillips MR, Piccinelli C, Pilz TM, Piradov MA, Pirsaheb M, Plass D, Polinder S, Polkinghorne KR, Pond CD, Postma MJ, Pourjafar H, Pourmalek F, Poznanska A, Prada SI, Prakash V, Pribadi DRA, Pupillo E, Quazi Syed Z, Rabiee M, Rabiee N, Radfar A, Rafiee A, Raggi A, Rahman MA, Rajabpour-Sanati A, Rajati F, Rakovac I, Ram P, Ramezanzadeh K, Ranabhat CL, Rao PC, Rao SJ, Rashedi V, Rathi P, Rawaf DL, Rawaf S, Rawal L, Rawassizadeh R, Rawat R, Razo C, Boston S, Reiner RC, Reitsma MB, Remuzzi G, Renjith V, Renzaho AMN, Resnikoff S, Rezaei N, Rezaei N, Rezapour A, Rhinehart PA, Riahi SM, Ribeiro DC, Ribeiro D, Rickard J, Rivera JA, Roberts NLS, Rodriguez-Ramirez S, Roever L, Ronfani L, Room R, Roshandel G, Roth GA, Rothenbacher D, Rubagotti E, Rwegerera GM, Sabour S, Sachdev PS, Saddik B, Sadeghi E, Sadeghi M, Saeedi R, Moghaddam SS, Safari Y, Safi S, Safiri S, Sagar R, Sahebkar A, Sajadi SM, Salam N, Salamati P, Salem H, Salem MR, Salimzadeh H, Salman OM, Salomon JA, Samad Z, Kafil HS, Sambala EZ, Samy AM, Sanabria J, Sanchez-Pimienta TG, Santomauro DF, Santos IS, Santos JV, Santric-Milicevic MM, Saraswathy SYI, Sarmiento-Suarez R, Sarrafzadegan N, Sartorius B, Sarveazad A, Sathian B, Sathish T, Sattin D, Saxena S, Schaeffer LE, Schiavolin S, Schlaich MP, Schmidt MI, Schutte AE, Schwebel DC, Schwendicke F, Senbeta AM, Senthilkumaran S, Sepanlou SG, Serdar B, Serre ML, Shadid J, Shafaat O, Shahabi S, Shaheen AA, Shaikh MA, Shalash AS, Shams-Beyranvand M, Shamsizadeh M, Sharafi K, Sheikh A, Sheikhtaheri A, Shibuya K, Shield KD, Shigematsu M, Shin JI, Shin MJ, Shiri R, Shirkoohi R, Shuval K, Siabani S, Sierpinski R, Sigfusdottir ID, Sigurvinsdottir R, Silva JP, Simpson KE, Singh JA, Singh P, Skiadaresi E, Skou ST, Skryabin VY, Smith EUR, Soheili A, Soltani S, Soofi M, Sorensen RJD, Soriano JB, Sorrie MB, Soshnikov S, Soyiri IN, Spencer CN, Spotin A, Sreeramareddy CT, Srinivasan V, Stanaway JD, Stein C, Stein DJ, Steiner C, Stockfelt L, Stokes MA, Straif K, Stubbs JL, Sufiyan MB, Suleria HAR, Abdulkader RS, Sulo G, Sultan I, Szumowski L, Tabares-Seisdedos R, Tabb KM, Tabuchi T, Taherkhani A, Tajdini M, Takahashi K, Takala JS, Tamiru AT, Taveira N, Tehrani-Banihashemi A, Temsah MH, Tesema GA, Tessema ZT, Thurston GD, Titova MV, Tohidinik HR, Tonelli M, Topor-Madry R, Topouzis F, Torre AE, Touvier M, Tovani-Palone MR, Tran BX, Travillian R, Tsatsakis A, Car LT, Tyrovolas S, Uddin R, Umeokonkwo CD, Unnikrishnan B, Upadhyay E, Vacante M, Valdez PR, Van Donkelaar A, Vasankari TJ, Vasseghian Y, Veisani Y, Venketasubramanian N, Violante FS, Vlassov V, Vollset SE, Vos T, Vukovic R, Waheed Y, Wallin MT, Wang YF, Wang YP, Watson A, Wei JK, Wei MYW, Weintraub RG, Weiss J, Werdecker A, Westerman R, Whisnant JL, Whiteford HA, Wiens KE, Wolfe CDA, Wozniak SS, Wu AM, Wu JJ, Hanson SW, Xu GL, Xu R, Yadgir S, Jabbari SHY, Yamagishi K, Yaminfirooz M, Yano Y, Yaya S, Yazdi-Feyzabadi V, Yeheyis TY, Yilgwan CS, Yilma MT,.Yip P, Yonemoto N, Younis MZ, Younker TP, Yousefi B, Yousefi Z, Yousefinezhadi T, Yousuf AY, Yu CH, Yusefzadeh H, Moghadam TZ, Zamani M, Zamanian M, Zandian H, Zastrozhin MS, Zhang YQ, Zhang ZJ, Zhao JT, Zhao XJG, Zhao YX, Zhou MG, Ziapour A, Zimsen SRM, Brauer M, Afshin A, Lim SS, Factors GR (2020) Global burden of 87 risk factors in 204 countries and territories, 1990–2019: a systematic analysis for the global burden of disease study 2019. Lancet 396(10258):1223–1249

Mycoo MA (2018) Beyond 1.5 degrees C: vulnerabilities and adaptation strategies for Caribbean Small Island developing states. Reg Environ Change 18(8):2341–2353

Myers CA (2020) Food insecurity and psychological distress: a review of the recent literature. Curr Nutr Rep 9(2):107–118

Nabatanzi M, Ntono V, Kamulegeya J, Kwesiga B, Bulage L, Lubwama B, Ario AR, Harris J (2022) Malaria outbreak facilitated by increased mosquito breeding sites near houses and cessation of indoor residual spraying, Kole district, Uganda, January-June 2019. BMC Public Health 22(1):1898

Naderi L, Karamidehkordi E, Badsar M, Moghadas M (2024) Impact of climate change on water crisis and conflicts: farmers' perceptions at the ZayandehRud Basin in Iran. J Hydrol-RegNal Stud 54

Nguyen TT, Grote U, Neubacher F, Rahut DB, Do MH, Paudel GP (2023) Security risks from climate change and environmental degradation: implications for sustainable land use transformation in the Global South. Curr Opin Environ Sustain 63

Nnadi NE, Carter DA (2021) Climate change and the emergence of fungal pathogens. PLoS Pathog 17(4):e1009503

Nosrat C, Altamirano J, Anyamba A, Caldwell JM, Damoah R, Mutuku F, Ndenga B, LaBeaud AD (2021) Impact of recent climate extremes on mosquito-borne disease transmission in Kenya. PLoS Negl Trop Dis 15(3):e0009182

Nwanaji-Enwerem JC, Casey JA (2022) Naming civic health in environmental justice discourse: the Jackson water crisis. Lancet RegNal Health-Am 14

Ogunbode CA, Doran R, Hanss D, Ojala M, Salmela-Aro K, van den Broek KL, Bhullar N, Aquino SD, Marot T, Schermer JA, Wlodarczyk A, Lu S, Jiang F, Maran DA, Yadav R, Ardi R, Chegeni R, Ghanbarian E, Zand S, Najafi R, Park J, Tsubakita T, Tan CS, Chukwuorji JC, Ojewumi KA, Tahir H, Albzour M, Reyes MES, Lins S, Enea V, Volkodav T, Sollar T, Navarro-Carrillo G, Torres-Marín J, Mbungu W, Ayanian AH, Ghorayeb J,. Onyutha C, Lomas MJ, Helmy M, Martínez-Buelvas L, Bayad A, Karasu M (2022) Climate anxiety, wellbeing and pro-environmental action: correlates of negative emotional responses to climate change in 32 countries. J Environ Psychol 84

Okaka FO, Odhiambo BDO (2018) Relationship between flooding and out Break of infectious Diseasesin Kenya: a review of the literature. J Environ Public Health 2018:5452938

Oliveira GM, Vidal DG, Ferraz MP (2019) Urban lifestyles and consumption patterns. Sustainable cities and communities. Encyclopedia of the UN Sustainable Development Goals. In: W Leal Filho, A Azul, L Brandli, P Özuyar, T Wall, Springer, Cham

Oliveira GM, Vidal DG, Ferraz MP (2020) Urban lifestyles and consumption patterns. Sustainable cities and communities. Encyclopedia of the UN sustainable development goals. In: W Leal Filho, A Marisa Azul, L Brandli, P Gökçin Özuyar, T Wall. Switzerland, Springer, Cham, 851–860

Parsons ES, Jowell A, Veidis E, Barry M, Israni ST (2024) Climate change and inequality. Pediatr Res

Paz S (2024) Climate change: a driver of increasing vector-borne disease transmission in non-endemic areas. PLoS Med 21(4):e1004382

Perkins-Kirkpatrick SE, Lewis SC (2020) Increasing trends in regional heatwaves. Nat Commun 11(1)

Peters A, Schneider A (2021) Cardiovascular risks of climate change. Nat Rev Cardiol 18(1):1–2

Program USGCR (2024) Recent U.S. temperature trends. Retrieved 26 July 2024, 2024, from https://nca2014.globalchange.gov/report/our-changing-climate/recent-us-temperature-trends

Rakib MA, Sasaki J, Matsuda H, Fukunaga M (2019) Severe salinity contamination in drinking water and associated human health hazards increase migration risk in the southwestern coastal part of Bangladesh. J Environ Manage 240:238–248

Raoelison OD, Valenca R, Lee A, Karim S, Webster JP, Poulin BA, Mohanty SK (2023) Wildfire impacts on surface water quality parameters: Cause of data variability and reporting needs. Environ Pollut 317

Ratter-Rieck JM, Roden M, Herder C (2023) Diabetes and climate change: current evidence and implications for people with diabetes, clinicians and policy stakeholders. Diabetologia 66(6):1003–1015

Redvers N, Aubrey P, Celidwen Y, Hill K (2023) Indigenous peoples: traditional knowledges, climate change, and health. PLOS Glob Public Health 3(10):e0002474

Ricciardi W, Marcheggiani S, Puccinelli C, Carere M, Sofia T, Giuliano F, Dogliotti E, Mancini L, Agrimi U, Alleva E, Busani L, De Castro P, Gaudi S, Michelozzi P, Rezza G, Testai E, Vella S (2019) Health and Climate Change: science calls for global action. Ann Ist Super Sanita 55(4):323–329

Robine JM, Cheung SLK, Le Roy S, Van Oyen H, Griffiths C, Michel JP, Herrmann FR (2008) Death toll exceeded 70,000 in Europe during the summer of 2003. CR Biol 331(2):171-U175

Rocklov J, Dubrow R (2020) Climate change: an enduring challenge for vector-borne disease prevention and control. Nat Immunol 21(5):479–483

Rockstrom J, Steffen W, Noone K, Persson A, Chapin FS 3rd, Lambin EF, Lenton TM, Scheffer M, Folke C, Schellnhuber HJ, Nykvist B, de Wit CA, Hughes T, van der Leeuw S, Rodhe H, Sorlin S, Snyder PK, Costanza R, Svedin U, Falkenmark M, Karlberg L, Corell RW, Fabry VJ, Hansen J, Walker B, Liverman D, Richardson K, Crutzen P, Foley JA (2009) A safe operating space for humanity. Nature 461(7263):472–475

Rodrigues GP, Brosinsky A, Rodrigues IS, Mamede GL, de Araújo JC (2024) Impact of reservoir evaporation on future water availability in north-eastern Brazil: a multi-scenario assessment. Hydrol Earth Syst Sci 28(14):3243–3260

Roque A, Wutich A, Brewis A, Beresford M, Garcia-Quijano C, Llorens H, Jepson W (2021) Autogestion and water sharing networks in Puerto Rico after Hurricane Maria. Water Int 46(6):938–955

Roy P, Pal SC, Chakrabortty R, Chowdhuri I, Saha A, Shit M (2023) Effects of climate change and sea-level rise on coastal habitat: vulnerability assessment, adaptation strategies and policy recommendations. J Environ Manag 330

Ruiz-Páez R, López-Bueno JA, Padrón-Monedero A, Navas MA, Salvador P, Linares C, Díaz J (2024) Short-term effects of fine particulate matter from biomass combustion and Saharan dust intrusions on emergency hospital admissions due to mental and behavioural disorders, anxiety and depression in Spain. Sci Total Environ 946

Rupasinghe R, Chomel BB, Martínez-López B (2022) Climate change and zoonoses: a review of the current status, knowledge gaps, and future trends. Acta Tropica 226

Salam A, Wireko AA, Jiffry R, Ng JC, Patel H, Zahid MJ, Mehta A, Huang H, Abdul-Rahman T, Isik A (2023) The impact of natural disasters on healthcare and surgical services in low- and middle-income countries. Ann Med Surg (Lond) 85(8):3774–3777

Sarkar S (2022) Pakistan floods pose serious health challenges. Bmj-Br Med J 378

Schilperoort M, Hoeke G, Kooijman S, Rensen PC (2016) Relevance of lipid metabolism for brown fat visualization and quantification. Curr Opin Lipidol 27(3):242–248

Schneider A, Ruckerl R, Breitner S, Wolf K, Peters A (2017) Thermal control, weather, and aging. Curr Environ Health Rep 4(1):21–29

Scortichini M, De Sario M, de'Donato FK, Davoli M, Michelozzi P, Stafoggia M (2018) Short-term effects of heat on mortality and effect modification by air pollution in 25 Italian cities. Int J Environ Res Public Health 15(8)

Semenza JC, Paz S (2021) Climate change and infectious disease in Europe: Impact, projection and adaptation. Lancet RegNal Health-Eur 9

Semenza JC, Ko AI (2023) Waterborne diseases that are sensitive to climate variability and climate change. N Engl J Med 389(23):2175–2187

Sever MS, Kati YA, Ozkaya U (2023) Destructive disasters, trauma, crush syndrome, and beyond. Acta Orthop Traumatol Turc 57(6):305–314

Shafique M, Khurshid M, Muzammil S, Arshad MI, Malik IR, Rasool MH, Khalid A, Khalid R, Asghar R, Baloch Z, Aslam B (2024) Traversed dynamics of climate change and one health. Environ Sci Eur 36(1)

Shaposhnikov D, Revich B, Bellander T, Bedada GB, Bottai M, Kharkova T, Kvasha E, Lezina E, Lind T, Semutnikova E, Pershagen G (2014) Mortality related to air pollution with the moscow heat wave and wildfire of 2010. Epidemiology 25(3):359–364

Sheridan SC, Allen MJ (2018) Temporal trends in human vulnerability to excessive heat. Environ Res Lett 13(4)

Skendzic S, Zovko M, Zivkovic IP, Lesic V, Lemic D (2021) The impact of climate change on agricultural insect pests. Insects 12(5)

Song XP, Guo XY, Hu XP, Zhang YJ, Wei DD, Hu Y, Jiang LZ, Zhang Y (2023) Human exposure risk assessment for infectious diseases due to temperature and air pollution: an overview of reviews. Environ Sci Pollut Res 30(38):88272–88280

Stanke C, Kerac M, Prudhomme C, Medlock J, Murray V (2013) Health effects of drought: a systematic review of the evidence. PLoS Curr 5

Symonds ME, Farhat G, Aldiss P, Pope M, Budge H (2019) Brown adipose tissue and glucose homeostasis—the link between climate change and the global rise in obesity and diabetes. Adipocyte 8(1):46–50

Tarolli P, Luo J, Park E, Barcaccia G, Masin R (2024) Soil salinization in agriculture: mitigation and adaptation strategies combining nature-based solutions and bioengineering. Iscience 27(2)

Tavares MPF, Oliveira GM (2023) Climate change and transmissible diseases. climate change and health hazards. climate change management. In: W Leal Filho, DG Vidal, MAP Dinis, Springer, Cham

Tchonkouang RD, Onyeaka H, Nkoutchou H (2024) Assessing the vulnerability of food supply chains to climate change-induced disruptions. Sci Total Environ 920

Thornton P, Nelson G, Mayberry D, Herrero M (2022) Impacts of heat stress on global cattle production during the 21st century: a modelling study. Lancet Planetary Health 6(3):E192–E201

Tran HM, Chen TT, Lu YH, Tsai FJ, Chen KY, Ho SC, Wu CD, Wu SM, Lee YL, Chung KF, Kuo HP, Lee KY, Chuang HC (2022) Climate-mediated air pollution associated with COPD severity. Sci Total Environ 843

Venegas RM, Acevedo J, Treml EA (2023) Three decades of ocean warming impacts on marine ecosystems: a review and perspective. Deep-Sea Res Part Ii-Top Stud Ocean 212

VER. from https://climate.copernicus.eu/climate-indicators/temperature

Vicedo-Cabrera AM, Sera F, Guo YM, Chung Y, Arbuthnott K, Tong SL, Tobias A, Lavigne E, Coelho MDZS, Saldiva PHN, Goodman PG, Zeka A, Hashizume M, Honda Y, Kim H, Ragettli MS, Roosli M, Zanobetti A, Schwartz J, Armstrong B, Gasparrini A (2018) A multi-country analysis on potential adaptive mechanisms to cold and heat in a changing climate. Environ Int 111:239–246

Watts N, Amann M, Arnell N, Ayeb-Karlsson S, Belesova K, Boykoff M, Byass P, Cai W, Campbell-Lendrum D, Capstick S, Chambers J, Dalin C, Daly M, Dasandi N, Davies M, Drummond P, Dubrow R, Ebi KL, Eckelman M, Ekins P, Escobar LE, Fernandez Montoya L, Georgeson L, Graham H, Haggar P, Hamilton I, Hartinger S, Hess J, Kelman I, Kiesewetter G, Kjellstrom T, Kniveton D, Lemke B, Liu Y, Lott M, Lowe R, Sewe MO, Martinez-Urtaza J, Maslin M, McAllister L, McGushin A, Jankin Mikhaylov S, Milner J, Moradi-Lakeh M, Morrissey K, Murray K, Munzert S, Nilsson M, Neville T, Oreszczyn T, Owfi F, Pearman O, Pencheon D, Phung D, Pye S, Quinn R, Rabbaniha M, Robinson E, Rocklov J, Semenza JC, Sherman J, Shumake-Guillemot J, Tabatabaei M, Taylor J, Trinanes J, Wilkinson P, Costello A, Gong P, Montgomery H (2019) The 2019 report of The Lancet Countdown on health and climate change: ensuring that the health of a child born today is not defined by a changing climate. Lancet 394(10211):1836–1878

Watts N, Amann M, Ayeb-Karlsson S (2020) The 2017 report of The Lancet Countdown on health and climate change: from 25 years of inaction to a global transformation for public health (vol 391, pp 581, 2017). Lancet 395(10239):1762–1762

Weatherly C, Carag J, Zohdy S, Morrison M (2023) The mental health impacts of human-ecosystem-animal relationships: a systematic scoping review of Eco-, Planetary, and One Health approaches. One Health 17

White BP, Breakey S, Brown MJ, Smith JR, Tarbet A, Nicholas PK, Ros AMV (2023) Mental health impacts of climate change among vulnerable populations globally: an integrative review. Ann Glob Health 89(1):66

Williams AP, Cook BI, Smerdon JE (2022) Rapid intensification of the emerging southwestern North American megadrought in 2020–2021. Nat Clim Chang 12(3):232

Wu CS, Zhang YQ, Wei J, Zhao ZH, Norbäck D, Zhang X, Lu C, Yu W, Wang TT, Zheng XH, Zhang L (2022) Associations of early-life exposure to submicron particulate matter with childhood asthma and wheeze in China. Jama Netw Open 5(10)

Xu R, Yu P, Abramson MJ, Johnston FH, Samet JM, Bell ML, Haines A, Ebi KL, Li S, Guo Y (2020) Wildfires, global climate change, and human health. N Engl J Med 383(22):2173–2181

Yao M, Hu Y, Zhang A, Ji JS, Zhao B (2022) COPD deaths attributable to ozone in 2019 and future projections using the WHO AQG 2021 in urban China. Eco Environ Health 1(4):251–258

Yari A, Ostadtaghizadeh A, Ardalan A, Zarezadeh Y, Rahimiforoushani A, Bidarpoor F (2020) Risk factors of death from flood: findings of a systematic review. J Environ Health Sci Eng 18(2):1643–1653

Yuan Y, Wang K, Sun HZ, Zhan Y, Yang ZM, Hu KJ, Zhang YQ (2023) Excess mortality associated with high ozone exposure: a national cohort study in China. Environ Sci Ecotechnology 15

Zhang P, Wiens K, Wang R, Luong L, Ansara D, Gower S, Bassil K, Hwang SW (2019) Cold weather conditions and risk of hypothermia among people experiencing homelessness: implications for prevention strategies. Int J Environ Res Public Health 16(18)

Zilbermint M (2020) Diabetes and climate change. J Community Hosp Intern Med Perspect 10(5):409–412

Zinsstag J, Crump L, Schelling E, Hattendorf J, Maidane YO, Ali KO, Muhummed A, Umer AA, Aliyi F, Nooh F, Abdikadir MI, Ali SM, Hartinger S, Mausezahl D, de White MBG, Cordon-Rosales C, Castillo DA, McCracken J, Abakar F, Cercamondi C, Emmenegger S, Maier E, Karanja S, Bolon I, de Castaneda RR, Bonfoh B, Tschopp R, Probst-Hensch N, Cisse G (2018). Climate change and one health. FEMS Microbiol Lett 365(11)

Ziska LH (2022) Rising carbon dioxide and global nutrition: evidence and action needed. Plants-Basel 11(7)

The manufacturer's authorised representative in the EU is Springer Nature Customer Service Centre GmbH, Europaplatz 3, 69115 Heidelberg, Germany. If you have any concerns regarding our products, please contact ProductSafety@springernature.com

Printed and bound by CPI Group (UK) Ltd, Croydon, CR0 4YY
30/09/2026
02225596-0010